X-Ray Contrast Agent Technology

X-Ray Contrast Agent Technology
A Revolutionary History

Christoph de Haën

CRC Press
Taylor & Francis Group
Boca Raton London New York

CRC Press is an imprint of the
Taylor & Francis Group, an **informa** business

CRC Press
Taylor & Francis Group
6000 Broken Sound Parkway NW, Suite 300
Boca Raton, FL 33487-2742

First issued in paperback 2020

© 2019 by Taylor & Francis Group, LLC
CRC Press is an imprint of Taylor & Francis Group, an Informa business

No claim to original U.S. Government works

ISBN 13: 978-0-367-73056-7 (pbk)
ISBN 13: 978-1-138-35164-6 (hbk)

Visit the Taylor & Francis Web site at
http://www.taylorandfrancis.com

and the CRC Press Web site at
http://www.crcpress.com

Contents

Preface

Some time in the 1990s I was invited to contribute to a work in preparation an article on the history of X-ray contrast agents. I concluded that the literature contained already enough such reports and that they reflected a reasonable consensus. I therefore declined my participation. Then I read a book written by my friend Meinolf Dierkes and his collaborators, Ute Hoffmann and Marz Lutz, entitled *Visions of Technology. Social and Institutional Factors Shaping the Development of New Technologies*. Therein they presented a particular sociological model for the genesis and shaping of new technology and submitted it to plausibility tests through application to the historical cases of the Diesel engine, the typewriter, and the mobile telephone. Performing a similar analysis for X-ray contrast agent technology promised to be unprecedented. It could offer a suitable response to a future invitation of the kind I had declined. Possibly lessons for the future could be learned. The rich historical literature appeared to contain all the necessary raw material. Alas, I soon became puzzled by the marginal role into which patents, as well as the industry and its representatives, were relegated. The situation failed to reflect my research and development experience in the contrast agent industry. But it also reminded me of my own similar failure of appreciation in my earlier working life as academic.

The envisioned sociological analysis could at best be as meaningful as the historical substrate. Thus, I embarked in my leisure on a reanalysis of the entire history of X-ray contrast agent technology. From the beginning, it was clear that I would need new and overwhelming evidence if I felt the necessity to challenge engrained historical interpretations. With time, the task and associated documentation took on monstrous dimensions. It became clear that the scope of the effort had to be reduced and even then, the subject would explode the frame of a journal article. Thus was born the plan for this book.

I decided to focus on uro-angiographic contrast agents, and in particular, on two periods in their development of special interest. The first, representing the technology genesis phase, had brought saltatory innovation, and the second, representing a technology shaping phase, had brought improvement innovation. The two case studies could be imbedded in a complete technology life cycle, wherein the technology finally reached its consolidation/economization phase, in which innovation stagnates.

I assembled new and old evidence, profiting gratefully from the magnanimous help provided by authors of earlier historical efforts, protagonists of the history and some of their relatives and collaborators, as well as archivists and librarians of public institutions and industry. I scrutinized the collected material with juridical rigor. The results called for revisions of the controversial case of Moses Swick and the first uro-angiographic contrast agent, and the fascinating tale that Torsten Almén told about the first nonionic one. The insight allowed replacement of the prevalent academia- and hospital-focused accounts with a revolutionary one that incorporated the role of industry and its patents. Prevalent historiography of contrast agents, i.e., great man history and chronicle of products, was replaced by a narrative that recognized technological innovation as a social process. The elaborated history now allowed plausibility tests for the sociological model under investigation and suggestions for minor modifications thereof.

My book aims at sensibilizing the reader to the social aspects of technological innovation, as well as to the importance of the industry and patents. In addition, I hope it to be entertaining with its web of ingenious and ambitious individuals acting in stories laced with human foibles. If my book were successful in both, then it would have achieved its objective.

Acknowledgments

I feel profound gratitude toward Michael F. Tweedle for his incessant encouragement, critique, and editorial assistance. Special thanks for particularly substantial contributions to my historical effort in form of documents and personal interactions go to Uno Erikson and Peter Rathert, as well as the deceased Torsten Almén, Ernst Felder, and Ronald G. Grainger. Many years ago, my friend Meinolf Dierkes gave me the opportunity to participate in a social science project, for which I am immensely grateful. Only with this experience could I dare the foray into sociology of the present effort. I am very obliged for help with information and documents to Max D. Adams, Christine Berghausen, Lars Björk, Jakob Björk, Erik Boijsen, Donatella Bonati, Franco Bonati, Bruno Bonnemain, Uwe Busch, Lee Clarke, Peter Dawson, Burton B. Drayer, Lars Fondberg, Michael Frings, Kirsti A. Granath, Togny Greitz, Thore Grimm, Tore Grönmo, Anders Grönwall, Johan Haavaldsen, Walter Habicht, Sadek K. Hilal, Geo. Brooke Hoey, Constance Holtermann, Björn Ingelman, Hitoshi Katayama, Aubry A. Larsen, Franz von Lichtenberg, Sergio A. Lugo, Hans-Joachim Maurer, Dominique Meyer, Richard F. Mould, Hans Müller, Reed M. Nesbit, Göran Nylander, Sigrid Oelze, Simona Pitrè, Michael Pohlenz, Ingunn Possehl, Carl Richter, Kathleen D. Rieger, Knut Sogner, Milos Sovak, Ulrich Speck, Susan B. Sterling (-Swick), Thomas Suter, Piero Tirone, Gunnar Törnell, Fulvio Uggeri, Eva-Maria Wartenberg (-von Schickh), and Hans Zutter.

A contribution by Bracco Diagnostics Inc., Monroe Township, NJ, to the fees for copyright permissions is gratefully acknowledged.

Author

Christoph de Haën, born September 9, 1940 in Switzerland, studied chemistry and biology in Lausanne, Zurich, and Rome, earning a master's degree in chemistry from the Swiss Federal Institute of Technology (ETH Zurich), followed by a PhD in molecular biology at the same institution. After a postdoctoral fellowship in biochemistry at the University of Washington, Seattle, he became research faculty at the same institution, jointly appointed in the Departments of Medicine (Division of Endocrinology and Metabolism) and Biochemistry. There he moved up the ranks to Research Full Professor and Graduate Faculty. In mid-career he joined the Bracco company, a worldwide active pharmaceutical corporation, headquartered in Milan, Italy. There he held various leadership positions in global research and development, including Head of the Milano Research Center and Scientific Advisor to the Bracco Group of Companies. The principle field of activity was contrast agents for various diagnostic imaging modalities. His publications span a wide variety of fields, including social science ones. He retired in 2005 and lives in Switzerland.

1 Introduction

"We use contrast agents like water," meaning, with the same insouciance. With this casual remark in the 1990s, an eminent radiologist referred to the then newest generation of uro-angiographic X-ray contrast agents or contrast media. Such a contrast agent is a sort of injectable ink, transparent to ordinary light but opaque to X-rays, and is used for radiological diagnosis in man. Excretion urography is the visualization of the whole urinary system by X-rays achieved with a contrast agent that enters the urine from the blood, typically after intravenous administration. Angiography is the visualization of the major blood vessels by X-rays after intravascular injection of a contrast agent. The characterization of a contrast agent as "uro-angiographic" refers to the property of being useful in both applications. In investigators involved in research and development of such products, the mentioned insouciance can stir feelings of satisfaction at having come so close to the technological *vision of the radiologically effective and totally innocuous contrast agent*; the term vision is being used here in the sense of guiding principle, ideal, myth, dream, and as the exact equivalent of the German term "Leitbild".[1] Yet at the same time, the statement causes concern that vision and reality may get confounded, leading to overconfidence and potentially dangerous misuse. Indeed, radiologists sometimes substantially exceed recommended doses for uro-angiographic contrast agents, even doubling them, not always realizing that this overstepping may only be done at the cost of increased risk of adverse reactions. The lax attitude regarding dose easily spreads to adjacent issues. Certain habits of handling contrast agents, such as that at one time widely practiced, of opening vials and pooling contents, are contrary to good practice for handling sterile injectables. Good practice norms reflect the need to maintain the pharmaceutical presentation sterile and particle-free, quality and safety characteristics, to which so much attention is devoted during manufacturing. Patterns of ill-advised behavior by radiologists and their assistants are expression of overconfidence. The overconfidence may be rooted in the perception that the products with their supposed lack of pharmacological effects need not be considered on a par with drugs.

The confusion between vision and reality alluded to is facilitated by the fact that contrast agents are remarkably well tolerated––an important property given the high doses required for diagnosis. In fact, no other substance is injected in a short time intravascularly to human beings in such enormous quantities, i.e., recommended doses of up to 142 g per 70-kg body weight within a few minutes for abdominal computer tomography or 107 g in 30 s in spiral computer tomography. Remarkable as their tolerability may be, it should never be forgotten that they are still pharmaceuticals, compounds foreign to the body, and for this reason inevitably endowed with residual toxic potential, especially if given at excessive doses. No pharmaceutical substance, not even a natural one, can ever be made totally innocuous. The vision underlying the expression "we use contrast agents like water" will, at best, be ever more closely approximated in the course of the evolution of the technology.

Here is neither the occasion to criticize professionals nor the occasion to teach proper use of contrast agents. Rare questionable behaviors of individuals or whole professional groups are of present interest solely insofar these anomalies provide evidence of a blurring between the vision and reality[2] and thereby serve as supporting evidence for the existence of a vision in the first place.

The existence of a vision has been considered a fundamental prerequisite for any collective human endeavor beyond survival (Böhler 1965), and thus of development of new technology. A technological vision, as intended here and discussed in detail by Dierkes, Hoffmann, and Marz (1996), for the duration of its spell singles out an objective collectively considered attainable and furnishes it with an aureole that makes its achievement desirable. On the temporal horizon of a vision, lines of extrapolation from collective past and present experience regarding desirability meet with corresponding lines of extrapolation regarding perception of realizability (Figure 1.1). A vision influences processes of perception, reflection, and decision. It exercises a powerful synchronizing function of a multiplicity of actors and actions in the processes of genesis and shaping of technologies. It is part and parcel of the carrying and enforcing power that, according to the French historian Fernand Braudel, alone determines the importance of an innovation (Braudel 1982). Note also that in this work the concept of a vision is conceived in a slightly narrower sense than in Dierkes, Hoffmann, and Marz (1996), mostly by requiring that a kernel of a vision has already infected a decisive group of human and institutional actors.

A vision is frequently captured in a concise formulation, which facilitates making reference to it. But a lengthy discursive format is not excluded (Kahlenborn et al. 1995). Familiar and concisely formulated visions effective today are as follows: the information superhighway, green industrial production, quantum computing, the hydrogen fuel economy, or specifically in the medical arena, minimally invasive surgery, magic bullet pharmaceuticals, personalized medicine, and organ transplantation banks. In addition to these visions on a macro scale, visions exist equally for

The Technological Vision

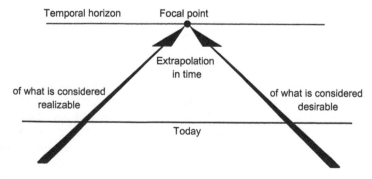

FIGURE 1.1   The vision of a technology—a pictorial metaphor.

much smaller technologies, as exemplified presently for contrast agents, but obviously they do not involve the same size of public. Moreover, effective visions must prove sustainable; otherwise they may lose out to competitors before having produced significant impacts. The vision of the "car-friendly city" yielded to that of the "car-free city center" before the former was realized in a major way. If what has been said is accurate, then the emergence of contrast agent technology too must have been guided by a vision. The mentioned misuse and mishandling of products betrays the specific content of the vision, namely that of *the radiologically effective and totally innocuous contrast agent*, captured too as "We use contrast agents like water" by the radiologist I have cited.

Technology is a broad notion. The case of contrast agent technology falls into the category defined by marketable human-made objects. The observations and theoretical considerations that will be presented apply primarily to such technology. But to various degrees they may be pertinent even for other categories, such as special knowledge and skills inherent in a circumscribed human activity, e.g., a surgical procedure.

Popular wisdom on the emergence, affirmation, and evolution of a technology, shared by the majority of exponents from scientific and technical disciplines, see a technology as the result of a process composed of rate-limiting discovery and implementation steps. Its success and failure are interpreted exclusively as evidence of its inherent technical performance merits or flaws, respectively—so similarly most contrast agent scientists and users, marketing people much less, hold such an objectivist worldview. Distinct from this view, scientists professionally studying technological innovation as a phenomenon have learned to appreciate its strong social process component. New technology involves both technical and social changes. In some publications, social aspects may have been emphasized excessively, but overall there can be little doubt that the so-called social constructivist perspective has substantially enriched the in-depth understanding of technology genesis and shaping. Definition of a technology is accompanied by attribution of meaning to it. This attribution is an eminently social process yet decisive for its affirmation and constructive shaping (Sturken, Thomas, and Ball-Rokeach 2004). Discoveries, inventions, functional designs, and the like are recognized as necessary prerequisites for a new technology, but they are not believed to suffice for its success. Obviously its price is an important factor for success and, as economy teaches, it is clearly a social factor. Apart from it, there exist numerous other social aspects to a technology. Who will predominantly benefit from it in terms of power, security, comfort, etc., often determines its adoption, independent on whether it fulfills its more strictly functional promises. The institutional context in which the technology arises can facilitate or hinder its success. Numerous technologies that began their careers at universities failed because inadequate patent protection did render unjustifiable the investment by industry necessary for continuation. It is here not the place to elucidate this issue in all of its facets; there exist many excellent articles and books (Winner 1986; Joerges 1989; Pool 1990; Jasanoff et al. 1995; Dierkes, Hoffmann, and Marz 1996; Spann, Adams, and Sounder 1995; Arns 1997). For actual purposes, it suffices to say that contrast agent technology was also generated and became reshaped multiple times in social processes.

Accordingly, this book addresses the social dimension of the technology. It focuses on people and institutions together with mental and social processes behind the innovation. It does so at the price of covering the more strictly natural science, technical, and medical perspectives somewhat superficially. It is satisfying to know that the latter perspectives are covered in reasonable depth by various authoritative texts.[3] The history of contrast agents offers favorable circumstances for the intended analysis, because of the existence of a precise and unequivocal starting point, the discovery of X-rays by Wilhelm C. Röntgen in 1895. This simplifies substantially at least one aspect of the project, namely the assessment of preceding periods. In addition, it allows to cover the progress of technology from the starting point onward to the present stage, in which its products have aged and further innovation stagnates.

The history of technology genesis and shaping is frequently envisioned by analogy to biological evolution as creation of variation followed by selection (see, e.g., Constant 1980). Technical artifacts that found significant application are arranged into a succession of the winners *ad interim*. A pipeline of candidate solutions provides for quasi-continuous progress. For technology in its entirety this characterization may be valid. But for an individual technology and artifact thereof, the situation differs. Economists characterize it by a technology or product life cycle. Its life time is divided into a research and development period without business gain, and a limited vital life offering financial return. Here the term of technology life cycle will be used in a slightly modified version to structure the history of contrast agents at a macroscopic level. First, the genesis or conception phase results in a fundamental innovation (see, e.g., de Haën 2001). Then, there are shaping phases that bring improvement innovations. Some of them are gradual, others saltatory in character. A last of them culminates in products that have become commodities. One may speak of a consolidation and economization phase. Further significant product innovation ceases. The technology may even have entered a frozen state and persist for a long time. In that case, it takes exceptional circumstances to break out of the old mold and start a fresh cycle of innovation. More often the technology gets replaced by an entirely different one.

This study examines at the microscopic level the advent of a technology, mainly that of uro-angiographic contrast agents, in terms of the analytical model of technology genesis and shaping[4] proposed by Dierkes, Hoffmann, and Marz (1996), as slightly modified in Kahlenborn et al. (1995). This version will be referred to as the original analytical model. The version with slight modifications, applied here, will be referred to simply as the analytical model. The authors of the original analytical model criticized the aforementioned traditional models of technological progress as having too little explanatory power and substituted it with their model. Here the analytical model is seen principally as a tool for the elucidation of the events within individual phases of the technology life cycle. It will be exposed in greater detail in a subsequent chapter. This introductory chapter offers just an orienting glance at it. In the chosen model, technology genesis and shaping is likened to the phenomenon observed when the expanding wave trains from multiple stones thrown into a pond meet. By a phenomenon called interference, the ridges and valleys of the waves emanating from different sources add constructively or destructively, thereby creating novel patterns that are not at all prefigured in any of the individual wave trains (Figure 1.2).

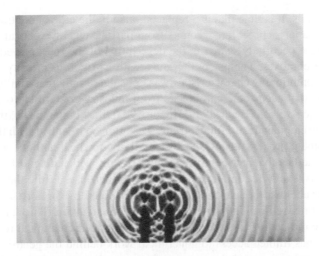

FIGURE 1.2 Interfering wave trains as pictorial metaphor for interfering knowledge cultures. The simultaneous agitation of the surface of a liquid by two wire tips creates a new pattern, not prefigured in the patterns created by individual wire tips. (Reprinted by permission from Springer Nature Customer Service Centre GmbH of lower figure in Kemp, M. 1998. Abbott's Absolutes. *Nature* 391 (6666): 445.)

In the case of technology genesis and shaping, the sources are identified with both actions and actors from distinct knowledge cultures and the wave trains are identified with the potential contributions of different sources.[5] The interference pattern created by wave trains from multiple sources stands for the knowledge and artifacts that constitute novel technology.[6] Unless otherwise specified, with interference here is meant a phenomenon productive for technology. A vision is thought to synchronize the interference phenomena. The model requires that traces of technology genesis and shaping be identifiable in technical artifacts as such pattern of interference, although not necessarily in the perfected final products. It rejects implicitly simplistic linear chains of cause and effect as satisfactory explanations and resists one-dimensional reductionisms.

The authors of the model just outlined have evaluated the plausibility of their analytical framework in three cases, i.e., the technology geneses and shapings of the Diesel engine, the mechanical typewriter, and the mobile telephone (Dierkes, Hoffmann, and Marz 1996). Others have applied the model to additional cases (Hellige 1996). Technology genesis and shaping in the case of uro-angiographic contrast agents offers a first plausibility test from the pharmaceutical and medical field.

The sociological analysis at hand will use history as its substrate. Numerous more or less thoroughly researched histories of contrast agent technology have been published, some of them by authors who were protagonists of the described events.[3] Although a few of these are meritorious, most were found to be seriously wanting. Some of them ignore the human beings involved. Others follow the mold of "great man history." Only a few laudable exceptions (e.g., Cannon 1914) have found an appropriate balance. "Great man history" has already for a long time been overcome by historians but continues to be cultivated in niche history writing by nonprofessionals. It is time that it is replaced, also in the case of contrast agents, by an up to date approach. So far, accounts from participants of events and their interpretations were peculiarly scarce in the description of the roles played by industry and its representatives. The patent literature with its rigorous priority dates and its technology conditioning characteristics was considered only marginally, if at all.[7] It is regrettable that even some chroniclers from industry conformed to the canon of their academic paragon (Hecht 1939; Hoppe 1963). Last but not least, the latest step in product development, i.e., clinical trials, was often confounded with the discovery and development process as a whole, especially by medical authors. This situation is rooted in the fact that new contrast agents typically widen diagnostic possibilities and reduce risks and discomfort associated with their use. While the latter is welcome, real excitement in academic circles is generated by the former opportunity. A plethora of new clinical possibilities calls for investigation. Through this activity the academic radiologist experiences innovation mostly as a clinical research activity, while he has limited access to nonclinical work behind the genesis of the novel article. No wonder, the result is a picture of progress biased toward academic contributions, above all those by university radiologists.[8] Of course, historical prevarications due to subtle misrepresentations by industry, more or less intentional, may be encountered too, pre-eminently in promotional materials.[9]

Creation of a well-balanced history, which is a precondition for a meaningful sociological analysis, encounters some obstacles. They arise on the one hand from the difficulty of obtaining access to company documents, and on the other hand, from industry's management of information exchange with academics and the general public, aimed more at a desirable image than at rendering its operation transparent. Additionally, the subject is clouded by a substantial secondary literature frequently tainted by chauvinism, suffering from distortions brought about by illiteracy of the authors in the languages of the original documents and ignorance of the patent literature. Propagation of misquotations, as well as distortions in quotation chains, further undermines reliability.

Thus, before distilling sociological processes from history, a revised version thereof is elaborated. Previously exhausted primary sources concerning selected periods of contrast agent technology genesis and shaping, the one regarding uro-angiographic ones in particular, are critically reinterpreted and the documentation is substantially integrated. The role of industry is reassessed based on previously unused material evidence where possible or with circumstantial arguments. A chronicle, in the better part drastically revised, is the result. Even if its interpretations were not shared completely by the reader, this work will deepen the understanding of the periods, persons, and events of interest. Besides, it provides fresh source material for alternative historical and sociological interpretations. With the present approach, I intend to heed the call of Charles Tilly (1981) for a sociology that does not illustrate its model expectations with the help of results from historians but makes its own historical research.

This study does not aim at fair coverage of the full history of contrast agents, all their classes, all areas dealing with mechanisms of contrast generation and toxicity, all areas of clinical applications, procedural advances and radiological insights, and parallel developments in the sector of imaging equipment and recording material.

Even less does it aspire to cover the history of medical diagnosis in fields in which contrast agents play a role. Rather, it focuses on the technology life cycle of contrast agents as products and on selected periods and technological innovation that promise to be enlightening in terms of social dynamics yet remain representative of the field of technology genesis and shaping as a whole. Given the specific scope, it seemed legitimate to pay particular attention to two periods that were rich both in terms of progress and documentation and in which large-scale political events, such as world wars or economic depressions, were not dominant factors. Other time spans and areas with important contributions to overall progress will regrettably, and somewhat arbitrarily, be glossed over. The synergies between developments in the contrast agent field and those in instrumentation will similarly be touched upon only superficially.

Chapter 1, the introductory one, explains the scope of the book. Second, it gives a gross overview of theoretical concepts that will be dealt with in detail more in what follows.

Chapter 2 deals with methodological aspects of research approaches used here in the two fields of interest, history, and sociology of technology.

Chapter 3 describes in an extremely cursory manner the history of X-ray contrast observation and manipulation from the discovery of X-rays up to the beginning of the first of the two periods onto which the principle investigation focuses. Its sole scope is to provide the reader with the minimal background historical knowledge necessary for being able to follow the main part of the study.

Chapter 4 is dedicated to the highly particular histories of the two periods of special interest and their sociological analyses. The first period essentially coincides with the technology genesis phase for uro-angiographic contrast agents. The second period encompasses the last portion of the technology shaping phase. The history begins with the advent of contrast agents that allowed visualization of the entire urinary system after intravenous injection, i.e., excretion urography, as well as contrast agents that allowed safe visualization of vasculature, i.e., angiography. It describes how in the late 1920s and early 1930s chemicals capable of both functions, so-called uro-angiographic agents, came to be seen as the Holy Grail and how successful products were found. The analytical model introduced in Chapter 2 is then evaluated for how well it serves to interpret dynamics, events, and outcomes in this empirical field, a field that encompasses a basic innovation (Mensch 1979). The model's concept map is richly ramified. Admittedly, here only elements of the model considered quintessential are considered.

A cursory description of the contrast agent evolution in the next three decades follows, before attention is turned to the second historical period focused upon, which lasted from the 1960s to the 1980s. That period encompassed the advent of low-osmolality and nonionic uro-angiographic contrast agents. In a structure that parallels the investigation of the first period, a thorough chronicle again forms the beginning. Only thereafter is the utility of the analytical model in interpreting this period scrutinized and ameliorations are suggested. This period stands for one of improvement innovation (Mensch 1979).

The chapter closes briefly recounting the consolidation and economization phase of contrast agent technology, a period in which low-osmolality and nonionic versions conquered the markets and eventually became commodities. It ends with a look at the present and its technological innovation challenges, abstaining from a sociological analysis.

Chapter 5, the concluding one, criticizes the work of lay and professional historians of contrast agent technology alike and attempts to justify why this author's opus constitutes an improvement. Recommendations for historiography of technology are derived. The merits of marrying historical with sociological perspectives are pointed out.

The reasons are given for declaring the analytical model to have passed the plausibility test posed by the history of contrast agent technology, even after taking some weaknesses into account. The question is explored whether besides scientific insights, the conclusions also offer recommendations for action.

2 Methodological Considerations

HISTORICAL AND SOCIOLOGICAL PERSPECTIVES

To date, the history of contrast agent technology has been dealt with almost exclusively by researchers trained in medical or natural sciences, often participants of the very history. The accounts typically focused on the chronology of successful products and applications, passing over anything between their advent. This focus allowed to streamline history and helped conservation of printing space. Histories typically glorified deeds and biographies of "great men," foremost clinicians. Such history risks of being abused to support claims of primacy. Reaching for primacies is rooted in individual self-interest and in the interests of social groups, such as groups of persons sharing a training pedigree, universities, hospitals, national professional societies, or social minorities. Primacy affords prestige and this in turn affords power. It accommodates the human romance with heroizing history, with believe in inevitable progress and its accompanying simplification and linearization. The corresponding kind of historiography, deprecatingly termed "whiggish," has rightly received heavy criticism (Staudenmaier 1985). The present history of the technology genesis strongly argues that only a complex network of communicating actors and actions can satisfactorily render the processes intelligible. Only examination of such networks in various instances of genesis and shaping of technology may reveal underlying shared patterns.

Professional historians of science and technology have so far shunned the field of contrast agent technology, to the extent that an otherwise most interesting contemporary history of Röntgen rays and their role in medicine barely mentions contrast agents (Dommann 2003). An exception is the book by Holtzmann Kevles (1997), in which the contrast agent technology is treated as part of the history of medical imaging. Historians of business, some of them by profession, have retrieved the histories of a couple of major contrast agent producing companies, including that of their radiological products (Amdam and Sogner 1994; Bonnemain 2014). Despite these exceptions, at the outset of the current effort social science perspectives on how contrast agent technology arose and was subsequently shaped were minimal. The author of this book has extensive experience both in academia and as an industrial scientist and manager but lacks formal training in social sciences. He may suffer from professional blinders in much the same way as his preceding nonprofessional writers of history. Fortunately, intense interactions with social scientists, although on themes different from those dealt with here, allowed interference of knowledge cultures akin to that described essential for technology innovation. This rendered possible a conscious effort to escape from all too trodden paths, and to engage the contrast agent history in plausibility tests for a particular sociological model of technological innovation.

Among the creators of the original analytical model, the principle author, Meinolf Dierkes, was the only one with whom I had direct interactions. The interactions were intentionally kept very limited regarding the subject of this book. I hoped thereby to strengthen through independence the persuasiveness of the plausibility test for the analytical model. The price paid for this was the lost opportunity to profit in full measure of the model creator's vast experience in the field.

Among many others, the British historian Arnold Toynbee asserted that history is only comprehensible to the extent that similarities and connections between temporally and spatially separated situations are uncovered (Dennett 1991). Without such similarities and connections it would picture complete disorder, unworthy of study. As it happens, historical situations are not like boulders, at a given time existing as isolated and rounded entities. Rather they are the fluid response to questions that codetermine the answers. Already the process of gathering raw information resents the hypotheses behind the inquiry. Hypotheses pilot both the nature and selection of what is considered relevant.

Even some type of raw historical records, foremost communications of all sorts, cannot be adduced in favor of any arguments without resenting our interpretation of their content. This content in turn may be seen as composed of what the author chose to express, and what he chose to conceal. The former is more or less immediately intelligible. In contrast, the latter requires reconstruction, taking into account a larger context, inevitably an activity with substantial subjectivity. As will be explained further down, the present project required sometimes reconstruction of so-called deleted modalities (Latour and Woolgar 1986) in scientific communication. Moreover, to distance oneself from the streamlining historiography also meant that dramatically more detailed facts needed consideration at the cost of a requirement for increased communication space. The potential reproval of persnickety history can be countered pointing out the necessity of juridical rigor, when a widely divulged and accepted historical consensus that is muddled by self-serving interests, calls for an extensive revision.

Here a particular sociological perspective on history was adopted, to be specific that of the analytical model. History was macroscopically structured accordingly. Macroscopic structuring occurred through the emphasis on selected periods and through coverage of certain episodes from more than one perspective. The result promised to provide a field plausibility test of the sociological model, combined with occasions for its refinement. It is admitted that the structuring does limit the validity of the case studies as an independent plausibility test for the model. But there is compensating merit in a congruent historical and sociological perspective. At the microscopic level, historical elements were cast as narratives, i.e., within the limits imposed by the macroscopic structure, individual chapters and paragraphs were organized as chronologically ordered, coherent stories, involving specified actors, actor-networks, and specific deeds linked to outcomes. Linguistic and national affiliations, as they existed at pertinent times, found use merely as an ordering principle of minor importance. In this form, the history should remain useful even to those who are not willing to embrace the associated sociological interpretations. With such readers in mind, the two perspectives have generally been put into separate chapters.

Chemical formulas, patents, research articles, pharmaceutical presentations, tables of prices, clinical protocols, and radiological pictures are the traces of activity left behind by networks of actors. These traces speak with different clarity to the various readers, according to their specific knowledge cultures. Some readers readily recognize homologies and analogies in chemical formulas, some can appreciate the radiological image quality *per se* or in the appropriate historical frame, others find it easy to put costs into economic perspective, and yet others allow pictures to speak more than a thousand words. Similar culture-specific messages are frequently found, for example, in physicochemical parameters or clinical indications. The inevitably knowledge culture-dependent intelligibility of certain messages should not discourage the reader, since distinct aspects are chronicled with enough redundancy to guard against shortage of evidence for the proposed sociological analysis.

A rewarding technique was the construction of trees of branching timelines of all documented events that had a chance, as slim as it may at first appear, of being historically relevant, followed by judicious pruning.[10] The resulting diagrams being still quite unwieldy, their lessons were then transformed into narratives. As it turned out, solely by meticulous reconstruction of timelines it became possible to demonstrate that histories of contrast agent technology to date have ignored crucial players and have distorted chronologies, to the detriment of opportunities to detect generalizable patterns of technology genesis and shaping.

The present history is distinguished from earlier efforts through a substantially enlarged collection of primary sources, such as original publications, documented events, dates, and artifacts, comprising appointments, encounters, data in archives of professional, industrial, governmental institutions and other groups or individuals, newspaper articles, theses, patents, laboratory notebooks, affidavits, awards, medals, employment records, and letters. Whenever possible, oral narratives were gathered in open-ended interviews with important protagonists of the events.[11] Any future alternative rendering of the history will at least have to take the new documents into account.

Principle sources of historical information were original publications in professional journals, book chapters, and other documents in all relevant languages. The secondary literature offered a gross overview of the field and partial support for the present endeavor, but given the frequent errors therein, an effort was made to check all cited documents, events and circumstances independently, unless noted otherwise. The consulted scientific literature encompassed in addition to the leading professional journals of the times, many minor and local ones. While articles published in minor and local journals may not have had immediate large impact, their inclusion here as primary sources helps best to capture the rich fabric of activities that nurtured progress. Furthermore, their message was frequently disseminated at congresses, where they found an ample audience. In fact, repeatedly so-called opinion leaders at prestigious academic centers and in technologically leading countries picked up with delays innovative approaches from minor players in less conspicuous environments. After all, even Röntgen's first two papers on X-rays were initially published in a journal of truly secondary stature and reports in newspapers played a dominant role in spreading the message.

The early literature is often very difficult to find, because of incomplete bibliographic citations, incoherent styles of referencing, and errors that have crept in during serial recitation. In numerous cases, exact dates are crucial for correct chronologies. For these reasons, in this publication a fully verified and unusually detailed bibliography has been included. The literature additionally abounds with biographical confusions. Thus, all actors are unequivocally identified through their full particulars, to the extent these could be retrieved. Beyond present scopes, such data serve quantitative social science (Schich et al. 2014). Unless there is a potential confusion of names, only on the first mention a person's full name is given.

The named protagonists of the stories reflect the selection of publications and other documentation. This selection leads to the alignment of persons who made merely a onetime contribution to contrast agent technology with others who invested a lifetime into it. This may appear unjust, but once in a while one publication is all that is needed for having a significant impact. Moreover, I focus on people involved in the emergence of novel products, to the detriment of those, equally meritorious, who developed protocols for the extension of clinical application of existing agents, pioneered interpretation of contrast-enhanced images and clarified mechanisms of agent-related adverse events, to list just some of the many other important activities. If the specific goal of the book is kept in mind, the conspicuous absence of some famous names associated with these other activities may hopefully be excused.

For selected protagonists of the two historical periods that are examined in depth, a short, mostly professional-style *curriculum vitae* and a portrait are provided. The former helps gauge diversity of knowledge cultures and circumstances of employment among important protagonists. The latter reflects my conviction that they facilitate the reader's engagement with the subject matter of this book by offering emotional access and memory aids involving also the visual sense. Biographies and portraits are not intended as tributes to the individuals, without thereby wanting to deny anyone merited recognition by society. They should not be seen as relapse into "great man historiography."

Institutions, both public and private, play an eminently important role in technology genesis and shaping. They are part of the knowledge culture of human protagonists, they condition human activity, and they act as protagonists themselves. For this reason, human protagonists are introduced together with their institutional environment and their function therein. Institutions are called by their names at the historical moments of interest and, where instructive, their evolution is dealt with.

Decisions by companies are the result of complex processes involving many participants. For this reason, when information was sufficient and available, individual participants in industrial decision-making were identified. Otherwise, the entire company was treated as an actor on the same par. Such treatment is justified by the fact that the process by which a given company decision is taken is not always revealed, democratic, authoritarian, or otherwise.

While industry's role in technology genesis and shaping is seriously addressed, it has to be accepted that the retrieved documentation reached a satisfactory consistency only for the research and development area. To a more limited measure entrepreneurial argumentation could be identified. For lack of sufficient data, marketing ones concerning individual products and company strategies regarding mixes have been almost completely neglected. Data in this area are difficult to obtain and to interpret. It is simply admitted that marketing considerations and decisions beyond question have played a role in the genesis and shaping of contrast agent technology, chiefly by accelerating, decelerating, or abandoning altogether certain developments. The same may hold true for the political and economic environment. Neglect of these aspects remains a regrettable deficiency of the study. In the case of one producing company in the field, a splendid and unusually exhaustive historical inquiry, comprising marketing strategy, has been published (Amdam and Sogner 1994), but an industry-wide equivalent has yet to be made. It is hoped that future investigations in this and other directions would integrate the analysis, without invalidating its essential parts.

TECHNOLOGICAL INNOVATION

Technology should be thought of as the union of marketable technical artifacts with prescriptions for use that satisfies specific practical needs and wishes. Based partially on the understanding of the economist Joseph A. Schumpeter (Schumpeter [1912] 1997, 1939), technological innovation shall be specified as the affirmation in practice of a novel technology, while substituting old ones in a process called "creative destruction." It is equivalently specified as the reliable replication of a technological invention or a combination of multiple such inventions on a societally significant scale and at an acceptable cost.[12] Technology genesis is the creation of basic novel technology that can bring fundamental technological innovation. Technology shaping is the process of modifying existing technology in order to improve performance and/or cost, leading to improvement innovation. The distinction between genesis and shaping cannot be rigorously defined. It has something to do with the magnitude of the difference in nature and performance of old and new technology.

At the beginning of a technological innovation, there may well stand a discovery or an invention. The latter may be defined as the demonstration of experimental validity of a technological idea. It is usually the feat of only a limited number of actors from a small group of knowledge cultures. But its usefulness often requires integration with other inventions (Graham and Senge 1980; Graham 1982), which adds a few actors. Röntgen's role in the emergence of X-ray technology offers a shining example. The discoverer was not alone suspecting that gaseous electrical discharge tubes produced radiation of yet unidentified nature, but only he made an ample set of steps toward a technological innovation. It comprised specification of the apparatus for the generation of X-rays and their isolation, identification of means for their detection, in real time on a fluorescent screen and permanently captured on photographic plates, determination of their power to penetrate matter, as well as their application in the visualization of usually invisible objects, the human skeleton in particular. No matter how close to the feat others may have come and how essential their work may have been in preparing the ground for Röntgen's breakthroughs, nobody had foreseen the outcome of his endeavors. Röntgen made a discovery in addition to an invention, although without asking for a patent. Yet he still made only the first steps in technology genesis. Transformation of a laboratory experiment into X-ray technology with its combination of tubes, electrical generators, films, intensifying screens, fluoroscopes, and semeiotics required a collective effort for more than a decade by a vast number of individuals and organizations.

Genesis and shaping of technology involves conspicuous actors who in their thinking and communication transcend boundaries of professional knowledge cultures. At the same time, it relies heavily on numerous specialists who act mostly within those boundaries. Besides persons from strictly technological and scientific disciplines, many from outside those realms, such as entrepreneurs, marketers, sales persons, purchasers, regulators, insurers, consumers, etc., may be essential. The lone actor is never a suitable key to understanding the ways in which new technology arises. In spite of this, it is fully justified to pay special tribute to Röntgen and similar figures, not just for their scientific achievement but also for their instigation of technological innovation.

New technologies typically grow on a soil characterized by sharing of knowledge among numerous and diversified actors and imbibed by a common vision. At least in the scientific community, there is a widespread recognition that controlled atomic fusion could solve the world's energy problem. A solution to that problem is obviously desirable. Nobody to date has attained such fusion in the laboratory resulting in more energy than was needed to get the fusion going, but the billions of dollars spent over the years in pursuit of this goal testify to the widespread belief that it ought to be feasible. A multitude of discoveries and inventions, together with other creative contributions from people in diverse roles will be necessary to transform experiments into a technology. The American physicist Richard Feynman in 1982 conjectured that quantum mechanical computers would deal with quantum mechanical problems more efficiently than normal computers (Feynman 1982). When many researchers endorsed his idea, and let their activities be oriented by it, an idea had advanced into a vision. To deal with quantum mechanical calculations more efficiently was a longstanding desire. Feynman's track record of scientific accomplishments (authority), together with existing experience, provided the basis for extrapolating feasibility. There is now a great deal of fervor in the field, even though it remains unclear whether such a computer will ever be practical. Almost everything has still to be discovered, invented, and developed—hordes of investigators still have the chance of contributing uniquely enabling steps.

A similar situation furnishes the reasons why it makes no sense to talk of the inventor of the atomic bomb or the discoverer of moon rocks. Experts long considered it possible to make these items available before they actually became so. In connection with figures like J. Robert Oppenheimer, the technical leader of the Manhattan Project and Wernher von Braun, the director of the Marshall Space Flight Center that developed the rocket for the first manned moon mission, the term "father" of a technology or innovation is sometimes used. This choice of words captures quite well the seminal, motivating, coordinating, and caring role these individuals played in projects that, by necessity, involved many contributors over an extended time period. Still more significant is the accepted use of the curious plural "fathers" of a technology and innovation, which implicitly admits the pluralistic origin of all technology and innovation[13] (Figure 2.1). The role of fathers of technology can readily be demarcated from that of the many actors participating in the events who brought to bear merely their proper professional knowledge culture.

THE ANALYTICAL MODEL OF TECHNOLOGY GENESIS AND SHAPING

If the history of such a narrow field as that of uro-angiographic contrast agent technology is not to remain another "Just-so story," at least it must capture in exemplary form a widespread pattern of human activity and thereby contribute to the understanding of the emergence of new technology in general. Better, it would offer a lesson for future developments in similar areas. This expectation presupposes the attainability of such generalizable understanding and its robustness across differing societal and historical conditions. At present, no consolidated understanding exists of the way in which new technology comes about and continues to be shaped. Attention of management practitioners and consultants is usually directed at organizational

"But you *both* can't be the 'Father of Ammonium Pentoxide Phosphate'."

FIGURE 2.1 Cartoon by Sidney Harris, New Haven, CT, USA. (Reprinted with permission from the author.)

aspects, especially those pertaining to the late phases of development of a technology or of products. These late phases are dealt with by management science, typically in the form of system engineering, project management, and marketing. Earlier phases are still in need of a convincing conceptual framework. Analysis of the role played by the combination of a limited number of universal precepts, by the nature of preceding technical platforms and patterns of evolution thereof, as well as by the fertilization of one field of innovation by answers successful in another one, has yielded promising means for accelerating the identification of possible technical solutions (Altshuller 1988; Fey and Rivin 1997) but have nothing to say about the human and societal side intrinsic to technology genesis and shaping. A great deal of literature delights in praising the solitary hero who attained a breakthrough against obstacles in an uncomprehending or worse, a hostile environment. While such situations may have existed, singled out in this way, their descriptions cater more to the human predilection for heroizing historiography than to a deeper understanding of the processes preceding the general availability of a new technology. Isaac Newton's famous pronouncement about having seen further by standing on the shoulders of giants deserves complementation by one about giants having seen further by standing on the human pyramid formed by a host of less conspicuous forerunners and contributors.

By now, it is widely admitted that the subject of technological innovation requires focus on its social dimension (Winner 1986; Joerges 1989; Pool 1990; Jasanoff et al. 1995; Dierkes, Hoffmann, and Marz 1996; Spann, Adams, and Sounder 1995; Arns 1997; Sturken, Thomas, and Ball-Rokeach 2004). The requirement would continue to hold if the subject could be understood solely in terms of responses to needs of society. It becomes still more imperative, realizing that there exist no teleological necessities for current technology being just the way it is—alternative technologies could have satisfied societal needs just to the same extent.[14]

A given social configuration may favor one solution over another. This may be the case as early as at the level of conception and invention or as late as at the stage of adoption of the technology by society at large. A yet unresolved case may illustrate this. Presently X-ray imaging, magnetic resonance imaging, nuclear imaging and ultrasonography are heavily competing in some applications where all four offer valid approaches and a promising future. It is naive to believe that in each application the better solution will win, or even that the word better in such a context has real meaning. The outcome will be influenced by many additional factors, social ones in particular, e.g., the strength of the aversion of patients and insurance companies to the damaging aspects of ionizing radiation, the intensity with which medical care insurers are focusing on cost containment measures, the strategic directions, and investment decisions of the major producers of equipment, the size of the capital invested in installed facilities and the cost of their up-dating, the relative contribution of personnel and equipment to the cost of a diagnostic procedure, profit opportunities for physicians and healthcare providers, and not to underestimate, the research interests of medical opinion leaders.

Within the scope of understanding *technology genesis and shaping* as a social process, a number of approaches have been formulated, including the Social Construction of Technology Approach (Bijker 1987, 1995), the Techno-Economic

Network Approach (Callon 1992), and the Socio-Technical Network Approach (Elzen, Enserink, and Smit 1996). A deeply thought-provoking attempt to capture conceptually the process components of technology genesis and shaping has been put forth by Dierkes, Hoffmann, and Marz (1996) in the form of the original analytical model. The present study takes a slightly modified version of the model as a basis for selecting and ordering historical records in the specific area of uro-angiographic contrast agents. The ease with which history seems to comply with the model and the frequency with which exceptions have to be postulated or modifications have to be proposed are taken as an indication of its goodness. This way of perceiving the relationship between history and sociological model is found in the works of Dierkes, Hoffmann, and Marz (1996) and Bijker (1987, 1995). The model is under primary scrutiny, history providing the means of scrutinizing.

Another way of seeing the relationship would put history into the center and would use the model just as a chosen lens through which to explore history. Competing models could be called upon case by case according to their explanatory power. Deeper understanding of the history would be the primary aim. While accepting the legitimacy of both undertakings, I have opted in this publication for the former approach to the relationship. The separation between historical and sociological analyses into separate chapters is intended to help the reader who would like to follow the second approach on the same database.

Given the centrality of the analytical model of Dierkes, Hoffmann, and Marz (1996) with the presently proposed modifications, it deserves the following short, paraphrased description. To help understanding by readers not trained in social sciences, the major concepts are illustrated by examples. Terms with model-dependent meaning are written in italic type when first occurring in association with their explanation and exemplification. In anticipation of some weaknesses found in the model when applied to the specific case of contrast agent technology, certain well-circumscribed aspects have been modified and the modifications are identified as such. Moreover, especially in its application to contrast agent technology, I found myself unable to deal with all ramifications and components of the model's concept map and hence focused on those I considered most illuminating. Thereby I failed to do full justice to the work of the model's authors but believe not to have watered it down excessively or misrepresented it.

As already alluded to, the analytical model's most basic tenet is that the genesis and shaping of new technology happens as a result of the *interference* of different *knowledge cultures*, more often than not empirical knowledge cultures.[15] I would like knowledge cultures to be understood here as structures in the sense of Sewell (1992), i.e., as constituted by mutually sustaining cultural schemas and sets of resources that empower and constrain social action and tend to reproduce by that action. The triplet formed by cognitive knowledge, know-how (and skill), and "know how to behave," which is accumulated in the domain defined by a conventional designation for a knowledge culture, such as physics, pharmacology, radiology, regulatory affairs, forensics, or entrepreneurship, may serve as illustration of what is a resource linked to knowledge cultures as structure (de Haën 2009). Participants of a knowledge culture form a thought community in the sense of Fleck (1981; 1983), and they share emotional and/or subconscious elements of a worldview. Shared tacit knowledge (Polanyi 1966) and technical jargon are particular aspects of cultural schemas cementing actors of a knowledge culture together. The career ladder, professional societies, and other categorical associations are further cultural schemas linked to knowledge cultures as structures. The role of the latter in establishing and maintaining power is prominent. Females and males may in certain circumstances belong to different knowledge cultures despite sharing a label.[16]

Knowledge cultures may give rise to subcultures. Regarding the biology of the proteins (histones) that occur together with nucleic acids in chromosomes, for a long time two highly competitive knowledge subcultures within biology as mother culture coexisted, corresponding to schools of thought (Turner 2001). The two were in disagreement about whether histones participated actively in the regulation of gene expression, and this served to create and maintain separate power bases, with their manifestations in independent scientific journals, associations, and congresses. Belonging to one or the other knowledge culture meant, among other things, being at home in the corresponding institutional environment. Indeed, both the knowledge the individual has of its institutional environment and the institution's knowledge of its constituency are part of a knowledge culture. Similar parallel cultures exist even in mathematics (Gowers 2000). Any size area of endeavor may form structures in the sense intended here.

Knowledge cultures are not restricted to those expected to be found in engineers, natural scientists, and physicians of all varieties but comprise all possible others, mostly empirical categories, such as those of craftsmen, politicians, entrepreneurs, lawyers, military men, sportsmen, artists, photographers, journalists, etc. Lastly, institutions not only reflect knowledge cultures of their members, they possess knowledge cultures on their own. These are of an organizational nature and encompass such scopes as, traditions, facilities, territories, economic resources, empowerments, and means of communication.

In describing the differences among knowledge cultures, Dierkes, Hoffmann, and Marz (1996) concentrate initially on differences in the mode of regular knowledge acquisition that encompasses *production and reproduction of knowledge*. In the corresponding processes people necessarily relate to objects, to other humans (*actors*) and to themselves. In doing so they place their action in a three-dimensional framework whose reference dimensions are the objects, the actors, and the self. Knowledge cultures may be distinct in their differential emphasis of these otherwise mutually integrating and essential dimensions. They may be characterized by the coordinates of *object-reference*, *actor-reference*, and *self-reference*.

These abstract concepts may be easier to understand if they are exemplified. In his production and reproduction of knowledge, the natural scientist tends to pride himself in the emphasis on objective data collection (object-reference), often wanting to forget the importance, even for him, of networking among peers, cultivating the interests of his sponsor (actor-reference) and fostering one's image (self-reference), not to speak of the obvious self-reference in personal curiosity. In philosophical terms, his object-reference may be associated with science's epistemic values of consistency, coherence, truth, conformance, predictive power, and fertility, whereas actor- and self-reference may be associated with science's non-epistemic values, embodied in belief systems, and desire for status and rewards. Other knowledge cultures (e.g., entrepreneurship, management, marketing) involved in technological innovation cultivate knowledge production and reproduction with different object references, such as return on investment, risks, profitability, sustainability, growth potential, legality, transparency, etc. They also differ in the weight they attribute to object-reference relative to the weights given the other reference dimensions.

To the surgeon's profession a strong theatrical element has been attributed (Rosen 1997; Porter 2001), and the term "surgical theater" for the main place of his activities captures that perception. As the surgeon and pioneer of heart catheterization with contrast agents, Werner Forssmann, noted, young surgeons are prone to boasting[17] (Volkmann 1955; Forssmann 1972), a form of self-enacting. Accordingly, the knowledge culture of the surgeon can be described as closer to that of the performing artist than to that of the natural scientist or engineer in the emphasis on showmanship (mix of actor- and self-reference), a phenomenon perhaps linked to the similitude in short-term performance stress during a stage performance or an operation. Yet for self-protection the surgeon needs to experience the human body on the operating table primarily as an object (object-reference) and reserve his capability to relate to other people (actor-reference) for surgical team building, visits to pre- and postoperative patients and interactions with patient relatives.

A dominant actor-reference may be encountered for example in the knowledge culture of social workers and actors. But even among scientists and engineers some belong to knowledge cultures with a strong actor-reference. Think of professionals who dedicate themselves to the various activities of professional associations, including the publication of journals, the organization of congresses, and the assignment of awards. Sales engineers too tend to fall into this category. A dramatic illustration of pronounced combination of actor- with self-reference is found when the stage actor Sir Laurence Olivier answered to the question of his colleague Dustin Hoffman of why we do what we do with "Look at me, look at me, look at me, ...!" (Lipton 2006).

No doubt a healthy measure of self-reference is a necessity for highly creative activities, leadership, and certain kinds of performances (sports, musical instrument playing). Knowledge cultures excessively dominated by self-reference may be found among misunderstood artists, scientists and inventors, business and financial tycoons like Howard R. Hughes Jr., military leaders like George S. Patton, religious zealots, politicians, and sportsmen. Although the activities of such individuals require elevated skills in dealing with certain objects (object-reference) or with a public (actor-reference), crucial evaluations and decisions can be excessively influenced by self-reference.

The reference dimensions of Dierkes, Hoffmann, and Marz (1996) for the characterization of the modes of production and reproduction of knowledge in different knowledge cultures did not seem to capture the full richness of diversity encountered in this study. Therefore, a number of additional distinctions among modes are introduced here, without claiming thereby to have exhausted the possibilities. A first useful additional differentiation of modes involves preferential use of some of the five classical physiological senses. Given their multiplicity the preferences of a person or a collective may best be captured by a normalized radar chart. When dealing with some diseases the early radiologist with his characteristically visual orientation had to contend with the knowledge culture of the specialist in internal medicine who traditionally reconstructed clinical reality mostly from anamnesis and physical examination, i.e., inspection (vision), manipulation, palpation (touch), auscultation (hearing) and percussion (coordinated touch and hearing) (Lachmund 1997), as well as body temperature sensing (thermoception). The clash in paradigms has been exemplified perceptively in the case of lung tuberculosis with its disputed site of initiation and course of progression (Pasveer 1993; Lerner 1992). Regarding the sense of seeing and visualization, it is worth noting that before the advent of X-ray diagnosis, physicians paid scarce attention to visualization and corresponding communication of their findings. Apparently, it was the advent of X-ray diagnosis that triggered some of them to introduce marked drawings of human torsos into patient medical records (Howell 1986, 1989; Grigg 1965). Eventually this led to the illustrated Röntgen report as a separate document, prepared by the radiologist (Grigg 1965). Pertaining to the sense of hearing, it has been argued that the noise of machines to the adept is a form of music, and as such a source of inspiration of technology (Pacey 1999). The preferential use of senses manifests itself also through preferential choices of communication media such as the voice, the written word, the gesture, body contact, the image, dressing, or music. Even the choice between presenting an argument while writing in real time on a blackboard, flipchart, or overhead projector transparency, a modality today still common among physicists (Ball 2017), versus projecting ready images elaborated using sophisticated software, the dominant manner of physicians, may reflect a difference in the preferred use of senses. Although certainly exceptions, some scientists feel attracted to communicate their PhD work in dance format on occasion of the yearly contest organized by the American Association for the Advancement of Science, a highly respected scientific organization (Bohannon 2010).

In this context, it seems worthwhile to be reminded of the important role his preferred physiological senses in general, and sight above all, have played in the research of the experimentalist, Röntgen. His discovery of the photoacoustic effect in gases[18] (Röntgen 1881) relied completely on acute hearing, and that of X-rays came about by astute physical perceptions during playful manipulation of devices in the context of well-trodden ground. He abhorred elaborate hypotheses and theories,[19] although being fully aware of using both of them in judicious doses in his experimentation. He trusted his senses and cultivated them. On vacation in Pontresina in the Engadin valley in the Swiss Alps, he was noted for his habit of spending much time under the black cloth of a photographic camera. What he sensed he valued as facts. Given his typical visual mode of knowledge acquisition, he was convinced that a set of X-ray images would be the most direct means for quickly convincing the world of his discovery. He therefore personally spent during the Christmas vacation of 1895 an inordinate amount of time in preparing positive prints for his mailing of the news of his discovery to colleagues (see Chapter 3). With his attitude, he personifies the acme of a development that in the 18th and 19th centuries saw a gradual gain in epistemological status of sensuous experience in science, especially in physics (Sibum 2004).

Different knowledge cultures are further distinctive in whether their mode of production and reproduction of knowledge involves preferentially the following data collection means alone or in combination: observation, description, measurement, collection, excavation, dissection, contemplation, conceptualization, construction, synthesis, creation, etc. Related to these modes are knowledge culture specific preferences for means of data analysis, e.g., statistics, pattern recognition, comparison of objects among themselves or of artifacts with aesthetic norms, analysis in terms of threats and opportunities, etc. Knowledge cultures have traditions in these respects and members join such cultures because they like the type of activity, apart from the interest in the subject matter *per se*.

Some means of data collection that are characteristic for the production and reproduction modes of a specified knowledge culture may usefully be assigned a position on a continuum defined by a pair of pole modalities, the continuum defining a kind of axis of differentiation. Multiple such axes of differentiation may be constructed and used simultaneously. Their usefulness as descriptive tools is increased the more they capture mutually independent dimensions of the object of interest; in mathematical terminology, if they are orthogonal dimensions.

Observational-analytical and manipulative-constructive modes in knowledge acquisition are exemplary poles defining an axis of differentiation. Their relative importance differs in various knowledge cultures. This can hold even for similar subject areas, as illustrated by the various approaches to the study of animal behavior. They span the whole spectrum of modes, from those by the almost purely observational naturalist, Charles Darwin, the experimental naturalist, Niko Tinbergen, and the laboratory experimental psychologist, Frederic Skinner.

Partially aligned with the axis of differentiation just dealt with is that which on the one pole has the method of the abstractionist–theoretician–mathematician, and on the other, that of the tinkerer–gadgeteer–craftsman–experimentalist–phenomenologist (Jacob 1977; Lightman 1996; Galison 1997), the one working with sealing wax and strings (Price 1984), in possession of practical know-how and skills. This dimension can be gauged pondering the history of how the problem of determining geographical longitude at sea was solved. The Board of Longitude in early 18th century Britain presided over a parliamentary prize for who would find a practical, useful, and precise solution to this problem. This board was loaded with astronomers and mathematicians for whom an acceptable method needed necessarily to be rooted on the former pole of the axis. This prejudice deriving from their knowledge culture erected major obstacles to finally winning the prize for John Harrison, a master carpenter and clockmaker who actually solved the problem by building a sea chronometer that corrected for ship motion, variation of gravity and temperature, and which withstood the humidity at sea (Sobel 1998).

The New Zealander–British physicist Ernest Rutherford is quoted to have differentiated himself as a man in search of novel principles and phenomena, rather than one characterizing known ones with ever increased precision, saying I don't quite class myself in the same order as Hugh L. Callendar, who was more of an engineering type than a physicist and who took more pride in making a piece of apparatus than discovering a new scientific truth (Heilbron 1999). Instead of Callendar, Rutherford might as well have chosen as his antipode Albert A. Michelson, the untiring measurer of the speed of light and constructor of precision instruments, or the metrologist with early involvement with X-rays, Johannes Pernet from the Polytechnikum (today ETH) in Zurich. If Callendar, Michelson, and Pernet are accommodated in a common knowledge culture, then the same may be done with Rutherford and Albert Einstein. In the case of the interaction between Pernet and his student Einstein, the opposing knowledge cultures led to a clash, when after a laboratory accident Einstein was allegedly told something like: "You're enthusiastic but hopeless at physics. For your own good you should switch to something else, medicine or maybe literature or law" (Fölsing 1993, 57). Recently, the term "discovery science" has been introduced for technology-driven research, as opposed to hypothesis-driven endeavor (Leroy Hoood, quoted in Thomas and Gilbert 2002). If these latest polarizations do not merit proper axes of differentiation, they may be seen as variation of the one illustrated in the last paragraph.

Yet another axis of differentiation has the drive for primacy, gratification in the short term and overt public recognition, on one pole, and on the other the protracted search for lasting truth and beauty at the risk of not being esteemed for it in one's own lifetime (Knorr 1977). Among people in science, medicine, and technology, the heart surgeon Christiaan N. Barnard (Hoffenberg 2001) and Thomas A. Edison may serve as exponents with the former emphasis, while again Darwin may serve as an exponent with the opposite inclination.

Knowledge production and reproduction is further differentiated along the axis that spans the imperatives of practicality and economic reasonableness on the one hand and absolute, unlimited pursuit on the other. Typically, politicians, lawyers, engineers, and practicing clinicians operate more in modes that satisfy the former imperative,[20] whereas ideologues, natural scientists, and medical investigators tend to respond more to the latter. On the various axes of differentiation, the physicist and trained mechanical engineer, Röntgen, fell definitively closer to the second-mentioned poles than to the first-mentioned ones.

Next, in the analytical model differences in knowledge cultures with respect to modes of *knowledge documentation and conveyance* are considered.[21] Knowledge is documented and conveyed through sets of signs that represent actions. In these signs reside the functions *symptom* (expression), *symbol* (description, representation), and *signal* (catalyst, appeal, exhortation) in an interwoven fashion but with knowledge culture specific proportions. The industrial knowledge culture in its relationship with the outside world typically uses signs that emphasize signal over expression and symbol. This behavior is evident in package design, product brochures, advertisement for products, slogans used to shape the image of a company or a whole industry, and the flashy annual reports for investors. In contrast, knowledge cultures from the natural sciences and engineering typically use signs that emphasize symbols (chemical formulas, mathematical expressions, graphical representations, tables of values) to the detriment of symptoms and signals. Yet, even for natural scientists and engineers, presentation skills (symptom, expression) are of importance. Only when scientific and technical capabilities are accompanied by sufficient presentation skills can a successful career as an independent scientist or engineer be envisioned. In the humanities expression (symptom) and appeal (signal) take a greater share of importance than in natural sciences and engineering. The importance of expression for science in general may be gauged by the acclaim received by scientists who excel in the art of presentation, such as the historians Fernand Braudel and Golo Mann, the paleontologist Steven J. Gould or the geneticist Richard Dawkins. The performing artists' and criminal defense lawyers' signs are typically dominated by the symptom-expression and signal-appeal functions.

Two knowledge cultures that are similar in terms of production and reproduction as well as modes of documentation and conveyance may nevertheless give rise to interference phenomena if they differ in the subject area. Chemists in pursuit of innovative drugs or insecticides may serve as example. Conversely, two knowledge cultures may cover the same subject area but differ strongly in the modes of production and reproduction as well as modes of documentation and conveyance of knowledge. For instance, theoretical hydrodynamics has been approached by two separate, contemporaneous communities: the classical one using the mathematical analysis tools of theoretical physics and the other those of computational physics. The two knowledge cultures remain sharply divided, because they differ strongly in their valuation of what constitutes an appropriate answer to a common question. The ones strive for elegant mathematical expressions (aesthetics) although not rigorously applicable to the real world, the others strive for successful prediction of real situations regardless of how cumbersome a process it requires (utility). Feynman (1965) has discussed the same distinction in a comparison between modern western (Einsteinian) science and that of the Mayas[22].

Furthermore, there exist differences in knowledge culture between academic institutions and industry, related to the dominant format of documenting, and conveying scientific and technical progress. The former cultivates predominantly publications in professional journals and books, while the latter relies on internal reports, standard operation procedures, patents, marketing authorizations from governmental institutions, quality attestation and demonstration models, and in the end, has sales of products to show.

Although, in legal terms, patents count as publications on a par with scientific articles, it is well known that the two are constructed with completely different rationales and follow different conventions. These rationales and conventions manifest themselves from gross rhetoric strategies down to the finest details of every written word (Myers 1995). Knowledge documentation and conveyance in scientific articles and patents deserves some differentiation. In a scientific communication, typically taking the form of a research article published by a professional journal, a chronological account of a meandering endeavor, that would reflect reality, is replaced by a reconstruction in terms of a rational and purposeful undertaking. Rigorously formulated rationales for undertaking the research effort are constructed with the hindsight of knowing the results, substituting for the real ones that often have been vague and triggered by chance observations. The resulting central thread of the report almost never follows the true chronology (Medawar 1963; Simmons 2001). In recording the process from initial observation through validation by replication of the experiment and variation thereof to final communication, laboratory scientists progressively strip away contextual information they judge impertinent, a process that has been termed "deletion of modalities" (Latour and Woolgar 1986). Besides experimental details, this deletion can involve interactions among investigators, funding sources, employers, financial interests, location, and times. If any of these are ethically problematic, the practice is especially tempting. Following initial publication "deletion of modalities" often expands. Only historical review articles occasionally offer some corrective accounts. "Deletion of modalities" is principally dictated by the necessity to minimize publication space and costs, at the same time guarding the reader against information overload. It is amply legitimized by the successes of the natural sciences and their essential communication and memory vehicle, the refereed journal article. Regrettably it opens the door to distortions of varying gravity. Instrumental "deletion of modalities" for selfish purposes, such as neglect to acknowledge collaborators and their contribution, is condemnable but by no means uncommon. While in the case of a publication the exemplified deletion almost never has serious legal consequences, in the case of patents certain kinds of "deletion of modalities" can easily lead to invalidation.

Another characteristic of the scientific publication is its proclivity to emphasize the understanding acquired, and to show how it relates to previous understanding. Preferably acquisition of understanding is perceived as incremental and as occurring in a sequence of actions that, even if revolutionary, were still logically anchored in much preceding understanding. This manner of presentation favors consensus building.

On the contrary, well-written patents emphasize how little previous knowledge had been helpful in making the invention, how surprising and "nonobvious to the person skilled in the art" the accomplishment was in the light of knowledge and practice available at the time of the invention. Being able to show that the invention defies deduction from antecedent understanding and practice is meritorious. When the reasons for why the invention works could have been known, or has been figured out *post hoc*, it is tactically unwise to touch on that in a patent application. This corresponds to a patent writing-specific form of "deletion of modalities." Yet, in order for an application to survive the scrutiny by the intellectual property authorities and courts, the writer has to walk the thin line that separates admissible "deletion of modalities" from obligation to disclose all relevant "prior art." To the detriment of science and technology the differences between the illustrated communication modes serve as a pretext for most academics to ignore the patent literature, while industrial scientists cannot ignore scientific journals and books.[11, 23]

Returning to the issue of knowledge culture characteristics, differences in those of women and man are conceivable (Jaggar and Bordo 1989; Harrison 1993). The differences may extend to all aspects of such cultures.

Now that it has been explained what is meant here by knowledge cultures and their differences, the problem of interferences of such cultures can be addressed. The phenomenon of interference of knowledge cultures exists both at an outer and at an inner level. The outer level is located external to the actors and involves *communication* processes. The inner level is located essentially within the individual and involves *individuation* processes comprising patterns of cognition and decision-making (ways of thinking and behaving, strategies of cooperation).

In this context, communication is different from the one-way process analyzed by information transmission theorists. It needs to be understood in its widest sense as the coordination of the sustained cooperation through exchange of signs (Krüger 1990) within networks of human actors (Manning 1989). Following the terminology of Dierkes, Hoffmann, and Marz (1996) the communication processes of present relevance require (a) that as a consequence of the exchange of signs, agreed upon actions are taken, i.e., there is *cooperation*, (b) that this cooperation is sustained and repeatable, i.e., there is *reproduction of cooperation* and (c) that the sustained cooperation from different communication channels is coordinated, i.e., there is *coordination of reproduction of cooperation*. Coordination shall be understood as a voluntary commitment on the part of the involved actors, as opposed to the involuntary *synchronization* among situations and events yet to be introduced.

At the inner, or individuation level, an analogous triad of processes mirroring what happens in the outer world may be postulated. (a) For the actor in a communication network to be able to achieve cooperation he must be able to slip to some extent and in a convincing manner into the knowledge culture of the partner; he has to adapt his frame of mind (Goffman 1974). This process is called *internalization* of the foreign knowledge culture. (b) Sustained internalization requires effort and perseverance, characteristics that are essential to guarantee stability of the communication network. This is called *reproduction of internalization*. (c) Lastly, if more than one communication line to partners, each from different knowledge cultures, is to be maintained active, the internalization processes need to be integrated or coordinated, and we may speak of *coordination of reproduction of internalization*.

If processes of interference of knowledge cultures, both at the communication and the individuation level, are not to present themselves at random and without consequence but are to materialize as new technology, various such processes need to occur in synchrony. Something must favor simultaneity of several events or conditions, temporal dispersion of which would constitute a hindrance to the genesis and shaping of a new technology. *Synchronization* will here be understood as the extrapersonal process that creates coincidences of events and conditions in different places of the network of actors and actions. The process is distinct from the voluntary coordination efforts of individuals, although of course synchronization may be a facilitator of coordination and vice versa. A stronger delineation between synchronization and coordination than that used in Dierkes, Hoffmann, and Marz (1996) promises to add clarity.

Dierkes, Hoffmann, and Marz (1996) speak of the need for threefold synchronization, namely that of communication processes by themselves, that of individuation processes by themselves and that between communication and individuation processes. These authors also examine the specifications of that something, which could exercise the synchronization. They conclude that the *vision* as defined above satisfies the specifications. They consider a vision for a technology to be a first crucial result of interference of knowledge cultures. Such a vision is able to capture the imagination of a critical mass of actors. It expresses a belief in the existence of a technical solution to a perceived need and offers incentives (prestige, career, wealth, fame, power and self-fulfillment) for those who make a significant contribution to find it. Under this vision the actors are spurred into focused action. Both actors and actions are the sources of continued interference phenomena of knowledge cultures and become thereby responsible for the shaping of the technology.

Originally, Dierkes, Hoffmann, and Marz (1996) distinguished in a vision a *guidance function* and an *image function* with each one divided into subfunctions. In the cases to be explored, only in a few instances did this conceptual dissection of a vision help understanding. Apparently, the authors too saw a need to revise their theory. A first revised theory distinguished three functions in a vision, a coordination function, an orientation function, and a motivation function. A second revision distinguished solely an *orientation function* and a *motivation function* (Kahlenborn et al. 1995). Here this latest revision of the theory was adopted because it did offer insight into general dynamics when applied to specific circumstances. Through the orientation and motivation functions of a vision, perception, thought and decision processes are influenced. Visions achieve *orientation* by providing a vanishing point on the temporal horizon where the projection into the future of the desirable fuses with that of the feasible (Figure 1.1). By staking out common ground on the interpretation of the present and the future and by emphasizing those aspects that transcend the differing worldviews of knowledge cultures, they supply an orienting framework of references. The shared framework of references releases tensions, reduces the surface of interpersonal friction, and helps avoid conflicts in processes of communication. In the end, orientation manifests itself as goal-directed joint action.

Visions achieve *motivation* because they speak simultaneously to people's brain and heart. To a large degree through the desirability component, they touch on deep-seated norms and values. In addition, they are conditioned by various emotions, such as sympathy, fear, prejudice, and greed. From these unconscious roots, visions derive their pronounced power to elicit engagement and motivate action.

Through the combined effects of the orientation and motivation functions, visions achieve *synchronization* of events and actions involving actors that may not even know of each other. In the same vision the entrepreneur may see an opportunity for business where the scientist sees one to reach fame. There is interpretative flexibility[24] (Pinch and Bijker 1984). Yet under a shared vision, diverse actors coming from different directions move toward a common vanishing point as if guided by an invisible hand.

Even within an individual a vision can exert its various functions. People jump over their own shadow to get in conformity with the vision. The scientist may discover the entrepreneur in himself. The politician betrays his traditional stand to join a vision. A vision becomes a mission; Saul becomes Paul (Bible: Acts 9:1–19 and 13:9). Besides, the "vision within" may prepare for the moment of discovery during sagacious observation of accidental situations or in the course of playful exploration of phenomena. In Louis Pasteur's translated words: ... in the fields of observation, chance favors only the prepared minds (Pasteur 1854). When a discovery appears unrelated to an original quest the process may be called serendipity.[25]

In the analytical model, little attention has been given to how a vision for a technology emerges, except for conceiving it already the result of interference phenomena between knowledge cultures. How a vision can simultaneously be the result of interferences between different knowledge cultures, and synchronize interference phenomena between such cultures, merits further thought. Here one way in which it could happen is proposed.

A useful starting point for the discussion is the considerations that the Austrian ethologist Konrad Lorenz made about the existence of homologies and analogies in the evolution of technologies (Lorenz 1974). In order to understand his arguments a short excursion to the concepts of homology and analogy as they are used by him and generally in evolutionary biology is indicated. Homologous evolutionary solutions (anatomical structures, physiological mechanisms, behavioral patterns) to challenges posed by the environment to different organismic species show similarities because they evolved through divergence from a unique precursor version in a common ancestor species. Distinctive traits (e.g., pectoral fin of fishes, bat wing and human arm, or highly varied courtship and mating behavior in insects) revealing residual similarities among species and with precursor species are interpreted as of homologous origin. Analogous evolutionary solutions instead show similarities because similar environmental pressures favored convergence toward similar, though not identical traits, from completely unrelated departure points (e.g., need to reduce hydrodynamic friction in fishes and aquatic mammals through similar shapes, or need to arrive at spatial vision in squids and vertebrates using two camera-type eyes). The corresponding similarities may be interpreted in terms of similarities of environmental pressures on distinct species and in terms of physical limitations dictating possible solutions, i.e., analogous ones.

Lorenz in his Nobel lecture has identified in technical artifacts the existence of similarities for reasons of both homology and analogy (Lorenz 1974). He underlined how typically an available technical solution is taken as point of departure for finding responses to altered and diversified requirements. In the responses, the common ancestry remains recognizable; they are homologous. Gradual shaping of the technology is the result of proceeding this way. The gradualism is associated with the risk of tunnel vision, a connection illustrated by Lorenz with the case of railroad carriage design.

In contrast, analogies play an important role in saltatory technological change, breaks with traditional solutions, major innovation. Quite disparate situations and objects are assimilated in a creative act that anticipates the commonness, which is expected to typify the resulting technical products. These expectations may go well beyond available evidence. Common rules of inference may be transferred from one system to the other in a great leap of confidence. The commonness may exist mostly in the eye of the beholders, who in our case may constitute the knowledge cultures participating in the technology genesis at stake. Note that the described use of the term analogy almost coincides with the use it has in developmental psychology. I shall use the terms *proceeding by analogy* and *analogical argumentation* in connection with situations in which analogies, as understood by Lorenz, played a role.

Analogies as learning tools are generally admitted. Lorenz went a step further and elevated them to sources of knowledge (Lorenz 1974). The Scottish physicist, James Maxwell, would have agreed with him (de Regt 1996). I believe that they can serve in knowledge production even if they remain tacit or surface at the boundary between the conscious and unconscious state. An example of the former situation, that of fruitful tacit analogies, is offered by the case of Benjamin Franklin's pronounced moral belief system about the balance of pain and pleasure, or the balance of savings and clear revenues, which by analogy prefigured his conservation law for electrical charge (Heilbron 1999). An example of the latter situation is found in the famous case of the conception of the structural formula for benzene by the German chemist, August Kekulé. In this case, a connection between a dreamt snake biting its tale, the alchemist's ouroboros, and the structure of a curiously hydrogen-deficient hydrocarbon molecule occurred in a half-awake state. The cognitive transposition between "snake" and "chain of linked carbon atoms," an elementary step in an analogical argumentation, led to the formulation of an atomic chain closing on itself, the cyclic structure of benzene (Del Re 1996).

The analogical argument is in part irrational, since in the beginning the individuals are generally only dimly aware of the analogies on which they rely. One could ascribe the process of proceeding by analogy to the psychological function of intuition, which Jung (1923) classified as a perceptive and irrational function because it requires experience but is distinct from deductive thought. It is a saltatory process that draws on the unconscious with its vast memory bank, its associative accessing system, its speed, and its ability to process multiple items in parallel (Isenman 1997). It thus appears as though intuition is the psychological function used to achieve interference of knowledge cultures within an individual. The intuition of individuals is of elevated importance in the genesis phase of technology but remains important in the following phases too.

Articulated analogies acquire a semantic dimension, they become metaphors. "The delight in the metaphor lies in the enjoyment of an unexpected meeting, a pairing of two things that have not previously known each other"[26] (Umbral 2001). But there is more to metaphors than esthetics. Metaphorical language as rhetoric device has long been recognized to offer not only one of the most important modalities of communicating about innovative ideas but also to provide a means of generating knowledge (Brown 2003), allowing reflective exploration of unknown territory in known terms (Black 1962; Hesse 1966), lay[ing] the tracks for our trains of association (Draaisma 2001) and making available a negotiation vehicle among knowledge cultures with different interests and beliefs (Ceccarelli 2001). Metaphors as components of visions are often obvious, i.e., the "highway" in the vision of "the information superhighway" or the "green" in that of "green industrial production." They may play an important role in the formation of a vision without necessarily appearing in its final expression.

Novel products, processes, or uses arrived at through invocation of unusual analogies may be surprising to "persons skilled in the art," as the patent law expresses it, i.e., the results are not deducible from experience in an obvious way. If these conditions are met, one of the crucial criteria for a patentable invention is fulfilled. It follows that patents are a typical indicator of successful proceeding by *analogical arguments*.

The invoked analogies typically touch upon objects and situations taken from or illuminated with angulations belonging to different knowledge cultures. For this reason, *proceeding by analogy* is presently proposed to constitute the elementary process of interference of knowledge cultures in the formative phase of a vision. Such elementary processes of interference may concatenate to form the multiple steps in the emergence of a vision, identified earlier (de Haën 1995) and reiterated in amplified form here (vide infra). But concatenation alone is not sufficient. Exclusively when the sequence of steps results in a linked and formulated assembly of concepts capable of triggering the imagination of a critical mass of protagonists into focused action, a vision for a technology has emerged. Once a vision has consolidated it catalyzes further interference of knowledge cultures by its synchronization function, as discussed by Dierkes, Hoffmann, and Marz (1996).

Social environments that appreciate phantasy, such as science fiction and humor, favor the emergence of technological visions. Consider two works of phantasy created before the discovery of X-rays and contrast agents that through their curious premonitions illustrate the mental preparedness for the new technologies and associated full-fledged technological visions.

Three years before the discovery of X-rays the physician, Ludwig Hopf,[27] under the pseudonym Philander, published widely appreciated medical and anthropological fairy tales. In one of them, entitled "Elektra," the country doctor Redlich was faced with a severely sick patient resisting a diagnostic biopsy. "Oh, he sighed and sat down exhausted on a bench, if only there were a way to render people transparent like a jellyfish!"[26] (Philander [1892]). Promptly a sphere glowing in bluish white light appeared to him, out of which stepped Elektra. She promised to fulfil his wish and gave him a can that, when opened, produced a light with which he saw first the innards of a live frog and then the causal pathologic agent inside a patient's abdomen. The news about this event spread like wildfire through the daily papers, and since the content of the can was easy to reproduce, the tool became soon widely available. Being a modest man, Redlich refused a one million prize granted by the state. This fairy tale reads like a premonition of the story of Röntgen and X-rays. Immediately after Röntgen's discovery, this curiosity was noted even by the lay press (The Roentgen Rays 1896).

Figure 2.2 and associated text present an extract from a book by Wilhelm Busch ([1884] 1965), the German forefather of modern comics, dated to 12 years before Röntgen's discovery and 20 years before Hermann Rieder's introduction of bismuth subnitrate-supplemented semolina porridge as gastrointestinal contrast agent. Therein he tells the satirical story of a painter who as child created the drawing shown. Busch was at the time widely appreciated, and the translation made by the Nobel Prize winning physicist, Max Born, proves that this continued to be so for many decades. It is therefore not preposterous to assume that the originator of the oral contrast agent called Rieder's meal (for description vide infra) had seen the cartoon. It would be an over interpretation to assign this fact any causative role. But technological utopias, pipedreams, or prolepses in the belletristic literature have been suggested to help preparing the public terrain for a technology's rapid reception (Dommann 2003).

Science fiction may not only be a phantasy product of an author but can also be the result of unintended over- or misinterpretation of experimental observations. On erroneous ground, discernible only by the specialist, the professor of physics at Tufts University, Amos Emerson Dolbear wrote a year before Röntgen's discovery in the popular magazine *Cosmopolitan*: "So it is actually possible to take a photograph of an object in absolute darkness, with the ether waves set up by working an electrical machine" (Dolbear 1894). Despite distortion of reality such news can condition an unprepared audience for things to come.

FIGURE 2.2 Wilhelm Busch's 1884 gastrointestinal contrast agent premonition in a comic that begins with the creations of a gifted child.

> Through practice grows his human ken.
> Soon he produces total men
> draws with industry and zest
> A stout old gentleman at rest.
> And not his outer looks alone,
> No, his interior is shown.
> Here sits the fellow in his room
> And eats some porridge, let's assume
> He lifts the spoon toward his lips
> You see, how down it runs and drips,
> And in the belly concentrated
> Is visibly accumulated
> Thus, through the pencil of this gifted
> Small infant Nature's veil is lifted.

(Wilhelm Busch, "Klecksel the Painter," translated by Max Born, © Frederick Ungar 1965, used by permission of Bloomsbury Publishing Inc.)

An important aspect of a completed vision is the stabilization of objectives and typology of technical solutions it provides. In fact, in the emergence of new technology frequently one observes an early imprinting process of technical qualities that become almost impossible to rescind (e.g., Canzler 1997). Multifarious early options, through a closure process, make place for a consensual narrow set of similar solutions. A technology can even become frozen in a state that accepts merely small gradual change and resists all fundamental change. This generalization is beautifully illustrated by the prolonged total dominance enjoyed by the mechanical typewriter design of the Underwood Typewriter Company dating from the end of the 19th century, i.e., a bundle of swivel levers, conceived by Franz X[aver] Wagner, connecting key levers to type bars acting on a movable paper carriage that allows the typed words to be seen. This design had emerged from a vast number of valid alternatives, comprising some promoted by an unsuccessful cartel, and it resisted change for at least five decades (Dierkes, Hoffmann, and Marz 1996). When the frozen state was finally overcome in the 1960s, an era of turmoil for users and producers followed, leading to modern word processing, with the QWERTY keyboard as the only remnant of past technology.

Neither a full-fledged vision nor a critical mass of actors needed to bring a technology to a functionally satisfactory result is sufficient to guarantee that the consumer at large espouses the technology. Additional processes of evaluation involving larger populations and their representative institutions usually led to the final determination of their acceptability. Public fund-raising campaigns, such as the "March of Dimes," an effort in favor of vaccination against poliomyelitis, are examples of expressions of the positive societal forces of technology espousal. Skepticism finds its expression in technology resistance movements (Bauer 1995).

The consumer rejected the Wankel engine for cars with its favorable space requirement, weight and environmental impact profile before it could ever be developed to a level of maturity comparable to that of the conventional Otto engine. It so happened in good part because the consumer would have been required to get used to a different driving style and feeling (Knie 1994). In addition, the number of carmakers espousing the Wankel concept never reached a critical number, and even those involved responded to the need for better fuel efficiency after the 1973 oil crisis by diverting investments from risky novel to less so, old technology. By no means just every new technology that satisfies a perceived need, i.e., reduced emissions, will gain acceptance. Only on the proving ground of the market do the ultimate decisions on the acceptability of new technology become consolidated. It is the result of the larger social evaluation that determines the final destiny of a new technology.

Before we address the first of the two phases of contrast agent technology genesis and shaping selected for detailed historical and sociological analysis and apply the discussed analytical tools to it, we need to provide some basic information. This takes the form of a short review of what had happened from the discovery of X-rays to the assembled vision for contrast agents in general and the identification of so-called uro-angiographic ones. Space limitations dictated this extremely complex part of history to be reduced to bare bones. I found it shocking to realize how in the process of reducing the details the historical narrative progressively linearized and naturally approached the very form of historiography here in the critique. In consideration of this, the reader is invited to refrain from viewing the narrative of this part of history through the same sociological lens as those of the two historical phases explored in detail.

3 Assembly of a Technological Vision for X-Ray Contrast Agents

The discovery of X-rays by Wilhelm Conrad Röntgen in late 1895 has been recounted innumerable times (Glasser [1931] 1993). Most spectacularly it encompassed the observation that the new radiation could traverse a hand and produce a shadow image of the bones on a fluorescent screen. Evidently bones attenuated the radiation more than flesh. Röntgen explored the attenuation properties at first simply by interposing between radiation source and screen, casual objects of his environment, such as stacks of playing cards, books, and rubber sheets. Not surprisingly for the avid photographer in him, he also discovered that the shadow images could be captured on a photographic plate enclosed in its cassette, i.e., radiography, as it became known later. He took numerous show-piece radiographs of most diverse objects. Then, he switched to semi-quantitative estimation of opacities of geometrically well-defined pieces made of materials of the kind typically used by physicists in the construction of scientific instruments, and distinct in their mass density and optical refractivity. He tried to explain the results in terms of bulk physical properties, not in terms of chemical composition. Röntgen's research focused heavily on X-ray attenuation properties of materials and little on X-ray production, detection, quantification, and characterization. In print Röntgen refrained from commenting on the application potential of his discovery even after others had done so extensively. Röntgen's approach reflected his knowledge culture as academic experimental physicist and hobby photographer. Within 18 months of his discovery, he published three articles (Röntgen 1895, 1896, 1897) and then left the field to others.

With the mailing date, first of January 1896, Röntgen sent a preprint describing his discovery to numerous colleagues worldwide. In some cases, including that of Franz Exner, physics professor at the University of Vienna, the preprint was accompanied by photographic positives of several of his radiographs. Through Exner and his visiting colleague from Prague, Ernst Lecher, the latter's brother and editor-in-chief of the Viennese newspaper, Die Presse, Zacharias Konrad Lecher, learned about the discovery very early. This allowed him to be among the earliest to publish the news and the first to do so in an in-depth article ([Lecher Z.K.] 1896). Thanks to his past professional experience in animal anatomy and dissection, he was able to recognize and explain competently before anyone else the enormous significance of radiography for medicine. The two-part news article he authored conquered the world press by storm, before Röntgen's first proper publication (Röntgen 1895) submitted to the journal on December 1895, in the first half of January 1896 became available to the general public.

The press reports sufficed to attract to the exploration of X-rays numerous investigators. They came from all sectors of the educated class, representing universities, high schools, private individuals and their associations, nonacademic public institutions, and industry. Most investigators let the quasi obligatory initial hand radiograph be followed by radiographs of diverse objects, without systematic objective. But already within a few weeks after the news on X-rays had broken, radiographs of medical interest were produced. Bone pathologies were detected and imaging of foreign objects in tissues facilitated their surgical removal. Much attention was given to the study of the opacity of materials, and this is also the most relevant aspect for the present endeavor. The choice of the materials, and the type of explanation for the observations offered, reflected the knowledge culture of the investigators. Physicists, like Röntgen, addressed the role of bulk physical properties of materials used in their laboratories irrespective of their chemical composition (glasses, brass, ebony, lead, steel, water, etc.). Chemists examined pure elements in series related by the periodic chart and pure chemical compounds containing these, including salts as solids and in solution. Pharmacists concentrated on therapeutic products and their elemental composition. Military physicians explored wound dressings, disinfecting agents, and plaster casts.

The following story illustrates the role played by professional knowledge cultures in the birth of the concept of medical imaging with the help of contrast agents (Haschek and Lindenthal 1896). This all happened within a couple of weeks after the discovery of X-rays had become known. In Exner's physics laboratory his assistant Eduard Haschek was charged with exploring the X-ray phenomenon. He decided to collaborate with Otto Lindenthal, a medical doctor candidate and auditor at the Pathoanatomical Institute, who had access to anatomical specimens. Lindenthal was familiar with anatomical preparations of macerated and dried extremities whose vasculature was rendered visible by injection of colored masses. Teichmann's red injection mass for the visualization of arteries was a linseed-based, cinnabar [mercury(II) sulfide] and calcium carbonate-containing glazier's putty rendered injectable by dilution with carbon disulfide (Teichmann 1880). Experience with bone suggested to the investigators that the calcium would impart X-ray attenuation. Thus, the collaborators took radiographs of an amputated hand injected intraarterially with Teichmann's red mass. Reasoning by analogy, they introduced the concept of an intentionally administered contrast agent, produced the first angiogram of a cadaver hand, and started the dream of achieving the same in the living human (Figure 3.1).

The next stories sketch how iodine became recognized as the X-ray-absorbing element with outstanding potential as constituent of contrast agents. In early February 1896, Maurice Meslans, associate professor in the School of Pharmacy in Nancy, addressed the opacity of chemicals and pharmaceuticals selected on the basis of their chemical constituents (Meslans 1896). He found iodine, even as part of an organic compound, to contribute strongly to the opacity. In particular, he emphasized the extraordinary opacity of iodoform, i.e., CHI_3. Iodoform was used topically, intramuscularly, and orally in several pharmaceutical formulations against various infections. A preferred formulation for injection into cold tubercular abscesses and tubercular joint cavities, commercially called JODOFORM-GLYCERIN (Guttstadt 1891), consisted of a 9.1%(w/w)[29] fine suspension of iodoform in glycerol (Trendelenburg 1926).

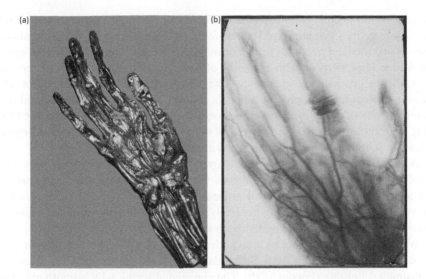

FIGURE 3.1 Manipulation of radiographic contrast—analogy at work in the first use of chemicals to create X-ray contrast—the first angiography *ex vivo*.

(a) Photograph of one of Ludwik Teichmann's original anatomical preparations. Because the arteries have been injected with a red putty, in reality the arteries are dark red on a light brown background formed by other tissue. (Courtesy of the Museum of the Anatomical Institute, Jagiellonian University Medical College, Krakow, Poland.)

(b) Radiograph produced on, or shortly before January 15, 1896, by Haschek and Lindenthal (1896) of a fresh anatomical specimen whose arteries have been injected with Teichmann's red putty and around whose index finger a copper wire has been wrapped to mimic a ring. Shown is a positive print made from the original radiographic glass plate.[28] (Josephinum, Ethics, Collections of History of Medicine, MedUni Vienna.)

Apparently, it was prepared on prescription by pharmacies, Chemische Fabrik auf Actien (vormals E. Schering) in Berlin[30, 31] being a major producer and supplier of iodoform.

Paul Geissler, military medical officer, resident and university lecturer at the Surgical University Clinic of the Friedrich-Wilhelms-University in Berlin described at the 25th Congress of the German Society of Surgery in late May, 1896, sharply contoured X-ray shadows in a joint of a patient with articular tuberculosis. Geissler correctly interpreted the shadow as being due to residual iodoform from earlier intramuscular therapy with JODOFORM-GLYCERIN (Geissler 1896). At the same congressional session, Hermann Kümmell, Assistant Medical Director of the Surgical Division of the General Hospital in Hamburg-Eppendorf, mentioned the X-ray opacity of iodoform impregnated wound gauze (Kümmell 1896). Not too surprising then, in the hands of Kümmell's assistants, chiefly Henry Graff, JODOFORM-GLYCERIN soon became a contrast agent for the radiological tracing of fistulas (Graff 1897/1898). It became the first clinically used iodine-based contrast agent in history.

Already on January 25, 1896, Giuseppe Vicentini, holder of the chair in experimental physics at the University of Padua in Italy and his then assistant, university lecturer Giulio Pacher, contrasted the stomach and the intestines of a dead rat with metallic mercury (Vicentini and Pacher 1896). By March Wolf Becher, general practitioner and medical historian from Berlin, had filled the stomach of a dead guinea pig for the same purpose with a solution of basic lead acetate, a very toxic product. He made the generalizing statement: "The properties of solutions that do not allow passage of Röntgen rays, provide a handle for taking also photographs according to Röntgen's procedure of inner cavernous organs of the animal body" (Becher 1896). This enunciation captured the first half of the vision of contrast agents that began to orient the search for further products and applications (Table 3.1).

Before contrast agents could be employed in man, low-toxicity responses to the problem had to be found. Three avenues were pursued. In one, contrast was generated by X-ray-absorbing surgical probes. In another, toxic contrast agents were enclosed in nontoxic enveloping materials. Eventually, suitable nontoxic chemicals were sought. In the end, the last approach won out.

While radiographing various wound dressing materials, physics professor Aimé Forster at the University of Bern, Switzerland, observed the strong X-ray absorption of bismuth subnitrate (Forster 1896). The elevated absorption by bismuth metal had previously been described (Battelli and Garbasso 1896). But now a new derivative was identified that could produce X-ray contrast and excelled with low toxicity, as documented by prolonged therapeutic experience in gastrointestinal disorders and in wound dressings. In early 1897, Theodor Rumpel, assistant medical director at the New General Hospital in Hamburg-Eppendorf, Germany, first demonstrated the clinical utility of oral bismuth subnitrate in form of a 5%[29] suspension for creating artificial roentgenological contrast in a living human being (Rumpel 1897). For several years, a diagnostic meal containing bismuth subnitrate, so-called Rieder's meal, helped establish gastrointestinal radiology (Rieder 1904). Beginning around 1910 it became replaced by products based on barium sulfate (Bachem and Günther 1910), special formulations of which are still today the contrast agents of choice in these applications.

Almost coevally with the establishment of Rieder's meal in gastrointestinal radiology, a first practical contrast agent for the retrograde examination of the urinary tract was found. This product was generically called what it actually was, namely a preparation of colloidal silver, after 1902 stabilized by protection colloid (Gehes Codex 1914, 130; Zaunick 1954). It obtained the trademark COLLARGOL™ (Chemische Fabrik von Heyden GmbH, Radebeul near Dresden, Germany). It found its application in the lavage of the infected bladder. Now Fritz

---

TABLE 3.1

The Stepwise Completion of the Vision for Contrast Agents in General

1st Half: Contrast agents consisting of X-ray-absorbing chemical substances for the radiographic visualization of hollow organs in anatomical preparations.

2nd Half: Such contrast agents that are totally innocuous when administered to man.

The complete vision:

   Contrast agents consisting of X-ray-absorbing chemical substances totally innocuous to man.

Voelcker, university lecturer at the Surgical University Clinic in Heidelberg, and his assistant, Alexander von Lichtenberg (Figure 4.20; see *curriculum vitae*), showed its utility in the radiographic visualization of urethra, bladder, ureters, and renal calyces (Voelcker and Lichtenberg 1905), here enclosed within the term pyelography. With JODOFORM-GLYCERIN, bismuth subnitrate and stabilized colloidal silver it became clear that chemical contrast agents with acceptable toxicities were possible. The search for improved agents was hence accelerated. The second half of the vision of contrast agents regarding their innocuousness completed the vision that would orient and motivate research up to the conception of excretion urography in 1923 (Table 3.1).

CURRICULUM VITAE

von Lichtenberg, Alexander (= Sándor) (b. January 20, 1880, Budapest, Kingd. of Hungary, Austro-Hungarian Empire//d. April 12, 1949, Mexico City, Mexico)

Königliches Katholisches Universitätsgymnasium, Budapest: High school dipl. [Érettségi] (1897)/Royal Hungarian Univ. of Sci. in Budapest: Stud. of med. (1897–1903); State med. lic. (March 1903); MD (1903); Anatom. Inst., Asst. (September 1902–November 1903)/Univ. Heidelberg, Germany: Surgical Univ. Clinic, Asst. (November 1903–March 1908)/Univ. Strassburg, Germany: Surgical Clinic (April 1908–December 1913 and May 1914–June 1917); *Venia legendi* [surg. and orthopedic surg.] (February 1910)/Surgical Policlinic, Budapest, Hungary: Chief Surgeon (December 1913–April 1914)/Wartime military serv., Royal Hungarian Army: Regimental Physician and Chief Surgeon, Pecs, Hungary (August 1914–March 1916); Director, Royal Hungarian Bureau of Invalides and Consulting Surgeon of the Military Command, Kassa, Hungary (March 1916–November 1918)/Univ. Strassburg, Germany: Titular prof. of surgery (June 1917–December 1918)/Bábakèpsö Hosp., Miscolc, Hungary: Chief surgeon (1919)/Private practice in orthopedic surg. within the private clinic of Rudolf Jahr, Berlin, Germany (1920)/Univ. Berlin, Germany: Recognition of *Venia legendi* and conversion of designation to surg. (September 1920); Assoc. Prof. [untenured] of Surgery (1922); Withdrawal of teaching permission by Nazi regime, followed by appeal (November 1933); Partial reinstatement (July 1934); Retroactive withdrawal valid for 1935 and extension of research permission until end of 1936 (July 1936)/St Hedwig Hosp., Berlin: Consultant urologist (1922–1924); founding chief of Urol. Dept. (December 1924–June 1936)/Univ. Pennsylvania: Honorary PhD (June 1930)/Escape to Hungary (June 1936)/Private urol. practice at Siesta Sanatorium, Budapest (1937–1939)/Escape to Mexico (1939); Private practice of urol. and urol. surg. at Clinica Londres and at Sanatorio Dr. Alfonso Ortiz Tirado (1939–1944), upon incorporation renamed Sanatorio Durango, Mexico City (April 1944–April 1949)/Hosp. Colonia de Ferrocarriles Nacionales de México, Mexico City: Consultant/Co-founder and co-editor, *Z. f. Urolog. Chir.* (1913–1936).

By 1900, Kümmell observed persistent X-ray shadows spreading from sites of therapeutic injections of the iodized oil, JODIPIN™ (E. Merck in Darmstadt).[32, 33] He attributed the shadows to the continued presence of the product (Kümmell 1901). By recovering original product from old injection sites, Max Landow, who in 1903 was assistant medical director of the surgical section of the Municipal Hospital of Wiesbaden, Germany, delivered the conclusive evidence for this disputed interpretation (Landow 1903). At Kümmell's institution visualization of cadaver arteries was pursued with various contrast agents (Hildebrand, Scholz, and Wieting 1901), including JODIPIN.

From 1899 to 1909, JODIPIN was a chloroiodized oil, whereas afterwards, without change in trademark, it became a purely a hydroiodized analog (Merck 1909, 156–159 & 350–351). The methods of preparation of durable chloro- and hydroiodized oils were covered by patents assigned to the producing company (*E. Merck in Darmstadt* 1897a,b). Contracts reveal that the methods were invented by the Austrian physician, Hugo Winternitz, a private scholarship holder working under the director of the Hygiene Institute of the Friedrich-Wilhelms-University in Berlin, Max Rubner (Merck and Winternitz 1897, 1900). The inventor early on published on his therapeutic experience with JODIPIN (Winternitz 1897).

As a reproduced page from his laboratory notebook documents, sometime in 1901 the method for producing hydroiodized oil was independently found by Marcel Guerbet, chief of practical chemistry exercises at the Superior School of Pharmacy of the University of Paris and co-founder with Laurent Lafay of the company today called Guerbet SA[34] (Bonnemain 2014). Whereas JODIPIN was based on sesame oil, that in its hydroiodized form was limited in practice to an iodine content of 20% (w/w), Guerbet's product, today generically named iodized oil viscous injection, and originally given the trademark LIPIODOL LAFAY™ (Société Guerbet & Cie, Paris, France)[34] was based on poppy seed oil, that allowed an iodine content of 40%(w/w). In any case, Kümmell's *in vivo* observation of the opacity to X-rays of this class of products prepared the ground for their future use as contrast agents (*vide infra*).

Karl Fritsch, in 1911 was an assistant of Hermann Küttner, the director at the Surgical Clinic of the Schlesische Friedrich-Wilhelms-University in Breslau.[35] He promoted the substitution of JODOFORM-GLYCERIN in the radiography of fistular tracts by the better tolerated and more practical hydroiodized oil, JODIPIN (Fritsch 1911). At the same institution during the World War I, Eduard Melchior, then university lecturer, together with the surgery assistant Maximilian Wilimowski, provided an extensive documentation of its use for obtaining radiological surveys of gunshot wounds and abscess cavities (Melchior and Wilimowski 1916). With these deeds, the era of hydroiodized oils as contrast agents was opened, but their affirmation as a major diagnostic tool and their conquest of increasing numbers of clinical applications had to await the arrival on the scene in 1921 of formidable clinical advocates. The most effective were Jean-Athanase Sicard, professor of medical pathology at the University of Paris and chief of service at the Hôpital Necker, and his young intern, Jacques Forestier (Sicard and Forestier 1932), as well as Carlos Heuser in private radiology practice in Buenos Aires (Heuser 1924, 1925, 1926a,b, 1927). An outstanding role was furthermore played by the company Société Guerbet & Cie, which could promote its product LIPIODOL LAFAY freely and on a worldwide scale, whereas E. Merck in Darmstadt, with its JODIPIN suffered from war-related economic sanctions.

In 1918 Don F. Cameron, on military duty as assistant surgeon at the Medical Research Center of the U.S. Navy in Fort Wayne, Indiana, replaced colloidal silver-based products in pyelography by solutions of alkali bromide or iodide, relying mainly on the anions for X-ray attenuation (Cameron 1918). The most widely used ones were sodium iodide (Cameron and Grandy 1918), sodium bromide (Weld 1918; Braasch 1925), and potassium iodide (Rubritius [1919] 1920). In any case, recommended concentrations varied with time and application. The contrast agent solutions were produced *ad hoc* by hospital pharmacies. The exception was lithium iodide, which after testing by Eugen Joseph, professor and chief of the Urological Division of the Polyclinical Surgery Institute of the United University Clinics of Berlin, in 1921 was brought to market as UMBRENAL™ (Chemische Fabrik C.A.F. Kahlbaum GmbH, Berlin-Adlershof, Germany). A role was played for a short while by a solution of sodium thorium dicitrate (THORIUM SOLUTION™, Hynson, Westcott & Dunning Pharmaceutical Laboratory, Inc., Baltimore, USA). It was introduced by J. Edward Burns, assistant resident urologist at the James Buchanan Brady Urological Institute of The Johns Hopkins Hospital in Baltimore (Burns 1915, 1916).[36]

Overlapping with the first historical phase of present special attention were two developments that will receive only marginal treatment. In 1923, Evarts A. Graham, a surgeon at the Washington University Medical School in St. Louis, USA, realized that the concentration in the bile of oral purgatives of the class of polyhalogenated phenolphthaleins (Abel and Rowntree 1909) offered an opportunity for biliary X-ray contrast generation (Graham 1931). Graham and his resident, Warren H. Cole, obtained different ultrapure chemicals from the director of organic chemistry research and development at Mallinckrodt Chemical Works, St. Louis,[37] Vernon H. Wallingford, and reported radiological success (Graham and Cole 1924). In 1924, the company brought iodophthalein sodium on the market under the trade name IODEIKON™ (Mallinckrodt Chemical Works 1931). With this product, synthetic organic chemistry in general and iodinated compounds in particular gained massively in importance for contrast agent research.

Around 1927, Theophil Blühbaum (*alias* for Czesław Murczyński) (Leszczyński 2000b), a radiologist postdoctoral fellow from Krakow, working with Karl Frik, the chief of the Werner Siemens-Institute for Röntgen Research at the Municipal Hospital Moabit in Berlin (Casper 1967) and his radiologist associate, Helmut Kalkbrenner, engaged in a search for improved contrast agents, returning to colloidal preparations. They discovered the clinical utility of plain colloidal thorium dioxide, a radioactive substance erroneously considered harmless, as contrast agent for the liver and the spleen (Blühbaum, Frik, and Kalkbrenner 1928). Siemens-Reiniger-Veifa Gesellschaft für Mediziniche Technik m. b. H, Berlin) and Chemische Fabrik von Heyden AG, Dresden, Germany, put the same colloidal thorium dioxide on the market, under the respective trade names TORDIOL™ and UMBRATHOR™.[31] Through the addition of protection colloid, the latter company achieved a product with improved stability in the blood, generically called stabilized colloidal thorium dioxide (THOROTRAST™). This product and its congeners allowed the acquisition of angiographic images of unsurpassable sharpness and resolution of various anatomical territories, the brain included but left a tragic legacy in the form of permanent deposition of the radioactive contrast agent in the liver, where in some cases, it produced tumors (Andersson, Juel, and Storm 1993).

The young physicians Joseph Berberich, assistant at the Senckenberg's Pathology Institute of the University of Frankfurt a. M., and Samson Hirsch, assistant to Walter Alwens, director of medical services of the Municipal Hospital Sandhof in Frankfurt a. M. and contrast agent researcher, joined forces in the search for angiographic contrast agents. Through Alwens' involvement in the clinical testing of the therapeutic candidate 10%–20% strontium bromide solution for injection, they came to the idea of using the product as angiographic contrast agent (Berberich and Hirsch 1923).[38] Contemporaneously with publication of the demonstration of the roentgenologic utility of strontium bromide by the end of 1923, it was brought to the market as DOMINAL-X™ (Chemisch-Pharmazeutische Aktiengesellschaft, Bad Homburg, Germany) (R. S. 1924). It was indicated for the visualization of arteries and veins, but its problematic performance allowed it to gain only marginal clinical acceptance.

With these gross historical and conceptual elements, the fundaments are now laid for delving into the detailed historical and sociological analysis of the first of the periods selected for special interest.

4 Uro-Angiographic Contrast Agents— The Holy Grail

TECHNOLOGY GENESIS PHASE

History

From Sodium Iodide to Uroselectan Sodium and Methiodal Sodium

Following in the footsteps of Cameron (1918), in 1919 Heuser tested improved formulations of potassium iodide for pyelography (Heuser 1919). In parallel, he explored the visualization of the vasculature after intravenous administration of the same. He visualized radiologically the leg veins of a dog by compressing the upper part of a leg and injecting into a vein of the paw nominally 10%, actually 9.1% (w/w) potassium iodide solution, corresponding to an iodine concentration of 74 mg(Iodine)/mL. He took radiographs of a patient's arm, into a dorsal hand vein of which, for therapeutic purposes, potassium iodide had been injected, and demonstrated opacification of the veins. In a syphilitic boy therapeutically treated with intravenous potassium iodide, he visualized the compound in the heart. In what can be perceived as an excessive leap of confidence he concluded: "To those involved in hospital services I point out that here there is a new way to examine the pulmonary artery and vein."[26] In part explainable by his superficiality in communication and his other peculiar publication habits,[39] his message did not resound in a timely manner, the Anglo-Saxon world being markedly unreceptive. In spite of this, in 1931 Heuser was awarded the gold medal of the Radiological Society of North America (RSNA) for his outstanding work in the development of the clinical application of the X-rays since the year following their discovery. These encompassed in addition to his many practical solutions to equipment problems, his contribution to the use as contrast agents of iodized oils in gynecology and his exploration of sodium and potassium iodides in retrograde pyelography and angiography (Anonymous 1932).

With or without having read the pioneering work of Heuser (1919) on angiography after intravenous injection of a potassium iodide solution, other clinicians began to experiment with iodide-containing solutions (Dünner and Calm 1923; Brooks 1924; Osborne et al. 1923; Moniz 1927a,b). As will be dealt with more in detail later, Osborne et al. (1923) studied excretion urography with intravenous 10% (w/v) sodium iodide [85 mg(Iodine)/mL] that has an osmolality of 1,320 mosmol/kg.[40,41] For the development of angiography most relevant is the addendum in their publication dated February 1923 that speaks of excellent venograms downstream from the site of intravenous injection and contains the prediction of the value of sodium iodide in this application.

Lasar Dünner, head of the Municipal Tuberculosis Care Unit Moabit and member of the Radiology Institute of the Municipal Hospital Moabit in Berlin, gave a presentation in March of 1923 describing the use of intravenous sodium iodide for radiologically distinguishing vessels from bronchi in the lung (Dünner 1923). In the corresponding publication with his junior colleague Adolf Calm, the authors claimed to have actually 2 years earlier injected 10%–15% (unspecified type of %) sodium iodide solutions intravenously, thereby obtaining a faint radiograph of the subclavian vein. But since their primary goal of visualizing the pulmonary circulation was missed, continuation of the project had been delayed. In any case, Dünner and Calm, most likely unaware of the results of Heuser (1919) and independent of the suggestion of Osborne et al. (1923), conceived the use of sodium iodide as an angiographic contrast agent and provided indications of feasibility.

Shortly after the publication by Osborne et al. (1923), and likely under the impression of their prediction, Barney Brooks, who held the position of Essential Teacher in the Department of Surgery at the Washington University School of Medicine in St. Louis, pioneered contrast-enhanced imaging of the arteries. Ongoing work in the same department by Graham and Cole, addressing contrast agents for the biliary tree, apparently had furnished another impulse to the work of Brooks (1924). In September 1923, he was the first to obtain human arteriograms *in vivo*. To this effect he injected intraarterially a 50% (w/w) sodium iodide solution. This was the same upper limit concentration experimented with before by Cameron for retrograde pyelography in animals (Cameron 1918). Brooks obtained excellent arterial angiograms of lower extremities, but the extremely elevated osmolality of the solution, i.e., about 21,000 mosmol/kg,[41] caused such strong pain that general anesthesia was required. He reported his results in the spring of 1924 (Figure 4.1) (Brooks 1924).

In 1920 or thereabouts high doses of intravenous sodium or potassium iodide were being promoted for the treatment of syphilis. For this reason, Earl D. Osborne, a fellow in the section on dermatology and syphilology at the Mayo Clinic in Rochester Minnesota, studied the pharmacology of sodium and potassium iodide in man. Since the earlier excretion studies of Lafay (1893), a substantial literature on the subject had arisen. Osborne integrated these by examining the speciation of iodide in the blood, and the pharmacokinetics of the appearance of iodide in blood and cerebrospinal fluid after administration of the above salt solutions by the intravenous, oral, or rectal routes. Among other observations, he noted that even after injection by the latter two routes (Osborne 1921, 1922) a significant fraction of the dose turned up in the blood. Sodium and potassium determinations were performed under the direction of Leonard G. Rowntree (Figure 4.2; see *curriculum vitae*), the head of the Division of Medicine at Osborne's institution. Rowntree was sensitized to contrast agents for at least two reasons. Given that at his institution retrograde pyelography with alkali halides was being experimented with, he was familiar with the state of the art (Weld 1918). Moreover, he had a few years earlier experimented with iodoform suspended in olive oil for bronchography (Waters, Bayne-Jones, and Rowntree 1917). Rowntree was a lateral-thinking academic with very broad experience in

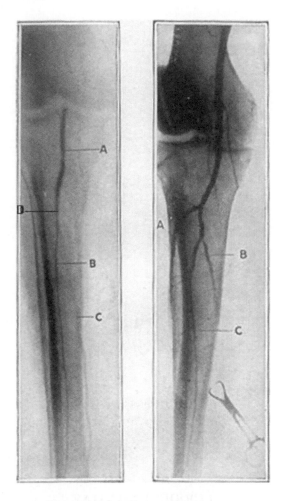

FIGURE 4.1 Sodium iodide as angiographic contrast agent.

Arteriographs of legs into whose femoral arteries, emptied of blood by a manipulation involving a tourniquet and an arterial clamp, Barny Brooks had injected 10 mL of 50% (w/w) sodium iodide.

Left: Right leg of a female patient suffering of foot ulcers. The patient was not anesthetized, causing some pain-induced motion during radiography. The popliteal (A), fibular (B), and posterior tibial (C) arteries are filled with contrast medium. Of the anterior tibial artery merely a faint outline at its origin is visible (D), evidence of occlusion (Figure 1 in Brooks 1924).

Right: Right leg of an anesthetized male patient suffering from diabetes-related gangrene in some tows. The popliteal artery is well visible, while the anterior (A) and the posterior (B) tibial arteries may be seen only for a short distance. The origin of the posterior tibial artery is markedly constricted and the lumen of the fibular artery (C) shows marked irregularities (Figure 2 in Brooks 1924).

FIGURE 4.2 *Leonard G. Rowntree.* (Reproduced with permission from Graner, J. L. 2005. Leonard Rowntree and the Birth of the Mayo Clinic Tradition. *Mayo Clin. Proc.* 80(7): 920–922.)

CURRICULAM VITAE

Rowntree, Leonard G[eorge] (b. April 10, 1883, London on the Thames, Ontario, Canada // d. June 2, 1959, Miami Beach, Florida, USA)

London Collegiate Inst., London, Ontario, Canada: High-school dipl. (1901)/Western Univ. London, Ontario, Canada: Stud. of med. (1901–1905); MD [Gold medalist] (1905)/Victoria Hosp., London, Canada: Internship (1905–1906)/General med. practice, Camden, New Jersey, USA (June 1906–1907 and summers of some following years)/Johns Hopkins Medical School, Baltimore, Maryland. Dept. of experimental therapeutics: Voluntary Asst., then Asst., then Instructor, then Assoc. (1907–1914), Assoc. Prof. (1914–1916)/Western Univ. London, Ontario, Canada: Honorary D.Sc., (1916)/Univ. Minnesota Medical School, Minneapolis: Prof. and Director of Dept. Internal Med. (1916–1920)/ Wartime U.S. military serv., finally as lieutenant colonel of the Army Medical Corps; Executive officer of the Medical Research Laboratories of the Air Serv. of the American Expeditionary Force (1917–1918)/Mayo Clinic, Rochester,

Minnesota: Head of a Sect. of Med. (April 1920–December 1921)/Mayo Foundation, Rochester: Chief of Dept. of Med. (January 1922–September 1932); Prof. of Med./Univ. Minnesota, Minneapolis: Member of the Graduate Faculty (April 1920–September 1932)/Philadelphia Inst. for Med. Research, Philadelphia: Director (September 1932–1940)/National Headquarters Office Selective Serv. System, Washington D.C.: Chief of Medical Div. (1940–1945)/ Presidential Medal of Merit (1945)/The American Legion: Chief Medical Advisor and Chairman of the Medical Advisory Board (1944–1953); Asst. Chief Medical Advisor (1953–1959)/Univ. Miami School of Med., Miami: Co-founder (1950–1952).

physiology, pharmacology, and aviation medicine, and thus ideally prepared to perceive the radiological opportunity opened up by Osborne's findings on metabolism, tissue distribution and excretion of sodium and potassium iodides. It was he who made the crucial connection between Osborne's observation of the rapid urinary excretion of potassium iodide and the use of sodium and potassium iodides as a contrast agent in retrograde pyelography.[42] He realized that alkali iodide, administered by any of the various routes, might allow urography after its excretion into urine, so-called excretion urography.[43] Radiographic visualization of the kidney and upper urinary tract in general was seen as a distinct possibility that would enable doctors to avoid the delicate procedure of ureteric catheterization.

Rowntree and Osborne decided to put the idea of using sodium iodide for excretion urography to the test. With the help of the radiologist Charles G. Sutherland and the urologist Albert J. Scholl they performed experiments on Osborne's patients. These were independently given iodides for therapeutic purposes. This circumvented the need for exploratory experiments in animals. By February 1923, the visualization of the urine in the bladder after intravenous or oral administration of sodium iodide was achieved (Figure 4.3, Left) (Osborne et al. 1923). Partial success with the visualization of the renal pelvis and ureters after intravenous administration followed.

There exists an analogy of the radiological approach of Rowntree and collaborators with various forms of kidney function examination developed much earlier that are based on the appearance of visible color in the excreted urine after intravenous injection of suitable dyes or precursors thereof (indigo carmine, methylene blue) (Heidenhain 1883). Used in combination with cystoscopy, these dyes facilitated localization of ureteric openings and assessment of individual ureteric functionality (Kutner 1892; Voelcker and Joseph 1903). The technique became known as chromocystoscopy. Incidentally, it was also attempted to use intravenously injected potassium iodide in combination with a bladder filling containing oxidants and starch, a filling that was to color in blue the ureteric jets of iodide-containing urine.

Immediately upon learning about the results with excretion urography achieved by the group around Rowntree (Osborne et al. 1923), the urological surgeon von Lichtenberg, by this time associate professor at the Friedrich-Wilhelms-University in Berlin and founding director of what was to become one of the most prestigious urological clinics worldwide, with 250 beds, reacted with his own experimentation.

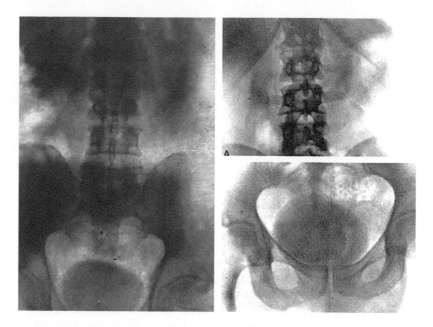

FIGURE 4.3   Staking out the new frontier—excretion urography.

Left: Excretion cystography in man performed by Leonard G. Rowntree and collaborators in 1923. Radiographs were taken one and one-half hour after slow intravenous injection of 100 mL of 10% (w/v) sodium iodide. Positive mode print of the radiograph prepared by black/white inversion from the negative mode print of Figure 2 in Osborne et al. (1923). (Adapted with permission from Osborne, E. D., C. G. Sutherland, A. J. Scholl Jr., and L. G. Rowntree. 1923, Roentgenography of Urinary Tract During Excretion of Sodium Iodid. *J. Am. Med. Assoc.* 80(6): 368–373. Copyright© 1923, American Medical Association. All rights reserved.)

Right: Excretion urographs in man achieved, purportedly with sodium iodide as contrast agents, by Johannes Volkmann in the spring of 1924. Right bottom: Bladder shadow. Right top: Kidney shadows. The two radiographs are reproduced from positive mode prints of the original radiographic glass plates that by now have been accidentally destroyed. The prints were kindly provided by P. Rathert, who had first published them as Figure 1 in Rathert, P., H. Melchior, and W. Lutzeyer. 1974. Johannes Volkmann, M.D., Pioneer in Intravenous Urography. *Urology* 4(5): 613–616 (Copyright Elsevier.)

Beyond reproducing the American experiments with sodium iodide, he searched for unspecified preparations that are excreted by the kidneys with elevated velocity and in high concentration, in addition to exploring improved radiographic techniques (von Lichtenberg [1928] 1929). He cultivated high ambitions of achieving a similar breakthrough for this procedure, as he had for retrograde pyelography with the introduction of stabilized colloidal silver (Voelcker and Lichtenberg 1905, 1906). To this end, he joined forces with Paul Rosenstein, the chief surgeon of the Hospital of the Jewish Community, Berlin. Together they explored the combination of excretion urography with pneumoradiology. While Rosenstein ([1924] 1925) communicated modest success, von Lichtenberg ([1928] 1929) limited himself to uttering his frustrations with the results during the discussion time at a congress. Noteworthy is the fact that in Rosenstein's group resided the resident Alfred Roseno ([1928] 1929), about whom much more will have to be recounted (Figure 4.5; see *curriculum vitae*).

The priority of Rowntree and collaborators of having conceived and demonstrated excretion urography with intravenous sodium iodide is undisputed. But independent achievement of the same results has been claimed, and efforts have been made to let the claimant share the merits with Rowntree. The following critical reconstruction of what happened led to an interpretation recognizing merit but for a different deed.

In 1922, the Surgical Clinic of the University of Halle a. S., Germany, was headed by Voelcker, one of the developers of chromocystoscopy (Voelcker and Joseph 1903) and of pyelography with stabilized colloidal silver (Voelcker and Lichtenberg 1905, 1906). At the time, his assistant, Hans Boeminghaus, attempted visualization of the urinary system in animals after oral administration of sodium bromide (Volkmann 1924c). The tolerated doses were too low for success, but the episode illustrates that in the early 1920s the idea of excretion urography was in the air.[44]

At Voelcker's clinic, working contemporarily with Boeminghaus was also Johannes Volkmann (Figure 4.4; see *curriculum vitae*), then a resident and university lecturer in the making. Much later Volkmann prided himself for having always experimented alone (Volkmann 1955). Indeed, there exist no signs for his

FIGURE 4.4 *Johannes Volkmann.* (Reprinted with permission from the University Archives, Halle-Wittenberg, Rep. 40/I, V 8)

CURRICULAM VITAE

Volkmann, Johannes [Allwill Max] (b. September 26, 1889, Waldheim, Kingd. of Saxony, German Reich // d. October 14, 1982, Hannover, State of Lower Saxony, Federal Republic of Germany).

Gymnasiumzu Zwickau; Germany: High-school dipl. [Abitur] (1909)/Univ. Leipzig (October 1909–June 1910; April 1912–March 1913; October 1913–December 1917); Univ. Erlangen (October 1910–June 1911); Univ. Grenoble, France (October 1911–March 1912); Univ. Kiel (March 1913–June 1913): Stud. of med./Peace-time voluntary military serv. (1910)/Municipal Hosp. Heilbronn: Unpaid medical trainee, Internal Med. Div. (July 1914–August 1914)/State medical exam., Dresden (August 1914)/Reservelazarett II, Stuttgart: Surg. training (August 1914–October 1915)/Univ. Leipzig: MD (December 1914)/Wartime military serv. in various medical duties, mostly as surgeon. Multiple military awards (October 1915–December 1918)/Landeskrankenhaus Braunschweig, Pathol. Bacteriol. Inst. (January 1919–October 1919)/Univ. Halle a. S., Surg. Univ. Clinic: Unpaid medical intern (November 1919–April 1920); Resident (May 1920–March 1923); *Venia legendi* in surg. (March 1923); First Asst. Surgeon (April 1923–November 1925); Asst. Medical Director (January 1926–December 1930); Untenured Assoc. Prof. (April 1928–December 1930)/Evangelisches Krankenhaus, Münster, Westphalia, Germany: Medical Director, Surgical Section and Chief of Staff (January 1931–May 1933)/Krankenhaus Bergmannstrost, Halle a. S.: Chief of Staff, and later Director (June 1933–1945)/Univ. Halle a. S.: Assoc. Prof. (November 20, 1939–October 1945)/Wartime military serv. in the German army's medical corps, last as lieutenant colonel (1941–1945), and in the Nazi party's SA organization, last as Obersturmführer (October 1941–October 1945)/Allgemeines Krankenhaus Schkeuditz, near Leipzig, German Democratic Republic: Chief Surgeon (November 1945–September 1950); Chief Medical Officer (October 1950–July 1952)/Univ. Greifswald, German Democratic Republic, Dept. of Surg.: Prof. and Chairman (August 1952–December 1955); Assoc. Prof. on temporary commission (January 1–January 31, 1956)/Prof. Emeritus, living in Hannover, West Germany (1956–1982). Member of Stahlhelm (Steel Helmet, League of Front Soldiers) (1922–1945)/Member SA (Storm Detachmnent) (1933–1945)/Member NSDAP (Nazi Party) (1937–1945)/Election to Deutsche Akademie der Wissenschaften, East-Berlin (1952)/Election to Deutsche Akademie der Naturforscher Leopoldina, Halle a. S., German Democratic Republic (May 1955).

co-involvement in the excretion urography experiments of Boeminghaus, although very likely he was aware of them. Up to the end of 1923, his research focus instead was on encephaloscopy as purely surgical technique (Volkmann 1923). But it soon extended to studies of the animal brain by radiological procedures requiring injection of a contrast agent directly into brain ventricles (Volkmann 1924b).

As agent, he preliminarily recommended tenfold concentrated Pregl's isotonic iodine solution for intravenous injection, supplemented with boric acid (concentrated PRESOJOD™, Chemische Fabriken Dr. Joachim Wiernik & Co. AG, Berlin-Waidmannslust, Germany) (Volkmann 1924b). Given its composition (Perutz, Siebert, and Winternitz 1930), this product acted as a contrast agent solution containing 3% (w/v) in iodine. Volkmann is credible when he hints that he arrived at this contrast agent and administration modality after examination of various other candidate compounds and administration modalities. More problematic is the impression he created later that these activities already represented efforts directed at excretion urography.

Before the submission deadline of March 12, 1924, possibly as early as 1923, Volkmann submitted a first short communication to the 48th Congress of the German Society of Surgery, to be held in Berlin on April 23–26, 1924. It exclusively dealt with encephaloscopy and ventriculography. It was regularly announced, presented, and published in the congress proceedings, as comment to a full-length talk on brain tumors by an unrelated investigator (Volkmann 1924b). In his comment, apart from recommending concentrated PRESOJOD as contrast agent, Volkmann suggested the creation of a new one, in which iodine was somehow bound to hexamethylenetetramine (urotropin), a chemical substance that, when introduced into the cerebrospinal fluid, was known to be rapidly excreted. With this proposal, he ventured into foreign knowledge cultures. He lamented that, regarding contrast agents, chemists and pharmacologists have let his profession down.

In later historical accounts of his involvement in excretion urography, Volkmann neglected to mention the communication on encephaloscopy and ventriculography. As will become evident, this deletion of modality later caused him to confuse the dates of two separate congressional abstract submissions.

In contrast to the straightforward story of Volkmann's involvement in brain studies, the story of his early efforts in excretion urography requires diligent analysis of incongruent claims and documented facts. The following chronology of Volkmann's urological research activities could be reconstructed heavily based on documents he had written personally, including some that were previously not considered, as well as meeting abstracts. On occasions where written recollections of the aging protagonist conflicted with his earlier descriptions of the events, the latter were privileged. The resulting chronology differs to some extent from that reconstructed by Peter Rathert and colleagues (Rathert, Melchior, and Lutzeyer 1974, 1975; Rathert 1992) and which has become adopted by several other authors. Their chronology considered in addition to written documents an interview with the 84-year-old protagonist, which resulted in some inconsistencies with earlier accounts.

In the course of his clinical duties, at one point Volkmann and coworkers were confronted with a patient needing a retrograde pyelography, which was rendered impossible by a severely restricted urethra. He has delivered an extensive late reconstruction of the events that followed (Volkmann 1966). Accordingly, the report of Berberich and Hirsch (1923) on strontium bromide as angiographic contrast agent led him to consider whether for urography "instead of the administration from below, excretion from above was feasible" (Volkmann 1924f, 1966). In October of 1924, he wrote that "injections into the bloodstream, with which, last year, following the communication of Berberich and Hirsch on vascular injections of strontium bromide, I resumed the earlier experimentation, had to remain initially unsuccessful, until the doses were augmented considerably"[26] (Volkmann 1924f). Available reports leave it unclear whether strontium bromide was actually tried on the aforementioned patient, but it appears unlikely.

Anyway, stimulated by the clinical case, Volkmann picked up systematic excretion urography studies where Boeminghaus had left off. His research notebook (Volkmann 1924a), in Kurrent handwriting, reveals the date he began recorded work on excretion urography. His experiments were numbered chronologically, the one of March 7, 1924, bearing the number 1. On that occasion, he explored the urographic utility in a patient of oral sodium iodide at two split doses for a total of 5 g. He observed weak X-ray shadows of the kidneys and the bladder. Besides the numbering of the experiments, also the progressive choices of contrast agents and administration protocols in the course of the investigation speak for March 7, 1924, as the date marking the beginning of Volkmann's systematic pursuit of excretion urography. Yet not until March 24, did experiment number 2 take place. It involved intravenous administration of 10 mL of 10% strontium bromide. The product was ineffective, suggesting that no previous experience, e.g., experience with the patient who had inspired him initially, had offered helpful orientation.

With experiment number 3, performed on March 25 with a 24-year-old male suffering from gonorrheal cystitis, began Volkmann's work with intravenous sodium iodide. He described that a 1 g oral dose of sodium iodide followed a day later by a 1 g intravenous dose of sodium iodide, i.e., 10 mL of a 10% (w/v) solution, sufficed to generate discernible shadows of the right kidney, the bladder, and left ureter.[45]

The next day, in experiment 4, he examined a female dog intravenously injected with the same dose of sodium iodide. Bladder and kidneys on radiographs were at best hinted at. This poor performance of a dose that in man of much larger body weight was more successful is puzzling, all the more so, as later at least tenfold higher doses of sodium iodide were considered essential for such success in humans.

Volkmann recounted that in pursuit of excretion urography by the end of April, 1924, he had experimented in animals and man administering orally, intramuscularly, subcutaneously, rectally, and intravenously, solutions of sodium iodide, potassium iodide, lithium iodide (UMBRENAL), sodium bromide and strontium bromide (DOMINAL-X), as well as other easily available chemicals and formulations, such as sodium rhodanide (thiocyanate) and tenfold concentrated isoosmolal Pregl's iodine solution (Volkmann 1924c,f). Some of these administrations in low doses for tolerability testing in animals were already ongoing within the earlier research program on ventriculography. They smoothly continued after March 7, 1924, now in pursuit of excretion urography.

Purportedly as early as fall of 1923 Volkmann submitted a communication on excretion urography to the 48th Congress of the German Society of Surgery in Berlin (Volkmann 1966, 1974). The recipient was the organizer of the congress, his uncle Heinrich Braun. Around March 25 the preliminary congressional program is claimed to have announced Volkmann's planned intervention,[46] but to the submitter's disappointment, only as comment to the ordinary full-length talk on renal diagnostics by Eduard Rehn (1924). By bad luck, a job interview in Freiburg i. Br. impeded the latter from delivering his talk. Only two other full-length talks on urological issues were scheduled and both also got cancelled. Opportunities for oral comments regarding urological diagnostics vanished.

The surgeon and radiology pioneer, Georg Perthes, was very disappointed about Volkmann's missed occasion to present his work at the surgery congress. He counseled him to immediately join the German Röntgen Society and offered to organize the opportunity for him to give a presentation at the immediately impending 15th Meeting of the German Röntgen Society, to be held in Berlin, April 27–29, 1924 (Volkmann 1966). It took again the form of a comment to an ordinary full-length talk, but this time it was presented and published (Volkmann 1924c). Short reports at a couple of regional meetings followed (Volkmann 1924d,e). Volkmann's urographic results generated sufficient interest for the *Münchener Medizinische Wochenschrift* to solicit a paper from him. In October 1924, he delivered a short, regular but unillustrated report (Volkmann 1924f), his only formal publication ever on the subject.

Despite not being presented at the surgery congress, but in accord with tradition, the talk of Rehn and the two others dealing with urological diagnosis were published in the congress proceedings but without comments. Ignorant of these facts, Volkmann felt wronged by his uncle, Braun. He blamed familial animosities for the situation (Volkmann 1966). Their description lets one suspect that his political extremism had played a role. Volkmann was an early and persistent member and officer of various extremely nationalistic and anti-Semitic organizations. From 1922 to 1945, he was a member of the ruthless Nazi organization, SA (Martin-Luther-Universität Halle-Wittenberg 2005) (Figure 4.4; see *curriculum vitae*). Actually, Volkmann's communication had not been treated by Braun differently from others in the same situation. Despite this fact, after retirement, Volkmann instrumentalized this story to foster the impression that Braun had caused a delay in communication, which deprived him of the opportunity to present excretion urography with sodium iodide before the achievement of Rowntree and collaborators had become known in Germany.

The account presented poses several intertwined puzzling questions. When did Volkmann submit his communication on excretion urography to the surgery congress, which results did he have in hand at that moment, what did he intend to present at the surgery congress and what did he actually present at the meeting of the German Röntgen Society.

At a medical meeting in Frankfurt a. M. on October 23, 1923, in form of an oral presentation, and on December 3, in print, Berberich and Hirsch (1923) had introduced strontium bromide as intravascular contrast agent for angiography. Right away the corresponding commercial contrast agent, DOMINAL-X, had become available (see Chapter 3). Volkmann's declaration that this contrast agent triggered his ideas on excretion urography fix to no earlier than late 1923 the time he had to deal with the difficult patient suffering from a restricted urethra. Thus, while his own initial dating of entry into the urological field to 1923 (Volkmann 1931), with further restrictions, may be accepted, his much later dating of the event to 1922 (Volkmann 1957, 1974), i.e., to a date preceding the publication of Rowntree and collaborators (Osborne et al. 1923), is untenable.[47]

The four earliest experiments on excretion urography in his notebook give the strong impression that they had not profited from preceding ones, at least not of ones aimed at the same objective. In particular, the claim of promising radiological signs after intravenous administration of 200 mL of 10% sodium iodide to the patient with the urethral restriction in 1923 (Rathert, Melchior, and Lutzeyer 1974) is incompatible with the experimentation in March of 1924, especially experiment 3. It needs to be rejected.

Based on Volkmann's research notebook as most reliable evidence, March 7, 1924 marks the beginning of his excretion-urographic experimentation. March 12 was the official deadline for submission of full-length communications to the surgery congress. By that time Volkmann had performed only experiment 1. It is unimaginable that his corresponding communication would have been accepted under such circumstances. We must conclude that short communications could be submitted also at a later time. The results of experiment 3 of March 25, dealing with intravenous sodium iodide, offered the earliest occasion. Whatever the real dates and contents were, it may be admitted that the manner with which the idea of excretion urography was addressed convinced Braun to accept a short remark at the surgery congress.

Moreover, until the beginning of the congress on April 23, there remained enough time for further experimentation. Herewith the claimed submission of a communication on excretion urography in 1923 (Volkmann 1974) is shown to be in error, probably because of the aforementioned confusion of two separate submissions. The fact is that objective documentation in support of the discussed submission to the surgery congress and its acceptance is at present missing. Only for the presentation at the subsequent meeting of the German Röntgen Society the documentation is solid.

Figure 4.3, right top & right bottom, shows some of Volkmann's purportedly earliest successful excretion urographs. Rathert, Melchior, and Lutzeyer (1974) obtained positive paper prints of the original glass plates from him and published them for the first time. Allegedly the urographs were obtained in 1923 and formed the basis for the aforementioned submission, still the same year, of a presentation to the surgery congress (Volkmann 1966; Rathert, Melchior, and Lutzeyer 1974). Disturbing, however, is that the 1923 date precedes March 25, 1924, when Volkmann's research notebook records the first experiment with intravenous sodium iodide (Volkmann 1924a). Thus, either the urographs were really produced with sodium iodide after March 25, 1924, or they were produced in 1923 with another contrast agent, namely strontium bromide. Against the latter possibility speaks the fact that the first experience with strontium bromide, described as experiment number 2 in the research notebook entry of March 24, 1924, was a total failure. The urographs do not completely exclude the possibility that they are the results of experiment number 3 in Volkmann's research notebook.

Unfortunately, Volkmann's urographs in Figure 4.3 lack contemporaneous descriptions of the exact date, the patient, the contrast agent, and the administration protocol. Later, they were said to have been obtained in 1923 on a patient with ure-thral stricture, but it is unclear whether this was the same patient as the one that had triggered Volkmann's entry into the field of excretion urography. Based on their interaction with Volkmann, Rathert, Melchior, and Lutzeyer (1974) give the intra-venous contrast agent employed as 200 mL of 10% sodium iodide, for a total 20 g of the salt. Being twice the one shown effective by Rowntree, this dose is plausible, but it is 20 times higher than the one Volkmann described in experiment 3, the first radiologically successful experiment with intravenous sodium iodide found in his research notebook.[45] A couple of months after his first presentation of urographs at the Meeting of the German Röntgen Society, Volkmann gave presentations at some regional meetings, where he recommended the following specific procedure. A blad-der shadow could be produced after preparatory oral administration of potassium or sodium iodide (day –2: 1 g; day –1: 3 × 1 g), followed on day 0 by intravenous admin-istration of 10% (w/v) sodium iodide (8–10 g) over 5 min and delaying radiography for at least ½ h. Kidney shadows were obtainable by the same procedure, except that the intravenous dose of sodium iodide was 15–18 g and the delay time was 2–2.5 h (Volkmann 1924c,d).

The order and timing of contrast agents, dosing protocols and administration modalities, initially experimented with in man, supports Volkmann's affirmation that up to shortly before the meeting of the German Röntgen Society at the end of April 1924, specifically at least until the experiment number 3 of March 25, he still had been unaware of the success in excretion urography with intravenous sodium iodide by Rowntree and collaborators (Volkmann 1966). However, by the time of his meeting presentation he did acknowledge the earlier achievement of unspecified American authors (Volkmann 1924c). This means that he could have produced his first really successful urographs with a 10–20 g dose of sodium iodide after having been informed about the paper of Osborne et al. (1923).

Volkmann's claim of having achieved excretion urography with sodium iodide independently of Osborne et al. (1923) can be accepted. But his ignorance of the work of American competitors in a leading journal more than 1 year after its pub-lication is not fully excusable, especially taking into account that the library at his university had at the time subscriptions to both the journal in which Osborne's work was published and *Index Medicus*, a compilation of all current medical publica-tions. Indeed, to von Lichtenberg the paper of Osborne et al. (1923) became known shortly after its publication. Was only Volkmann or were a significant number of German academics ignorant of what was happening in America? Given that fur-ther cases of European ignorance of American scientific literature exist, I favor the second hypothesis. One partial explanation for this situation may be found in the decade-long ostracism German and Austrian doctors suffered by their colleagues from allied countries after World War I. In fact, the former were not readmitted to a major international forum on X-rays until 1928 (Holtzmann Kevles 1997). Another partial explanation, I believe, must be found in a type of chauvinism among German-speaking researchers at the time, who had become used to perceive themselves so much at the center of important progress that they allowed themselves the luxury of paying scarce attention to what was happening elsewhere, obviously to their own disadvantage. Lastly, Volkmann's extreme nationalism may have blinded him to achievements in the USA.

Volkmann's account of his involvement in excretion urography reveals his aspiration to be recognized for having discovered the new diagnostic procedure independently of the American authors (Volkmann 1955, 1966). In a lecture to students in 1955 he even declared "I consider intravenous pyelography as my most important discovery" (Volkmann 1955), without mentioning anybody else. In interviews and letter exchanges some authors of historical papers have lent Volkmann a sympathetic ear (Rathert, Melchior, and Lutzeyer 1974, 1975; Rathert 1992; Moll and Rathert 2012; Hausmann 1990), but the story lost its persuasiveness when the herein described inconsistencies were eliminated through focusing on dates and chronologies in original documents.[48,49] Notable historians of contrast agent technology have ignored Volkmann altogether (e.g., Grainger 1982a; Pallardy, Pallardy and Wackenheim 1989), or minimized his role (Pollack 1996; Skrepetis, Paranichiannakis, and Antoniou 2004). As late as 1974 Moses Swick (*vide infra*) in a letter to Peter Rathert complained that "I could never find a publication of Professor Volkmann's work" (Swick 1974a). This is most curious given the fact that he cited the paper of Volkmann (1924f) in an early American publication of his (Swick 1930b).

In any case, Volkmann's noteworthy contribution to excretion urography was neither its conception nor the independent discovery of sodium iodide as prototypical contrast agent for this procedure, but the promotion of the new diagnostic approach combined with the realization that improved contrast agents had to come from an effort by industry. He not only called publically for its assistance (Volkmann 1924f, 1928) but, as will be described below, also utilized a personal contact to get a particular company successfully involved. No similar efforts and impact by the American investigators are known.

Declaredly stimulated by Volkmann's reports, additional studies with intravenous sodium iodide were performed at the University of Padua in Italy. Guerrino Lenarduzzi, assistant in the radiology unit of the department of clinical surgery and Renzo Pecco, assistant in that department, studied in intravenous excretion urography with 10% sodium iodide in dogs. With typical doses of 1 g/kg body weight, they exceeded by factors of more than two those explored by others. Despite the high doses, successful imaging of renal calyces and ureters required ligation of the latter (Lenarduzzi and Pecco 1927). In man, this meant external compression of the ureters (Ziegler and Köhler 1930). Notwithstanding this patient manipulation, combined with extensive patient preparation, such as oral doses of contrast agent on previous days and induced bowel voiding (Volkmann 1924f), or use of pneumoradiology (Rosenstein [1924] 1925; von Lichtenberg [1928] 1929), the diagnostic performance of simple salt solutions was never really satisfactory. In addition, there were significant adverse reactions (von Lichtenberg 1932).

With excretion urography, Rowntree and collaborators and Volkmann had independently established a next frontier for contrast agent research. Their merit was not so much having found a partial solution to the problem of excretion urography but to define it as a goal that was both worthwhile and considered attainable. They

TABLE 4.1

The Stepwise Emergence of the Vision of Ionic Uro-angiographic Contrast Agents

1) Intravenously injectable, mostly renally excreted contrast agents for excretion urography.
2) Intraarterially injectable, mostly renally excreted contrast agents for angiography.

The complete vision:

Uro-angiographic contrast agents whose low toxicity allows cardioangiography and cerebral angiography.

implicitly extended the earlier vision by adding the first part of a new one, excretion urography as a valuable and challenging clinical indication (Table 4.1). The vision maintained the fixed core requiring contrast agents as X-ray-absorbing chemical compounds totally innocuous to man.

Dissatisfied by the salts up to that date experimented with, Volkmann sought help from his elder student fraternity brother Georg Otto, a pharmacist and food chemist who, at that time, was head of the scientific division at the pharmaceutical company Gehe & Co. AG, Chemische Fabriken, Dresden-N[50] (Volkmann 1966; Pharmacist Georg Otto 1957). He must have described his disappointing excretion urography experiments with simple salts, including rhodanides (Volkmann 1924c,f). Peter Rathert, Düsseldorf, possesses a letter of response on company stationary to Volkmann from Otto, dated May 14, 1924, wherein he called attention to some literature on the substitution in therapy of iodides by rhodanides and mentioned related unspecified negative experiences gained in his company. The letter gives no evidence for studies made in response to solicitation by Volkmann. On October 16, 1924, company scientists released an internal memorandum in response to a letter by Volkmann (Wissenschaftliches Labor 1924). Most likely they referred to a second letter to Otto, who had passed it along to his scientific staff. According to the memorandum, the letter contained the very general proposal that the company search for better intravenous urographic diagnostics among inorganic salts, organic iodinated molecules and radioactive substances (Wissenschaftliches Labor 1924; Rathert, Melchior, and Lutzeyer 1975). With this proposal, he joined the earlier published calls by others for detoxified organic iodinated compounds (Schepelmann 1910; Cameron 1918). Noteworthy is the fact that in pursuit of the suggestions received, the company attempted in vain to render the long-known topical antiseptic tetraiodopyrrole (JODOL™, Kalle & Co. AG, Biebrich, Germany) (Ciamician and Silber 1885), water-soluble[51] by transforming it into sulfonamidic derivatives (Wissenschaftliches Labor 1924). As a result of this failure, an early and imaginative occasion for searching among iodinated organic molecules for urographic contrast agents was prematurely discarded. By the end of 1924, Otto left Gehe & Co. AG to join Pharmazeutische Abteilung Bayer-Meister Lucius of I. G. Farbenindustrie AG. His staff maintained interest in contrast agents, however.

FIGURE 4.5 *Alfred Roseno* around 1930. (Courtesy of Dr. Susan R. Fahrenholtz (-Roseno), Bloomfield NJ, USA.)

CURRICULAM VITAE

Roseno [in some early documents Rosenow], **Alfred** (b. July 31, 1896, Hamburg, City State of Hamburg, German Reich // d. January 29, 1965 New York, New York, USA)

Gymnasium St. Georg in Horn, Hamburg: High-school dipl. [Abitur] (1914), Univ. Munich (1914–1915); Univ. Berlin (1915); Univ. Rostock (1916–1917); Univ. Giessen (1917); Univ. Freiburg i. Br. (1917–1920): Stud. of med./Univ. Freiburg i. Br.: MD (May 1920)/Spital Hamburg-Barmbeck: Asst. of Surg. (1921–1922)/ Jüdisches Krankenhaus, Cologne: Assoc. Surgeon (1922–1925)/Jüdisches Krankenhaus, Berlin: Resident Surg. Div. (1925–1928)/Augusta-Hospital and Bürgerhospital, Cologne, Germany: Surgeon (1928–1930)/Jüdisches Krankenhaus, Cologne: Chief Surgeon (1930–1936)/Escape to USA (1936)/ License to practice in the State of New York (1937)/Israel Zion Hosp., New York City: Urol. (1937–1939)/Private urol. practice, New York City (1939–1965).

In 1929, Gehe & Co. AG commercially launched its first contrast agent, sodium iodide/urea, with the trade name, PYELOGNOST™ (Figure 4.5; Roseno [1928] 1929; Roseno 1929; New Specialties 1930). The concept behind the combination of sodium iodide with urea was the acceleration of the excretion of sodium iodide by the well-known diuretic effect of urea. It is ascribable to Roseno. While still working in

Rosenstein's clinic, he had explored the radiological semeiotics of the dog kidney and ureters contrasted with sodium iodide under low pressure conditions that mimicked the situation in excretion urography better than retrograde pyelography. To this effect the contrast agent was injected through an artificial kidney fistula (Roseno [1928] 1929). In 1928, after taking up his new position at the Surgical Clinic of the Augusta- and Citizens Hospital, associated with the University of Cologne, he conceived the idea of combining sodium iodide with urea. With the radiological help of Hans Jepkens, the idea was tested first in the dog, and then preliminarily in man (Roseno [1928] 1929; Roseno and Jepkens 1929).

Sometime in late 1928 Roseno and Gehe & Co. AG must have come into contact. The company had a prepared mind for Roseno's idea and a collaboration ensued. On February 3, 1929, a patent application for a particular pharmaceutical formulation of sodium iodide and urea as contrast agent was deposited (*Gehe & Co. AG and Roseno* 1929). Gehe & Co. AG and Roseno figured as inventors (Figure 4.6, Top). The patent application described a formulation as solid, ready for dissolution in water for injection, consisting of an inclusion compound of sodium iodide in urea with a 1 to 1 molar ratio.[52] This diagnostic was an analog to an oral therapeutic that was composed of the molecular inclusion compound of calcium iodide and urea (Greenbaum 1929) and formulated as a tablet for dissolution in water. Chemische Fabrik Arthur Jaffé, Berlin, had commercialized the latter preparation at least since 1918 as JODFORTAN™, with therapeutic indications for syphilis and atherosclerosis (Thoms 1928; Unger 1928). At the time, in Germany, neither the composition nor the clinical use of a mixture of sodium iodide and urea in solution as contrast agent could be patented. Roseno's idea per se was not patentable. Solely a novel process for the manufacturing of the product allowed it. Patenting of the process indirectly conferred to the product a proprietary status (*Gehe & Co. AG and Roseno* 1929). Some form of temporary exclusivity was an essential condition for the development of an economically viable drug then, just as today. Scientists at Gehe & Co. AG let the various requirements interfere with their encyclopedic awareness of available therapeutic drugs, their chemical composition, and pharmaceutical formulation.[53] They formulated a sodium iodide/urea inclusion compound, for which a production process could be patented. This was a highly ingenious solution to the multidimensional problem at hand. Upon dissolution, the two components of the inclusion compound lost their association, yielding the iodide ion that generated X-ray contrast and the urea that enhanced diuresis. The renal excretion in man of iodide after administration of sodium iodide/urea was described as very fast (Roseno [1929] 1930). Roseno's inclusion as inventor of the product recognized his technology enabling contribution. True to the tradition of Gehe & Co. AG, the company rather than any of its employees appeared as principle inventor (*Gehe & Co. AG and Roseno* 1929).

Naturally Gehe & Co. AG allowed Roseno to perform pivotal clinical studies with sodium iodide/urea. He reported substantial progress in intravenous excretion urography (Roseno 1929; Roseno [1929] 1930). Highly disturbing in his communications is the absence of a quantitative description of the contrast agent composition and concentration, combined with an extreme scarcity of information about injected doses. The initial clinical publications failed to acknowledge the fact that Gehe & Co. AG had provided the new contrast agent. He had to reveal this information in a *post scriptum* to one of them, wherein he furthermore provided previously deficient

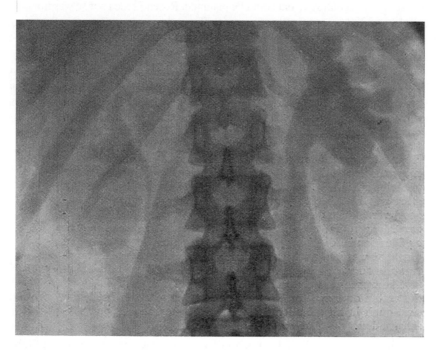

DEUTSCHES REICH

AUSGEGEBEN AM
7. OKTOBER 1930

REICHSPATENTAMT

PATENTSCHRIFT

№ 509265

KLASSE 12o GRUPPE 17ʟ

G 75543 IVa/12oᵇ

Tag der Bekanntmachung über die Erteilung des Patents: 25. September 1930

Gehe & Co. Akt.-Ges. in Dresden und Dr. Alfred Roseno in Köln a. Rh.

Verfahren zur Herstellung von Molekülverbindungen aus Jodnatrium und Harnstoff

Patentiert im Deutschen Reiche vom 3. Februar 1929 ab

Wenn man wässerige, alkoholische und andere Lösungen von Harnstoff und Jodnatrium erwärmt, dann mit einem Fällungsmittel, z. B. Chloroform, Äther usw., vermischt oder keine Abscheidung zeigt. In diesem Falle 35 handelt es sich sogar um ein molekulares Verhältnis von 4 Molekülen Harnstoff zu 1 Molekül Jodnatrium.

FIGURE 4.6 Sodium iodide/urea (PYELOGNOST™), the first commercial but in the end insufficiently effective contrast agent for excretion urography.

FIGURE 4.6 (CONTINUED)

Top: Title portion of front page of the patent by *Gehe & Co. AG and Roseno* (1929) for sodium iodide/urea, a molecular inclusion compound of sodium iodide in urea, which as a solid pharmaceutical formulation called PYELOGNOST, served to prepare solutions for injection containing 10% (w/v) sodium iodide.

Bottom: Copy of Figure 9 in Roseno (1929). Intravenous sodium iodide/urea at a dose equivalent to 38 g iodide ion was used to visualize the normal right kidney and ureter and the heavily enlarged left renal calyx and ureter, caused by a later operatively removed ureteric stone. (Reprinted by permission from Springer Nature Customer Service Centre GmbH of Figure 9 in Roseno, A. 1929. The Intravenous Pyelography. II. Communication. Clinical Results [in German]. *Klin. Wochenschr.* 8(25): 1165–1170 & (35): 1623.)

---

dosing information (Roseno 1929). It looks as though at first, he had tried to let the reader believe that he was the sole originator and producer of the contrast agent and was asked by the company to correct the situation.[54]

PYELOGNOST, prepared for intravenous injection, consisted of sodium iodide and urea, 0.3 mol each, dissolved in 200–250 mL of water [190–152 mg(Iodine)/mL]. The osmolality of the most concentrated solution should be of the order of 5,600 mosmol/kg,[41] although it was not measured. The instruction to take between 5 and 15 min for the administration reflected this elevated value, without specifically mentioning it anywhere. As dose for patients between 50 and 90 kg body weight, a single vial of PYELOGNOST containing 63 g dry product was recommended. This corresponds to 38 g of iodide ion, or 45 g of sodium iodide (Roseno 1929).

It is worth noting that in the case of intravenous injection to man, apart from the presence or not of urea, the major difference between the use of contrast agent by Roseno on the one hand, and Rowntree and colleagues or Volkmann on the other, was the total dose of iodide ion for a person of average weight. While the latter clinicians limited themselves to a maximum of 18 g iodide ion, Roseno and Gehe & Co. AG recommended sodium iodide/urea at a dose equivalent to 38 g iodide ion. Also, the recommended concentration in terms of sodium iodide was higher. With 20% (w/v) it reached around double that used by the others. No wonder sodium iodide/urea allowed better contrasted images of the bladder and kidney calyces and allowed visualization of the ureters for greater lengths than ever before (Figure 4.6, Bottom). But it is not clear how attribution of success could be partitioned between this increased dose of sodium iodide and the presence of urea.[55]

Sodium iodide/urea constituted at the time a decent solution to the problem of intravenous urography but did not allow visualization of blood vessels and suffered from substantial adverse reactions (Chwalla 1930). Although in 1928 von Lichtenberg was extremely skeptical of this contrast agent (von Lichtenberg [1928] 1929), he subsequently attested to its distinction as the first clinical success in excretion urography (von Lichtenberg [1929] 1930a; Binz, Räth, and von Lichtenberg 1930), and others agreed (Bronner 1929; Bronner and Schüller 1930; Ziegler and Köhler 1930; Swick 1930b). In addition, he and also Moses Swick (*vide infra*) attributed to it an important role in motivating further research, writing: "Roseno, through emphasis of physiological and pathophysiological aspects, was the first to achieve significant progress in this field and to restore good impetus to the already faltering idea"[26] (von Lichtenberg and

Swick 1929; von Lichtenberg [1929] 1930a). If the product failed to establish excretion urography as the up to date standard of patient care, it was primarily due to the almost contemporaneous introduction of the first iodinated organic contrast agents (*vide infra*).

Available documents support the conclusion that prior to Volkmann's suggestion to Gehe & Co. AG, the company had not been active in the contrast agent field. Thus, at least up to October 7, 1930, when the patent covering sodium iodide/urea became public, it is understandable that Volkmann felt wronged by the preference given to Roseno (Volkmann 1966; Rathert, Melchior, and Lutzeyer 1975; Rathert 1992). As an ardent anti-Semite, the Jewish descent of the latter must have disgruntled him doubly. In anger, he soon turned his back on excretion urography research altogether.

Gehe & Co. AG perhaps chose deliberately to avoid collaborating with Volkmann in order to make it clear that its own technology, realized in sodium iodide/urea, lay outside of the suggestions he had made. More enlightening is the fact that Roseno was co-inventor on the process patent covering the manufacturing of sodium iodide/urea (*Gehe & Co. AG and Roseno* 1929) (Figure 4.6, Top), a fact suppressed in interviews given by Volkmann (Rathert, Melchior, and Lutzeyer 1975; Rathert 1992) and acknowledged by him only in a much later, unpublished letter (Volkmann 1966).

In 1921 began the engaging scientific adventure of the medicinal chemist Curt Räth (Figure 4.7; see *curriculum vitae*) and the chemical engineer Arthur Binz (Figure 4.8; see *curriculum vitae*), an adventure that would eventually lead to uroselectan sodium (UROSELECTAN™, Schering-Kahlbaum AG, Berlin), the contrast agent that established excretion urography permanently.[56] This development was accompanied by a parallel one. In the Archives of Bayer AG, Leverkusen, it could be ascertained that beginning in 1922 in the chemical laboratories of Farbenfabriken vormals Friedrich Bayer & Co. [AG], Leverkusen,[57] the medicinal chemists Anton Ossenbeck (see *curriculum vitae*) and Ernst Tietze (see *curriculum vitae*) were involved in synthesizing a number of water-soluble halogenated compounds, most notably, iodinated ones[58] (Frings 1994). Among these methiodal sodium would later be identified by Gerhard Hecht (Figure 4.17; see *curriculum vitae*), a pharmacologist in the Elberfeld laboratories of the same company, as an eminently practical and general contrast agent and which, as ABRODIL™ [Pharmazeutische Abteilung Bayer-Meister Lucius der I. G. Farbenindustrie AG in Leverkusen], would in 1930 follow UROSELECTAN onto the market after just a few months of delay. What follows is the intertwined history of these two developments.

CURRICULAM VITAE

Ossenbeck, Anton [Heinrich Hubert Maria] (b. February 20, 1875, Köln, Rhine Prov., Germany // d. May 29, 1939, Köln-Mülheim, Rhine Prov., Germany).

Realgymnasium, Cologne, Germany: High-school dipl. [Abitur] (1893)/ Apprentice in pharmacy (1893–1894)/Univ. Bonn: Stud. of chem. (Spring–Fall 1894)/Technical Univ. Charlottenburg: Stud. of chem. (Fall 1894–Fall 1895)/Univ. Marburg: Stud. of chem. (Fall 1895–Spring 1898); PhD (February 1898)/Farbenfabriken Friedrich Bayer & Co., AG, Leverkusen, after 1925 I. G. Farbenindustrie AG, Werk Leverkusen: Chem. (August 1900–March 1935).

Tietze, Ernst [Adolf Wilhelm] (b. March 21, 1895, Köln-Ehrenfeld, Rhine Prov., Kingd. of Prussia, German Reich // d. August 24, 1971, Tübingen, State of Baden-Württemberg, Federal Republic of Germany).

Gymnasium in der Kreuzgasse, Cologne, Germany: High-school dipl. [Abitur] (Spring, 1914)/Königliche Eisenbahnwerkstätte, Cologne-Nippes: Unpaid trainee in mechanical engineering (April 1914–August 1914)/Wartime military serv., last as second lieutenant (August 1914–November 1918)/Univ. Bonn: Stud. of chem (December 1918–1920); PhD (August 1921)/Privat Asst. of Prof. Ludwig Claisen, Godesberg a. Rh. (April 1920–December 1920)/ Farbenfabriken vormals Friedrich Bayer Co. AG, Leverkusen, from 1925 to 1951 called Lower Rhine Div. of I. G. Farbenindustrie AG, Werk Leverkusen and from 1951 to 1974, Farbenfabriken Bayer AG: Chemist (1922–1963).

After Räth finally could complete his formal chemistry education, rendered very difficult by years of military service on different fronts and repeated severe wounding in World War I, he was formally promoted to PhD. on August 10, 1920, by the Friedrich-Wilhelms-University in Berlin. His dissertation had been performed under the guidance of Julius von Braun at the Chemistry Institute of the Agricultural College[59] and explored structure–activity relationships in alkaloids related to cocaine (Räth 1920). Räth himself synthesized the compounds and had their pharmacological activities determined by Julius Pohl, professor of pharmacological chemistry in Breslau.[35] Thus, he was trained as a medicinal chemist. On July 1, 1920, he was hired as von Braun's assistant at the institute at which he had performed his dissertation work (Personalblatt Curt Räth). At an astonishingly precocious stage in his career, he was allowed to conceive and direct doctoral dissertation projects (e.g., Walter 1922; Prange 1923), which concerned mostly his current field of interest, i.e., the synthesis of quinoline and indole derivatives. In general, he published the results, including those of his graduate students, as sole author (Räth 1924a,b). The choice of authorship reflects the dominant habit in the environment of the German universities of that time to emphasize strongly institutional position and intellectual input into project choice and organization, rather than the resolution of practical problems encountered during execution of the projects.

In 1921, Binz ended a 3-year stint as head of the chemistry division at the Georg-Speyer Haus für Chemotherapie in Frankfurt a. M., the former workplace of the physician Paul Ehrlich and a place with a long-standing tradition for the medicinal chemistry of aromatic compounds substituted by arsenic. It was there that he, emulating the development pathway of Ehrlich, had changed his research emphasis from the technology of dying and tissue printing to chemotherapy, becoming deeply involved in the development of neo-silver-salvarsan (Binz 1937a). This change in interests was also a return to the progressive pharmacology with which his famous father, the pharmacologist who pioneered chemical drugs, Carl Binz, had imbibed his son in his youth. Arthur Binz then returned to the Friedrich-Wilhelms-University in Berlin, where he had previously been, this time to occupy, as successor of von Braun, the chair of the Chemistry Institute of the Agricultural College. In order to

FIGURE 4.7 *Curt Räth.* (Reprint of figure on p. 81 in Schlenk, O. 1934. Chemische Fabrik von Heyden Aktiengesellschaft, Radebeul-Dresden, 1874–1934: Erinnerungsblätter aus 6 Jahrzehnten. Radebeul: Kupky & Dietze) (Present copyright owner unknown.)

CURRICULAM VITAE

Räth, Curt (b. December 1, 1893, Schöneberg [today Berlin], Prov. of Mark Brandenburg, Kingd. of Prussia, German Reich // d. November 29, 1944, Berlin[-Charlottenburg], State of Prussia, German Reich)

Premature termination of gymnasium, Berlin, Germany (1912)/ Landwirtschaftliche Hochschule [Agricultural College], Berlin: Stud. of agriculture and African languages (1913); Stud. of chem. (October 1913–July 1914)/ Chem. Association Exam. (July 1914)/German wartime military serv., last as second lieutenant of the reserve (August 1914–December 1917) with hospitalizations (September 1916–March 1917 and December 1917)/Kriegsmetall AG, Berlin: Steelwork supervisor (July 1918–January 1919)/Helmholtz Realgymnasium, Berlin: High-school dipl. [Abitur] make-up exam. (May 1919)/Landwirtschaftliche Hochschule Berlin: Stud. of chem. (February 1919–1920); PhD (August 1920); Asst. am Chem. Inst. (March 1920–September 1928); *Venia legendi* for chem. and chemother. (December 1925); Non-tenured Assoc. Prof. [and Veterinary Faculty, Univ. Berlin] (December 1929–November 1936)/I. G. Farbenindustrie AG, Werk Leverkusen, Germany: Head of Biological Inst., Pesticide Div. (December 1928–1929)/Chemische Fabrik von Heyden AG, Dresden-Radebeul: Head of Scientific Laboratories (April 1931–1936); Member of the Board of Directors (May 1933–1936)/Rütgers-Werke AG, Berlin: Manager of phenolate production site in Niederau, Saxony (1937–1942).

FIGURE 4.8 *Arthur Binz.* (©Universitätsbibliothek der Humboldt-Universität zu Berlin, Porträtsammlung: Arthur Heinrich Binz.)

CURRICULAM VITAE

Binz, Arthur [Heinrich] (b. November 12, 1868, Bonn, Rhine Prov., Kingd. of Prussia // d. January 25, 1943, Berlin, State of Prussia, German Reich)

Gymnasium Bonn, Germany: High-school dipl. [Abitur] (March 1887)/ Owens College, Victoria Univ., Manchester, and City and Guilds of London Inst. for the Advancement of Technical Education, London: stud. of math. & nat. sci. (April-Fall, 1887)/Peace-time military serv. 6 mo. (1887–1888)/ Univ. Bonn: Stud. of law (Fall 1887–Spring 1888); Stud. of chem. (Spring 1888–Summer, 1890)/Univ. Göttingen: Stud. of chem. (Summer 1890– March 1893); PD, (May 18, 1893)/Univ. Leipzig: Post-doctoral Fellow (November 1893–March 1894)/Chem. at Rolffs & Co., Siegburg, Germany (Spring 1893–Fall 1893)/Rhodes Works, Calico Cotton Printing Company, Manchester, UK: Head of Production (April 1894–July 1898)/ETH Zuerich, Switzerland, and Farbwerke vormals Meister Lucius und Brüning, Höchst a. M., Germany: Guest chem. (July 1898–November 1898)/Univ. Bonn, Chem. and Phys. Inst., investigator (November 9, 1898–1906); *Venia Legendi* [Techn. Chem.] (January 12, 1899); Freelance Lecturer (1899–1906)/Techn. Univ. Berlin-Charlottenburg, on sabbatical (1906)/Handelshochschule Berlin: Prof. and founding Director of the Chem. Inst. (October 1906–October 1918), including rectorship (Fall 1910–Summer 1913)/Georg-Speyer Haus

(Continued)

CURRICULAM VITAE (*Continued*)

für Chemotherapie, Inst. für Experimentelle Chemotherapie, Frankfurt a. M.: Head of Chem. Sect. (October 1918–March 1921)/Landwirtschaftliche Hochschule Berlin: Prof. and Director of Chem. Inst. (April 1, 1921–March 31, 1935), including rectorship (1925–1926)/Handelshochschule Berlin: Part-time Lecturer (Spring 1923–Spring 1925); Lecturer (Spring 1925–Spring 1929); Honorary Prof. (February 1929–July 1935)/Friedrich-Wilhelms-Univ. Berlin: Honorary Prof. (July 1929–July 1935); Prof. Emeritus (July 1935–1943)/ Dr.-Ing. h. c. Technische Hochschule Karlsruhe (July 10, 1929)/Philadelphia Franklin Inst. - Biochemical Research Foundation (February 1937–August 1938)/Guest, Deutsche Chem. Gesellschaft, Vice-Pres. (1931–1933); Gen. Secr. (1932–1937)/Chief Editor of *Z. f. Angew. Chem.* (1922–1932).

satisfy his environment, he ostensibly promised to apply chemotherapeutic methods to veterinary medicine, above all infectious diseases of domestic cattle (Binz and Räth 1927; Binz 1937a). Contradicting himself, in an alternative reconstruction of his goals, he reports on how he set out to attach arsenic and other elements such as iodine to heterocyclic aromatic nuclei in an attempt to intensify therapeutic efficacy in the treatment of syphilis (Binz 1930, [1930] 1931). Be that as it may, he did get permission to augment his Chemistry Institute by a chemotherapeutic department with the possibility to perform pharmaco-toxicological studies.

From among his predecessor's assistants he retained exclusively Räth, whom he judged the most gifted (Binz 1937a). The latter had the personality of a doer, with leadership experience both on the battlefield of World War I and as supervisor of a military-operated steel plant (Räth 1920). Binz wanted to put arsenic atoms onto heterocyclic molecules and Räth was already working on heterocyclic compounds. Beyond this the two shared a general interest in medicinal chemistry. Being his assistant, Räth became involved in all of his superior's research on arsenic and antimony-containing chemicals and for a decade all publications on the subject listed them as co-authors. The synthesized substances carried the internal codes B. R. followed by a number, reflecting their origin in the Binz-Räth collaboration (e.g., Räth 1927a; Dahmen 1927). Similarly, during that period all those patents concerning arsenical and antimony compounds, in which the inventor's names were revealed, bore both names (e.g., Binz and Räth 1923, 1924a, 1925). The situation did not change when Räth on December 17, 1925 was promoted to university lecturer in chemistry and chemical therapies (Personalblatt Curt Räth). It remained unaltered even when on October 1, 1928 he left the institute for industry and retained merely consulting and limited teaching activities at the university. But when he was appointed untenured associate professor on December 21, 1929 (Steiger 1929), Binz began publishing on arsenical compounds without his traditional co-author.

Räth's supervision of several doctoral dissertation projects is demonstrated by the acknowledgments he received for proposing the subject and for supervision of the work (Walter 1922; Prange 1923; Schiffmann 1925; Niculescu 1928; Schlottmann

1928; Garthe 1929; Hunold 1929). Binz, in most cases, was not even a member of the dissertation committee. Nonetheless, he counted all students as his by virtue of their work in his institute (Binz 1937a).

Räth early on convinced his superior to give him the chance to pursue his own academic career, which required research in an area distinct from that of the chairman of the institute and eventual presentation thereof in the form of a habilitation dissertation.[60] There is ample evidence that he was granted the necessary independence. Later Binz reported that his assistant was charged with guiding the group involved in veterinary chemotherapeutic exploitation of iodine derivatives of pyridine (Binz [1930] 1931, 1937a). A total of 73 such chemicals were synthesized (Binz [1930] 1931, 1937b). In contrast to the codification of arsenical and antimonial compounds, whose numbers were preceded by the letters B. R., the numbers for iodinated ones were preceded by only the letter R (e.g., Dahmen 1927), indicating that Räth after 1924 was alone running the iodine-based program. In some publications compounds were instead designated simply with *ad hoc* roman numerals, a common practice in chemistry journals. He was also permitted to patent some corresponding inventions by himself, an opportunity he, with the help of industry, did not allow to escape (Räth 1924c, 1925, 1927b).[61]

Unlike today, the German patent law instituted in 1877 did allow neither coverage of novel chemical substances destined to become drugs, nor their novel medical uses. Only manufacturing processes could get protection. Chiefly the medical community frowned upon the idea of patenting a medicinal product and industrial profits derived from people's illnesses[54] (Mann and Plummer 1991). This situation gave chemists and pharmacists a clear advantage over pharmacologists and clinicians. The situation in Germany changed on January 28, 1935, when contrast agents were declared commercial goods distinct from drugs and therefore patentable (Roentgen Contrast Agents 1935). In effect, the change in the German law was an incentive to the native industry to invest in the field. While similar situations existed in various other countries, the United States followed its proper path having a tradition of allowing composition of matter patents also for drugs, including contrast agents.

Räth (1924c) applied for a patent covering a process for the preparation of selectan (Figure 4.9, **1**)[61] and in so doing managed to beat by only a few months, the submission date for publication of the paper by Magidson and Menschikoff (1925) that described the synthesis of selectan in a scientific journal. He next achieved coverage of a potent purification process for 2-amino-5-iodopyridine (Räth 1925), a crucial intermediate in the synthesis of selectan by the Magidson and Menschikoff process. A number of iodinated relatives of selectan were synthesized by Räth's graduate students,[62] but it was he alone who invented a process for preparing chemical derivatives of halogenated oxypyridines with elevated water solubility. On May 12, 1927, he, as sole inventor and Schering-Kahlbaum AG as assignee, applied for a patent (Figure 4.10) covering the alkylation process used to obtain the highly water-soluble halogen substituted oxypyridines (Räth 1927b). The exemplifications included in particular selectan neutral (Figure 4.9, **2**), as well as uroselectan that became so named much later on von Lichtenberg's suggestion. It disclosed solubility results for the former product and for uroselectan sodium (Figure 4.11, **3**), the latter being the compound that would later bring about a breakthrough in excretion urography.

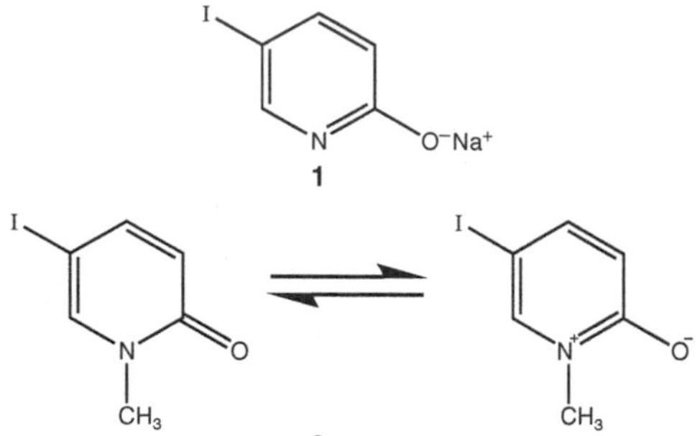

1

2

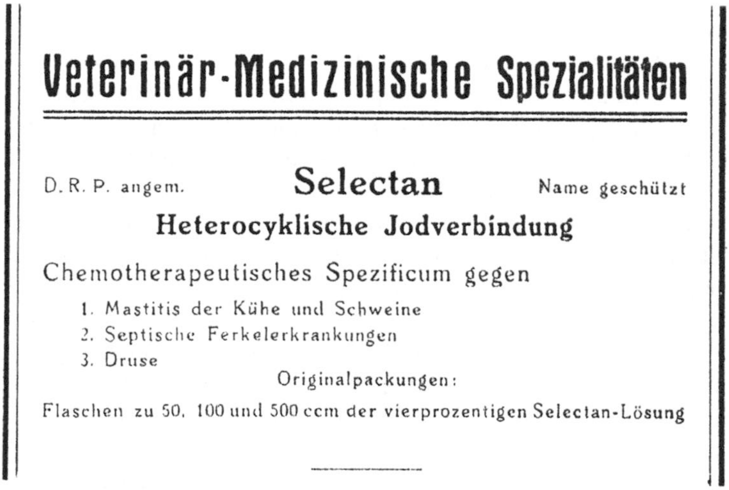

Veterinär-Medizinische Spezialitäten

D. R. P. angem. **Selectan** Name geschützt

Heterocyklische Jodverbindung

Chemotherapeutisches Spezificum gegen

1. Mastitis der Kühe und Schweine
2. Septische Ferkelerkrankungen
3. Druse

Originalpackungen:

Flaschen zu 50, 100 und 500 ccm der vierprozentigen Selectan-Lösung

FIGURE 4.9 Intravenous veterinary therapeutics as departure points for contrast agents:

1: Selectan sodium (SELECTAN™, Schering-Kahlbaum AG, Germany). *I/P* = 0.5.
2: Selectan neutral (SELECTAN NEUTRAL™, Schering-Kahlbaum AG, Germany). *I/P* = 1.0.

Top: Both SELECTAN and SELECTAN NEUTRAL were developed, and in 1928 put on the market as veterinary antibacterial agents.

Bottom: Advertisement of Schering-Kahlbaum AG for SELECTAN in 1928. The veterinary indications are mastitis of the cow and the pig, septic diseases of the piglet, and strangles, an infection of horses. (Reprinted with permission from Schering Archiv, Bayer AG, Berlin.)

One of the essential prerequisites of a patent is that the invention has an element of surprise to the "person having ordinary skill in the art." Yet that surprise quite often vanishes already before the initial reading of its description is finished. To an observer with a long historical perspective, the technological step forward captured by a patented invention sometimes seems astonishingly small. The one by Räth (1927b) fully exemplifies the situation. The synthetic process was new but little surprising. The elevated water solubility of selectan neutral qualifies as a surprise,

DEUTSCHES REICH

AUSGEGEBEN AM
5. SEPTEMBER 1930

REICHSPATENTAMT

PATENTSCHRIFT

№ 506425

KLASSE 12p GRUPPE 1/??

R 71177 IVa/12p

Tag der Bekanntmachung über die Erteilung des Patents: 21. August 1930

Schering-Kahlbaum A.-G. in Berlin*)

Verfahren zur Erhöhung der Löslichkeit von halogensubstituierten Oxypyridinen in Wasser

Patentiert im Deutschen Reiche vom 12. Mai 1927 ab

Die Erfindung bezweckt, in Wasser schwer lösliche Verbindungen der Pyridinreihe, die Hydroxylgruppen und außerdem noch Halogenatome, wie Jod, Brom, oder auch mehrere verschiedene dieser Elemente im Kern substituiert enthalten, in dem erwähnten Mittel leichter löslich zu machen und auch nach anderen Richtungen ihre Eigenschaften in therapeutischer Hinsicht zu verbessern, z. B. derart, daß Reizwirkungen bei ihrer Verwendung beseitigt oder vermindert werden.

Die in Betracht kommenden Pyridinverbindungen enthalten die OH-Gruppe in 2-Stellung und können die Halogenatome in 3- bzw. 5-Stellung oder auch in beiden Stellungen substituiert enthalten. Als schwer lösliche Ausgangsstoffe kommen vorzugsweise in Betracht 2-Oxy-5-jodpyridin, Dijodverbindungen usw.

Nach dem vorliegenden Verfahren wird die Löslichkeit derartiger Ausgangsstoffe dadurch erhöht, daß man sie mit alkylierenden Mitteln behandelt, wobei unter solchen Mitteln auch saure Gruppen, insbesondere die Carboxylgruppe am Alkylrest enthaltende, zu verstehen sind.

Man behandelt hierzu die Verbindungen, deren Löslichkeit verbessert werden soll, mit Alkylhalogeniden, Halogenfettsäuren oder anderen alkylierenden Mitteln vorteilhaft bei Gegenwart von säurebindenden Stoffen.

Als Produkte der Reaktion kommen beispielsweise Verbindungen nach folgenden Formeln in Betracht:

N-Methyl-2-oxo-5-jodpyridin:

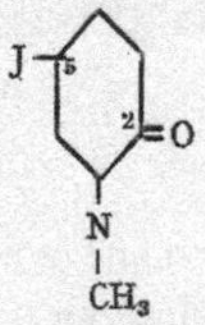

2-Oxo-5-jodpyridin-N-essigsäure

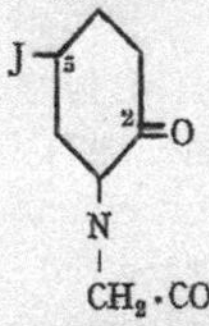

2-Oxo-3-brom-5-jodpyridin-N-essigsäure

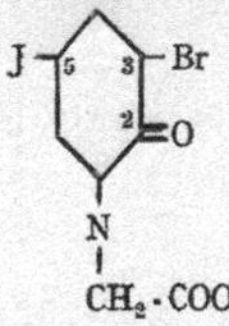

*) Von dem Patentsucher ist als der Erfinder angegeben worden:

Dr. Curt Räth in Rangsdorf, Kr. Teltow.

FIGURE 4.10 Front page of the patent of Curt Räth covering the synthesis process for uroselectan sodium, the first commercial contrast agent for excretion urography based on an iodinated organic molecule. The missing double bonds of the benzene rings can be confusing, but the nomenclature is unambiguous.

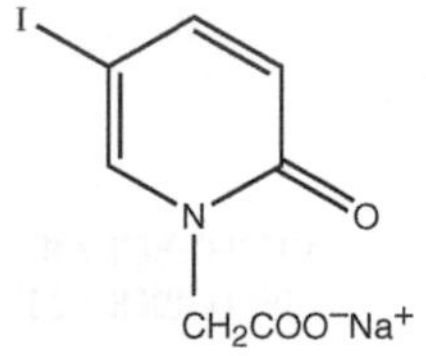

FIGURE 4.11 Uroselectan sodium (UROSELECTAN™), an iodinated organic compound that became the breakthrough commercial contrast agent for excretion urography, and its congeners.

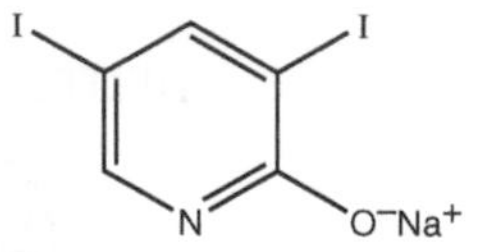

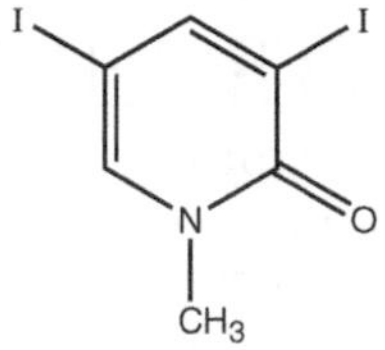

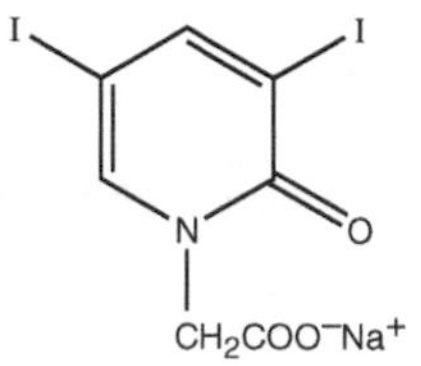

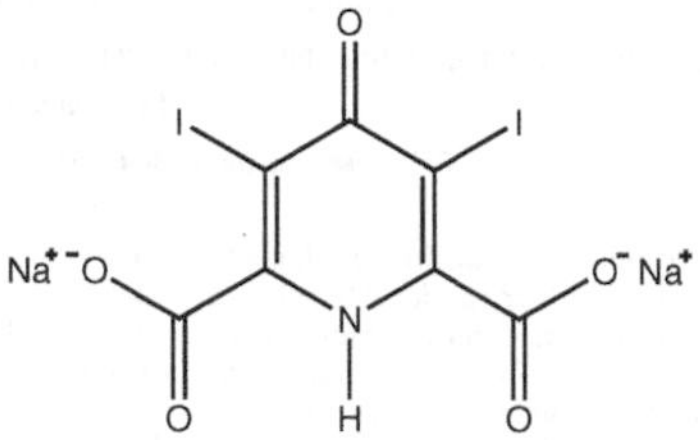

3: Uroselectan sodium (UROSELECTAN, Schering-Kahlbaum AG, Germany). $I/P = 0.5$.
4: Compound of Pfeiffer (1887), sodium salt. $I/P = 1.0$.
5: Diiodo-analog of selectan neutral (2). $I/P = 2.0$.
6: Diiodo-analog of uroselectan sodium. $I/P = 1.0$.
7: Diiodochelidamic acid sodium salt (Lerch 1884). $I/P = 1.0$.

less so that of uroselectan sodium. Yet evidently the surprisingly elevated water solubilities of the new selectan-type compounds did suffice to convince the patent examiners of the invention.

The patent stated that the invention besides teaching how to create halogen-bearing oxypyridine derivatives with elevated water solubility also aimed to improve their properties in therapeutic terms, e.g., such that irritating effects are eliminated or reduced in their use. Already on December 10, 1926, the surprisingly low toxicity[63] of uroselectan sodium (Figure 4.11, **3**) had been discovered (Binz 1937a, [1930] 1931).[56] However, given the legal regime the information was not helpful for getting a patent and was therefore withheld.

In terms of publications, Binz and Räth apparently reached a sophisticated agreement that was to satisfy simultaneously three prerequisites[64]: recognition of the existing long-term collaboration, their independence in the area of iodinated compounds, and well-separated printed acknowledgement of public and industrial research support. Two numbered series of publications were started: one in chemistry and one in biochemistry. Irrespective of the real author(s) and title of the article, all publications of these series carried as subtitles or footnotes one of the following lines: "Communication regarding derivatives of pyridines and quinolines by A. Binz and C. Räth"[65] or "Communication regarding biochemistry of pyridine derivatives." In self-citations of their work, these subtitles were provided or deleted to suit the purpose. The agreement documented the collaboration and allowed Binz from then on to refer to compounds of the Binz/Räth series, even when they had been reported just in publications for which Räth had signed as sole author. The agreement permitted Räth to publish a series of papers alone or at least without his superior as co-author (e.g., Räth and Prange 1928; Räth 1930, 1931a–d). Lastly, the agreement foresaw that all papers bearing both authors to list only public research support, whereas the papers authored by Räth alone acknowledged only industrial support (*vide infra*). Binz honored the agreement until long after his assistant had left the institute (Binz 1935b).

Binz specifically credits Räth for having discovered the anticoccal activity of some iodinated pyridine derivatives, probably sometimes in 1926. This means that their screening for anticoccal activity in infected rabbits (Räth 1927a) was the latter's initiative.[63] In his reports, Binz ([1930] 1931, 1937a) expressed high esteem for how his assistant had helped transform into reality the grand scheme with which he had boosted his research after returning to Berlin. Although one must concede that Binz during their collaboration had a genuinely magnanimous attitude toward his assistant, his later revised story does not do full justice to the contributions of the latter. Similar underrating of his involvement, even to the extent of having his name omitted from the acknowledgments, occurs in chronicles by other important participants of the developments (Swick 1975). Räth undeniably was instrumental in Binz's research on arsenical compounds. A similarly crucial scientific role of Binz in Räth's pursuits in the field of iodinated ones is much less evident. The fact shines through that Binz originally expected great therapeutic progress on syphilis, a human disease, mostly from arsenical chemicals and he left Räth to try his luck with iodinated ones and veterinary applications, especially after a group of the latter substances had failed to show antisyphilitic activity (Räth 1927a). There is little doubt that, contrary to the suggestion of much of the published historical literature, Räth deserves at least as much credit as Binz for having pushed the field of iodinated pyridine derivatives through innovative synthetic approaches and for having monitored the solubility and toxicity of these compounds and their therapeutic potential. The usual assignment of the greater credit to Binz is understandable, though not defendable, if one considers four aspects. First, he was the head of the chemistry institute at which Räth worked and thus the point of reference in the strongly hierarchical German university system. Second, Binz at some critical point assumed an important coordinating role in the events that in due course led to the breakthrough in excretion urography. Third, he knew how to sell himself and his role through publication as sole author of his version of history (Binz [1930] 1931, 1937a). His elevated position and his permanence in academia facilitated cultivation of his image. In contrast, Räth never put forth his version of history independently of Binz and, before Swick (*vide infra*) successfully evaluated his compounds as contrast agents in the clinic, he switched from academia to industry. Industry positions were then, and to a large degree are still today, less intellectually prestigious than academic ones, a fact possibly attributable to a perceived difference in idealism of purpose between the two environments.[66] Fourth, Schering-Kahlbaum AG, which was deeply involved in the research, development and distribution of their contrast agent, uroselectan sodium (*vide infra*), may have preferred the loyal consultant Binz over Räth, who already preceding the commercial launch of the product had become associated with I. G. Farbenindustrie AG and subsequently was employed by Chemische Fabrik von Heyden AG.[67]

Räth had been appointed university lecturer by the end of 1925 and stayed at the institute in Berlin until October 1, 1928 (Binz 1937a). Thereafter, for a 2-year trial period[68] he assumed responsibility for research and development in the section on pesticides at I. G. Farbenindustrie AG in Leverkusen (Binz 1928; Bayer AG 1930s, 97), while pro forma continuing to teach in Berlin, to consult with Binz and to publish work from his Berlin time. In 1929, he was promoted to untenured associate professor in Berlin. In 1932, he took a leave of absence from his duties at the university and in 1936 his academic appointment was withdrawn, because his industrial activity had prevented him from fulfilling the expected teaching duties (Vahlen 1936). In April 1931 Räth joined the Chemische Fabrik von Heyden AG in Dresden, the producer of the contrast agents COLLARGOL, UMBRATHOR, and THOROTRAST, to become its director of research and eventually also a member of the board of directors (Schlenk 1934). Apparently after Räth's departure from the chemistry institute in Berlin, Kurt Hillgruber, one of Binz's assistants, took over the task of synthesizing known iodinated compounds (Binz and Räth 1930a). Furthermore, he developed an improved analytical method for quantifying organically bound iodine in urine (Hillgruber 1930).

Based on acknowledgments in publications concerning arsenocompounds Binz's institute had research support in that sector from the German Ministry of Agriculture, Domains and Forests and from the Emergency Action Group for German Science.[69] In his account, Binz (1937a) recorded that in July 1924 he contacted a large chemical company, but he did not reveal its name. He succeeded in convincing it to financially supplement public research support at his institute. He reports that through this support he and Räth came to gain numerous patents in Germany and in certain foreign countries. This fact has allowed identification of the company as Deutsche Gold- und Silber-Scheideanstalt, vormals Roessler [AG], Frankfurt a. M.[70] The fit between its activities and those of Binz and Räth was perfect. Both were exploring arseno compounds (*Deutsche Gold- und Silber Scheideanstalt and Albert* 1924) and novel pyridines (*Deutsche Gold- und Silber Scheideanstalt* 1925), including iodinated ones, e.g., Kochendoerfer (1925) and Vieweg (1926).

It appears that company funding was dedicated specifically to Räth, who in turn in publications in which he was the principal author thanked the company by name for its help. He did so by thanking it for having supplied precursor substances for his chemical syntheses (e.g., Räth and Prange 1928; Räth 1930, 1931c), probably a euphemism for financial support. The company's name never appeared in papers that comprised Binz as author. Probably this pattern of acknowledging the help of industry reflected merely an accounting necessity in the face of otherwise incompatible funding sources.

In 1926 Deutsche Gold- und Silber-Scheideanstalt signaled to Binz its intention to abandon support of the research program. His negotiations with a first alternative industry failed, but in November of the same year, without having had support for their research interrupted, Chemische Fabrik auf Actien (vormals E. Schering), Berlin, took over sponsorship (Binz 1937a). Only a few months thereafter this company was to become Schering-Kahlbaum AG. The associations of Binz and Räth with this company hence began in 1926 and they lasted until about 1933[71] and October 1, 1928, respectively (Binz 1937a). Given the record of inventorship and patent assignments, Schering-Kahlbaum AG must have acquired from the former sponsor, Deutsche Gold- und Silber-Scheideanstalt, the intellectual property that the research program had procured.[70]

At the time Schering-Kahlbaum AG had a fanned-out activity with iodine-containing chemicals. Since 1887 one of its precursor companies traded iodoform manufactured by a patented electrolytic process (*Chemische Fabrik auf Actien* 1884; Wlasich and Berghausen 1996). As producer of iodoform it surely was informed about this chemical's previous application as a contrast agent when formulated as JODOFORM-GLYCERIN for injection. Other pharmaceuticals of the company included the iodinated hormone thyroxin purissimum that it isolated from natural sources (Schoeller and Gehrke 1927) and a 25% (w/w) lithium iodide solution (UMBRENAL) (see Chapter 3) (Joseph 1921) as contrast agent for retrograde pyelography. The iodinated oral cholecystographic contrast agent diiodophenylcinchonic acid (BILOPTIN™)[72] had earlier been introduced, but by the time the collaboration with Binz and Räth began, it had been taken off the market. Apart from having these articles, the company was working on the synthesis of polyiodinated phenol ethers (Schoeller and Schmidt 1926), the iodinated compounds thyroxin (Schoeller and Gehrke 1927), iodophthalein sodium (von Schickh 1926), monoiodooxindole (Schoeller and Schmidt 1924a), polyiodoisatin (Schoeller and Schmidt 1924b) and contrast agents related to diiodophenylcinchonic acid (Schoeller and Dohrn 1926), as well as various chemical substances bearing arsenic residues (Schoeller and Gehrke 1924). Pyridine derivatives, among them polychlorinated ones (Dohrn and Horsters 1922), were likewise under study. Schering-Kahlbaum AG must have seen a general strategic value in amplifying its presence in this field and closing out the competition through acquisition of the results of Räth and Binz.

In 1926 Räth had discovered the anticoccal activity of iodinated compounds of the selectan series. A consequence of the discovery was that the research at Binz's institute achieved at least temporary recognition in veterinary medicine, when in December 1927 the German Ministry of Agriculture, Domains and Forests authorized marketing in sequence of the iodinated compounds, selectan sodium (Figure 4.9, **1**), and selectan neutral (Figure 4.9, **2**), for the intravenous treatment of mastitis of the cow (Binz, Räth and von Lichtenberg 1930; Binz 1937a). Although a clear, yellow 4% solution of selectan sodium was sold to veterinarians for a short while in 1928, its effectiveness was too variable (e.g., Diernhofer 1928; Neumann-Kleinpaul and Pessinger 1929) to find widespread use.[73] Apparently, the veterinary clinical studies organized by Schering-Kahlbaum AG, although extensive, had been too superficial to clearly establish efficacy (Dahmen 1927).

While the veterinary trials were still ongoing clinical trials in man became desirable. Binz's brother-in-law, Carl von Noorden, who at Frankfurt a. M. clinically tested arsenical compounds, put him in contact with his friend Leopold Lichtwitz (Figure 4.12; see *curriculum vitae*), who at the time was professor at the University of Hamburg and chief of the Section of Internal Medicine of the Municipal Hospital in Altona, today a part of Hamburg. Binz being interested in clinically evaluating the compounds that had originated in his institute visited Lichtwitz on May 22, 1927 and reports having provided him shortly thereafter with selectan sodium and then with selectan neutral (Figure 4.9, **1** & **2**) for clinical trials regarding their anticoccal activity in human genitourinary infections (Binz 1937a). Lichtwitz instead maintains that Schering-Kahlbaum AG furnished the drugs (Lichtwitz 1930a), which is plausible given that the company had at that time both selectan sodium and selectan neutral under development for veterinary applications, and had much easier access to the necessary quantities of pure material.

Binz and Räth (1930a) have stated that through collaborating with Lichtwitz "they had entered unwittingly the area of urology."[26] There is no mention of radiology.

FIGURE 4.12 *Leopold Lichtwitz.* (Portrait provided by the Staatsarchiv, Hamburg, Germany.)

CURRICULAM VITAE

Lichtwitz, Leopold (b. December 9, 1876, Ohlau [today Oława, Poland], Prov. of Silesia, Kingd. of Prussia, German Reich // d. March 16, 1943, New Rochelle, New York, USA)

Gymnasium Ohlau, Silesia, Germany: High-school dipl. [Abitur] (Spring 1896)/Univ. of Breslau, Germany [today Wrocław, Poland]: Stud. of med. (Spring 1896–Spring 1897 & Spring 1898–Spring 1899)/Univ. Freiburg i. Br.: Stud. md. (Spring 1899–Fall 1899)/Univ. Munich: Stud. med. (Spring 1897–Spring 1898 and Fall 1899–Winter 1900); State medical lic. (December 1900)/ Private practice of his father and Hosp. in Ohlau: Asst. (January 1901–April 1906)/Royal Children's Clin., Berlin: Volunteer Physician (April 1901)/Univ. Leipzig: MD (April 1901); Stud. of chem. and phys. chem. (Spring 1906–Spring 1907)/Univ. Freiburg i. Br., Medical Policlinic: First Asst. (Spring 1907–December 1908)/Univ. Göttingen, Medical Clinic: Resident (January 1908–April 1910); Asst. Medical Director (April 1910–1930); Head of Medical Policlinic (October 1910–August 1914); *Venia legendi* for internal med. (December 1908); Titular Prof. of Med. (1913); Subst. Director of the Clinic and Policlinic (October 1914–November 1916)/Voluntary wartime military serv. as physician and chief of a red cross column (August–October 1914)/Städtisches Krankenhaus zu Altona: Chief of Internal Med. (December 1916–September 1931)/Rudolf-Virchow Krankenhaus, Berlin: Director (October 1931–March 1933)/Escape to USA (1933)/Montefiore Hosp., New York: Director Internal Med. Ward (1933–1943)/Columbia Univ., New York: Clinical Prof. of Med. (1935–1943).

This passage indicates that at least up to this time the two chemists were under the impression that Schering-Kahlbaum AG was interested exclusively in the antiinfective activities of their therapeutic candidates for the treatment of bovine mastitis and human sepsis. Evidently, they had spent no thoughts on candidates for radiological applications. In contrast, by 1927 Schering-Kahlbaum AG had accumulated experience in producing and merchandizing such goods, which included, besides the already mentioned lithium iodide (UMBRENAL) and diiodophenylcinchonic acid (BILOPTIN) also barium sulfate for gastrointestinal use (RÖNTYUM™) (Simons 1923/1924). The company therefore monitored very carefully the goals pursued by others, in order to develop potentially competing products. It must have considered the possibility that in the collection of iodo-compounds synthesized under its sponsorship by Räth and his students some might have potential as contrast agents. But, either the industry initially left its academic collaborators in the dark over the full range of its interest or these were too enthralled with the antiinfective activities for the contrast agent idea to register.

Parallel to the human antisepsis trials with selectan sodium and selectan neutral performed under Lichtwitz, Ernst Philipp of the Gynecology Clinic of the Friedrich-Wilhelms-University in Berlin, tested intravenous selectan sodium in human cases of severe puerperal sepsis (cited in Binz, Räth, and von Lichtenberg 1930).[74] In this context, Räth organized at his work place the study of the product's excretion in animals. Rapid excretion through the urine was demonstrated. As will be explained below, after October 24, 1927, Binz and Räth knew about Schering-Kahlbaum AG's interest in iodinated chemicals as potential urographic diagnostics. Indeed, Binz, Räth, and von Lichtenberg (1930) report having aired the urographic use of selectan sodium at that time but that the issue had not been insisted upon. It appears that Räth and Binz did not take the contrast agent application potential of their compounds seriously until about 1 year later.

On October 24, 1927 Binz mailed uroselectan sodium to the urological surgeon, Theodor Hryntschak (Figure 4.13; see *curriculum vitae*), then university lecturer at the University of Vienna and chief of the Urological Ward of the Wilhelminen Hospital. Binz reports (Binz 1937a) that the mailing occurred in response to a request by the following persons from Schering-Kahlbaum AG: Walter Schoeller (Figure 4.14; see *curriculum vitae*), the director of the main laboratory and Otto von Schickh[75] (Figure 4.15; see *curriculum vitae*), the inventive young chemical engineer involved in chemical process development for the syntheses of the cholecystographic contrast agent, iodophthalein sodium (von Schickh 1926) and iodinated pyridine derivatives (von Schickh 1927a,b; Allardt and von Schickh 1930).

FIGURE 4.13 *Theodor Hryntschak.* (Josephinum, Ethics, Collections of History of Medicine, MedUni Vienna.)

CURRICULAM VITAE

Hryntschak, Theodor (b. July 15, 1889, Vienna, Archduchy of Austria, Austro-Hungarian Empire // d. June 28, 1952, Vienna, Republic of Austria).

Piaristengymnasium, Vienna: High-school dipl. [Abitur] (1907)/Univ. Vienna: Stud. of med. (1907–1913)/MD (January 1913); Post-doctoral training in pathol, histol. and phys. chem. (1913–1914); Surg. training (1913–1914)/ Wartime military serv. as roentgenologist of mobile Austrian military surg. unit [first lieutenant] (1914–1918)/Sophienspital, Vienna: Urological surg. training with assistant position (1918–1927)/University Vienna: *Venia legendi* for urol. (April 1925)/Wilheminenspital, Vienna, Chief of Urol. Sect. (1927–1944)/Wartime military serv. at Standortlazarettt Rudolfspital, Vienna: Urol., Oberstabsarzt, (~1940–1944)/Univ. Vienna: Lecturer (September 1939); Titular assoc. prof. (November 1939, confirmed 1948)/Wiener Allgemeine Poliklinik: Chief of Urol. Serv. (1944–1952).

Elsewhere, Binz specified that the supply to Hryntschak encompassed "selectan-compounds among which was the later so called uroselectan"[26] (Binz [1930] 1931). According to him, the purpose was to test them for usefulness as a urographic contrast agent, without specifying the mode of administration (Binz, Räth,

FIGURE 4.14 *Walter Schoeller*, 1931. (Courtesy of Schering Archiv, Bayer AG, Berlin.)

CURRICULAM VITAE

Schoeller, Walter [Julius Viktor] (b. November 17, 1880, Berlin, Prov. of Mark Brandenburg, Kingd. of Prussia, German Reich // d. July 25, 1965, Konstanz, State of Baden-Württemberg, Federal Republic of Germany)

Wilhelms-Gymnasium, Berlin: High-school dipl. [Abitur] (1899)/Univ. Bonn (Fall 1899–Fall 1900); Univ. Berlin (Fall 1901–Fall 1906): Stud. of chem./ Chem. Association Exam. (May 1905)/Univ. Berlin: PhD (December 1906); Asst. of Organic Chem. (April 1907–August 1914); *Venia legendi* for chem. (December 1915)/Wartime military serv. as cavalry lieutenant, column commander and gas protection officer (August 1914–1917)/Planning of a German univ. in Bukarest, Rumenia (1917–1919)/Univ. Freiburg i. Br., Germany: *Venia legendi* in chem. [Nostrifikation] (April 1919); Asst. (April 1919–May 1919); Assoc. Prof. and Head of Medicinal Chemist Section, Chem. Inst. (May 1919–April 1924); on leave of absence (October 1923–April 1924)/Chemische Fabrik auf Actien (vormals E. Schering), after 1927 Schering-Kahlbaum AG, after 1937 Schering AG: Initially Substitute Member, later full Member of the Board of Directors (1923); Head of Principal Laboratories (1927–1944)/ Univ. Würzburg, Dr. med. h. c. (1932)/Kaiser-Wilhelm Inst. of Biochem., Berlin-Dahlem: External Scientific Member (1936)/Technische Hochschule Braunschweig: Dr. rer. nat. h. c. (1950)/Forschungs-Inst. Heiligenberg für Med. und Chem., Heiligenberg, State of Baden-Württemberg, West Germany: initially Founding Member and Scientific Director, later Honorary President (1946–1965)/West-German honor: Grosses Verdienstkreuz des Bundesverdienstordens (October 1955).

and von Lichtenberg 1930; Binz 1937a). Lichtwitz confirmed "that the preparations of Professor Binz were tested for the same purpose before Dr. Swick began his studies (without his knowledge) by a man of great experience in research work—without sufficient results" (Lichtwitz 1930d). Lichtwitz must have referred to Hryntschak. In one of his later publications von Lichtenberg admitted that Hryntschak had enjoyed limited success, while asserting that the candidate contrast agent he had used was selectan neutral (von Lichtenberg 1932). Maybe he was poorly informed. More likely he did not want to divulge the fact that the idea of testing uroselectan sodium in excretion urography had been examined before he ever got involved with it.

The choice of Hryntschak as collaborator for radiological experiments by the people from Schering-Kahlbaum AG was natural. He had a track record of publications about retrograde pyelography (Blum, Eisler, and Hryntschak 1920; Hryntschak and Sgalitzer 1921; Sgalitzer and Hryntschak 1921, 1924). Together with Max Sgalitzer, he had contributed substantially to the semeiotics of X-ray shadows of the kidney and the urinary tracts of both man and animals contrasted through retrograde administration of potassium iodide solution, as advocated at his university by Hans Rubritius (Rubritius [1919] 1920; Sgalitzer 1921). Since the patent of Räth (1927b) for uroselectan sodium was not published until August 21, 1930, Hryntschak in the

FIGURE 4.15 *Otto von Schickh*, about 1935. (Courtesy of Eva-Maria Wartenberg [-von Schickh], Puchheim, Germany.)

CURRICULAM VITAE

von Schickh, Otto (Edler) (b. March 21, 1899, Vienna, Archduchy of Austria, Austro-Hungarian Empire // d. August 7, 1971, Ludwigshafen, State of Rhineland-Palatinate, Federal Republic of Germany)

Austrian wartime military serv., last as standard-bearer (March 1917–November 1918)/Humanistisches Gymnasium, Znaim, Moravia, Austria-Hungary [today Znojmo, Czechia]: Wartime high-school dipl. [Kriegs-Abitur] (1918)/Technische Hochschule, Vienna: Stud. of chem. (Fall 1918–Spring 1923); Dipl. Ing. Chem. (May 1923)/Eduard-Spiegler-Stiftung, Vienna: Collaborator of Prof. Sigmund Fränkel (1923–1924)/Chemische Fabrik auf Actien (vormals E. Schering), Berlin, in 1927 fused to become Schering-Kahlbaum AG, Berlin: Engineer of chem. process development (November 1923–March 1932)/Technische Hochschule, Vienna: Dr. Ing. (July 1933)/

Landwirtschaftliche Hochschule Berlin: Extraordinary Asst. at Chem. Inst. (June 1933–October 1936)/Chemische Fabrik Tempelhof Preuss und Temmler, AG., Berlin-Tempelhof: Chem. (October–December 1936)/I. G. Farbenindustrie AG, Werk Ludwigshafen/Rhein, Germany, in 1951 becoming BASF AG, Ludwigshafen: Laboratory Chem. (January 1937–March 1964)/ German wartime military serv., last as lieutenant in the reserve, anti-aircraft defense (1939–1945).

autumn of 1927 could not have become aware of the compound on his own. We are forced to conclude that the person who first recognized potential contrast agents for excretion urography among Räth's iodopyridone derivatives and singled out uroselectan sodium for early testing was somebody at Schering-Kahlbaum AG.

The fact that Binz (Binz, Räth, and von Lichtenberg 1930; Binz 1937a) cites the 28-year-old von Schickh as jointly responsible with Schoeller for the initiative to test Räth's iodopyridine derivatives in excretion urography is highly significant. Unless their contribution had really been outstanding, young collaborators were mostly ignored at that time and in this respect von Schickh was singularly at risk, having at that moment not yet earned a doctoral degree in chemistry. Thus, his specific acknowledgement very likely means that he played a preeminent role. With an e-mail from Christine Berghausen, Scheringianum, Berlin, March 5, 2004, I was informed that von Schickh also perceived himself as having played a central role, and he ended even pursuing legal avenues to make his point.

Through his chemical process development activity at Schering-Kahlbaum AG on the well-known biliary contrast agent, iodophthalein sodium (von Schickh 1926), von Schickh had acquired, other than experience with iodination reactions, exposure to the concept of contrast agent. As a sole inventor, he patented an improved process for the synthesis of 2-amino-5-iodopyridine, a crucial intermediate in the production of the selectan family of compounds (von Schickh 1927b). It appears extremely likely that he was also working on the synthesis of the selectans themselves, uroselectan included. He certainly was ideally placed to make the connection between the application of the iodinated chemicals as anticoccal drugs and their potential utility as X-ray contrast agents. The rapid renal excretion of selectan sodium in animals demonstrated shortly before by Räth (cited in Binz, Räth, and von Lichtenberg 1930) could have played a role in the realization that substances of this class might serve as urographic contrast agents.

Apparently von Schickh acted as champion for the contrast agent application within the company and cultivated the contacts with Binz and Räth. Schoeller, as director of research at Schering-Kahlbaum AG, could not be passed over by Binz in his citation, but he probably acted foremost as senior advisor with strategic perspective and pharmacological experience during selection of the most promising candidate compound and the identification of suitable university collaborators. Schoeller's antecedent experience in the field of iodinated products is documented in a number of patents and papers on their synthesis and pharmacology (e.g., Schoeller and Schmidt 1924a,b; Schoeller and Gehrke 1924, 1927). The relative merits of von Schickh and

Schoeller will probably never be fully sorted out and hence we will have to cite them often together.[76]

Before ever receiving contrast agent candidates from the Berlin group, Hryntschak for the 7th Meeting of the German Society of Urology in 1926 in Vienna had announced an oral communication entitled "Regarding the question of the roentgenological representation of the kidney (animal experiments and demonstration)"[26] (Deutsche Gesellschaft für Urologie 1926). Later he claimed to have intended disclosure of his results on excretion urography with halogenated organic chemicals on that occasion but ended up not doing so because his studies were not yet complete (Hryntschak [1928] 1929). Nevertheless, through the printed announcement and through the grapevine Schering-Kahlbaum AG may well have known of Hryntschak's interest in the field. While his initial industrial supplier and the codification of his compounds made it possible to keep their structures secret, it would have been more difficult in a university environment to hide the fact that excretion urography was being pursued. This situation offers another reason for how Schering-Kahlbaum AG came to seek out this academic collaborator in pursuit of its urographic objectives.

Yet, an alternative scenario exists. Through the availability of diiodophenylcinchonic acid (BILOPTIN) (Pribram 1926), Hryntschak almost certainly realized the interest in iodine-containing organic contrast agents at Schering-Kahlbaum AG. The possibility can therefore not be excluded that selectans were provided to him in response to his request for brominated and iodinated compounds. He had been forced to look for a new source after the company he had collaborated with earlier in the autumn of 1927 had ceased as a supplier (Roseno [1928] 1929).

Hryntschak describes his involvement in excretion urography in the following way: He began animal experimentation on this procedure in the autumn of 1925, when he was university lecturer at the University of Vienna and intern in the Urology Ward of the Sophien Hospital in Vienna.[77] The experiments were performed in the Roentgen laboratory of the II. Medical University Clinic in Vienna (director, Gustav Singer) and later at the Roentgeninstitute of the Wilhelminen Hospital Vienna [director, Martin Haudek], the hospital at which he served as director of the urology clinic beginning in 1927. To start with Hryntschak repeated the work of others with sodium bromide and sodium iodide. Then, he switched to brominated and iodinated organic compounds that had kindly been made available by two large, unnamed industries (Hryntschak [1928] 1929; Hryntschak 1929). From what has so far been documented, we can assert that Schering-Kahlbaum AG was the second industry. Identification of the original sponsor was less straightforward.

Based on what has been said above about the origin of sodium iodide/urea at Gehe & Co. AG this company can be excluded. While numerous industries were active in the field of iodized oils and fatty acids, the number of companies patenting other iodinated or brominated molecules in the 1920s was rather restricted. Apart from the already aforementioned Deutsche Gold- und Silber-Scheideanstalt, Frankfurt a. M. and Schering-Kahlbaum AG, Berlin, and its predecessor companies, one of the few was Farbenfabriken vormals Friedrich Bayer & Co. AG, Leverkusen. After January 1, 1925, this company became the Lower Rhine Division of I. G. Farbenindustrie AG in Frankfurt a. M.

A case will now be made for the medicinal chemistry laboratories of I. G. Farbenindustrie AG in Leverkusen being the first supplier of brominated and iodinated compounds to Hryntschak. According to Frings (1994), no documents unequivocally proving such collaboration could be found in the company archives. Hence, the case will be built solely on circumstantial, though strong evidence.

The laboratory notebooks of the chemists Ossenbeck and Tietze document that the synthesis of iodinated chemicals at Farbenfabriken vormals F. Bayer & Co. in Leverkusen, Germany, had started in early 1922. Tietze (see *curriculum vitae*) terminated his activity the same year, but Ossenbeck (see *curriculum vitae*) continued with it at least until 1926, i.e., even after the company had become part of I. G. Farbenindustrie AG. The latter company and its precursor acquired externally, and came up internally, with improved and patentable processes for preparing halogen-containing alcohols (*I. G. Farbenindustrie AG* 1924; Callsen 1926). This situation, in part, prepared the ground for the discovery, development, and sale of tribromoethanol (USP) (Figure 4.16, **8**) (AVERTIN™) for rectal anesthesia[78] (Eichholtz 1927; von Schickh 1933). In 1925, the company patented a manufacturing process for some esters of diiodobehenolic acid, a particular form of iodized oil envisioned as contrast agents (Wingler 1925). Evidently, radiological products were in the company's field of view but did not yet receive high priority.

This changed when in 1929 the pharmacologist Hecht (Figure 4.17; see *curriculum vitae*) began testing various iodinated sulfodyes and numerous colorless halogen-containing chemicals for radiographic use. It cannot be excluded that the acquisition in October of the preceding year of Räth by a completely different division of the company but situated in Hecht's workplace, Leverkusen, has helped to stimulate increased interest in the field. Similarly, contacts to Otto, who by then had joined the same company, are possible. Efforts to identify a good candidate substance with biliary excretion, i.e., a potential competitor for the cholecystographic

$$CBr_3 - CH_2 - OH$$
8

$$CH_2I - SO_3^- Na^+$$
9

$$CHI_2 - SO_3^- Na^+$$
10

FIGURE 4.16 Methiodal sodium (ABRODIL™), the iodinated organic compound that became the first long-term commercially successful contrast agent, together with the historically related therapeutic agent tribromoethanol (AVERTIN) and the follow-up contrast agent dimethiodal sodium.

8: Tribromoethanol (AVERTIN, Bayer/I. G. Farbenindustrie AG, Germany).
9: Methiodal sodium (ABRODIL, I. G. Farbenindustrie AG, Germany; SKIODAN™, Winthrop Chemical Comp., USA). $I/P = 0.5$.
10: Dimethiodal sodium (TÉNÉBRYL™, Laboratoires André Guerbet & Cie, France; UROTRAST™, Nyegaard & Co. A/S, Norway). $I/P = 1.0$.

Among the commercial preparations only those of the inventor companies and their largest subsidiaries are listed. At different times dozens of other companies brought to market their own preparations of the same compounds. (Strain et al. 1964; Knoefel 1971b; Barke 1970.)

iodophthalein sodium, were unsuccessful. Instead, the prospects for renally excreted pharmaceuticals began to look good. From the collection of compounds available in the laboratory, one that became known as methiodal sodium (USNF) (Figure 4.16, **9**) was tried and immediately confirmed as being promising for excretion urography, both after intravenous and oral administration (Hecht 1930). Rather than being the result of a coeval experimentation in chemical synthesis, this candidate contrast agent belonged to a collection that predated by 3–7 years the initiation of their screening as radiographic contrast agents inside the company.

Within a year of the company's renewed research in the radiological field, the first of a series of process patents applications covering the preparation of the publicly never before described compound methiodal sodium were filed, which identified as inventors the chemists Ossenbeck and Tietze together with the pharmacologist Hecht (Ossenbeck, Tietze, and Hecht 1929a,b). The German patent was assigned to the corporate headquarters of the company they worked for, I. G. Farbenindustrie AG, Frankfurt a. M. (Ossenbeck, Tietze, and Hecht 1929a), and the American one was assigned to the Winthrop Chemical Company, Inc., New York (Ossenbeck, Tietze, and Hecht 1929b), of which the German group indirectly owned half.[79] A few months down the road the same inventors filed also a German patent application, tacitly aimed at covering the clinical use of iodomethanesulfonic acid salts as contrast agents (Hecht, Ossenbeck, and Tietze 1930a). They cleverly cast the invention into the form of a röntgenological process in order to get around the still reigning ban on patenting of medicinal compounds in their country (Roentgen Contrast Agents 1935).

FIGURE 4.17 *Gerhard Hecht*, 1957. (Reprinted with permission from Bayer AG, Corporate History & Archives, Leverkusen.)

CURRICULAM VITAE

Hecht, Gerhard (b. August 2, 1900, Hannover, Prov. of Hannover, Kingd. of Prussia, German Reich // d. February 22, 1981, Mölln, State of Schleswig-Holstein, Federal Republic of Germany)

Humboldtschule, Linden, Germany: Wartime high-school dipl. [Abitur] (June 1918)/Univ. Göttingen: Med. Stud. (January 1919–1924); State med. lic. (July 1924); Unpaid asst. of Prof. Wolfgang Heubner (May 1925–October 1925); detached to Univ. Inst. of Pharmacol. by I. G. Farbenindustrie AG, Elberfeld (October 1925–April 1926); MD (January 1926)/I. G. Farbenindustrie AG, Elberfeld, West Germany[57]: Pharmacol. (May 1926–1945); Head of Commercial Hygiene and Toxicol. (1945–1951)/Farbenfabriken Bayer AG, Leverkusen: Head of Commercial Hygiene and Toxicol. (1951–1954); Head of Pharmac. Div. (February 1954–End 1965); Authorized officer (January 1962–End 1965)/Univ. Cologne: Lecturer in Pharmacol. and Toxicol. (1951–Fall 1957); Untenured Assoc. Prof., Med. Faculty (Fall 1957–1965); Acting Chairman of Pharmacol. (1957–1958).

On the heel of the patent applications that were to cover methiodal sodium followed that, by the same inventors, covering the diiodinated analog, dimethiodal sodium (INN). In Europe, it sought coverage of a synthesis process and, in the USA, it also sought coverage of its clinical use (Figure 4.16, **10**) (Hecht, Ossenbeck, and Tietze 1930b). Dimethiodal sodium too must have been part of the pre-existing collection of compounds. Thus, I. G. Farbenindustrie AG, contemporaneously with Hryntschak's experimentation, had numerous iodinated and brominated chemicals available. With high probability, it was Ossenbeck who supplied them to Hryntschak with purely exploratory intentions, as was commonly done by industry, using code numbers for compound identification and keeping initially their structure a secret. The possibility that methiodal sodium and dimethiodal sodium were among those radiologically experimented with cannot be excluded; on the contrary, it appears highly likely. Hryntschak claims that the preparations he tested were brominated and iodinated aromatic molecules (Hryntschak 1929). One is justified to doubt this general statement given that he really never was in a position to reveal the chemical structures and I. G. Farbenindustrie AG lacked reasons to exclude nonaromatic bromo- and iodocompounds it already had in hand from screening.

In his chronicles Hryntschak gives the impression that the initiative for scrutinizing brominated and iodinated organic molecules for excretion urography had been his own. Specifically, he states that they were synthesized for him by two chemical industries in a meritoriously disinterested manner (Hryntschak [1928] 1929). The above-mentioned absence of records in the archives of the successor industries to I. G. Farbenindustrie AG concerning the collaboration is consistent with the relaxed, uninvolved atmosphere in which the venture was apparently conducted. Everything indeed speaks for Hryntschak to have taken the initiative to pursue excretion urography with organic bromo- and iodocompounds and for having solicited such products from industry. He claimed to have personally determined the toxicity of what he received (Hryntschak 1929). Taken for granted, this fact lends support to his account of himself as the initiator of the collaboration. But even if one were to accept his version of history, it is unlikely that any industry would have had 50 potentially suitable chemicals available in the necessarily high quantities needed for radiological experiments and have performed toxicological tests on them in rabbits, without beforehand having been engaged in a massive drug discovery program. Despite the possibility that the initiative had been coming from the academic corner, the supporting industry must have been committed to a project on iodinated and brominated compounds on a significant scale, though probably not primarily with contrast agents in mind. With drugs such as tribromoethanol (Figure 4.16, **8**) (AVERTIN) in development, I. G. Farbenindustrie AG must surely have been in such a situation. This circumstantial evidence strongly supports the conclusion that Hryntschak's first supplier of halogenated compounds was I. G. Farbenindustrie AG, Leverkusen.

In September 1928, Hryntschak reported in person to Roseno that at some point in the autumn of 1927 "he was forced to abandon his quest because the chemical industry, with which he had collaborated, could no longer produce the corresponding solutions for him"[26] (Roseno [1928] 1929). Since he collaborated with Schering-Kahlbaum AG only after that time, we can conclude that he was referring to I. G. Farbenindustrie AG as his first industrial supplier of chemicals. There are many ways to explain the situation. Here the interruption of his supplies is seen as unrelated to his performance or his activity. Rather, it appears that a certain research program at I. G. Farbenindustrie AG ceased when the industrial chemist, Ossenbeck, who was the local supporter of the activity, was assigned other tasks.

As earlier mentioned, Hryntschak had announced but not followed through, with a talk on excretion urography for the 7th Congress of the German Society of Urology in Vienna in 1926 (Deutsche Gesellschaft für Urologie 1926; Hryntschak [1928] 1929). In fact, he disclosed initial data not before September 28, 1928, at the 8th Congress of the German Society of Urology in Berlin, and solely as intervention in the discussion session (Hryntschak [1928] 1929; Roseno and Jepkens 1929). We know that he worked with uroselectan sodium after October 24, 1927, and on request by Schering-Kahlbaum AG., Räth and Binz supplied him with it. Therefore, in the aforementioned conversation with Roseno he must have referred to his past collaboration with I. G. Farbenindustrie AG. Moreover, he signaled that what he intended to present at the congress would be the results therefrom. Apparently, he had no permission by Schering-Kahlbaum AG to reveal information related to uroselectan sodium, and/or he had no interesting data obtained with it.

At the 1928 congress, after having pointed out that his work on excretion urography was not yet complete, Hryntschak explained his very ambitious goals. These were at least twofold: first, to opacify the renal parenchyma in order to obtain sharp outlines of the kidneys, to render tumors visible and to evaluate renal function, and second, to achieve visualization of the full urinary system under conditions of minimal disturbance. Then, he described having performed intravenous urographies on 150 rabbits with 50 unspecified compounds. According to the printed record of the intervention he did not even reveal that their X-ray contrast production was based on the presence of iodine or bromine atoms.

Two highly promising preparations, #13 and #27, were also studied in dogs, and preparation #13 even in man. Slides of the results with these candidate contrast agents were presented. Consider my literal translation[80] of some of Hryntschak's own, in part horribly awkward, German sentences describing his conclusions (Hryntschak [1928] 1929).

> Although some of these preparations proved quite useful, they did not entirely meet the conditions I desired: good visualization of kidneys, good pyelo-, uretro- and cystography after injection of correspondingly small quantities of substance.
> Demonstration of slides
> Two slides of rabbits, intravenous injection of preparation 27:
> Both kidneys are well visible; especially the left one presents itself as an extraordinarily dense shadow (The left kidney in the rabbit lies low, does not perform respiratory excursions and is therefore roentgenologically always better depicted than the right one that rests against the diaphragm). The renal medulla stands out clearly from the renal cortex; the renal papilla is especially conspicuous. The ureter, as so far in all animal experiments, cannot be seen, a fact probably explained by its extraordinarily narrow lumen. The shadow of the bladder is visible with particular clarity.
> Two slides on experiments in the dog: Preparations 13 and 27.
> Kidneys as well as the bladder are clearly delineated.
> Slides on intravenous pyelography in man with preparation 13:
> The right renal pelvis and the right ureter are very well represented, less so the structures on the left side. In contrast, the renal parenchyma is not visible or barely more visible than in control exposures.
> Thus, from these exposures it is already evident, that attainment of the aims delineated at the outset lies apparently within the realms of possibility. With this, then, a new and promising method for the diagnosis of kidneys, ureter and bladder would be gained

(Hryntschak [1928] 1929)

As revealed by a subsequent publication (Hryntschak 1929), the results in man shown by Hryntschak (Figure 4.18) were not fundamentally different from those that later convinced Swick (*vide infra*) that he was on the verge of success, initially working with selectan neutral. The doses used were in the range of 0.06–0.12 g(Iodine)/kg, i.e., doses not too dissimilar to those subsequently accepted as effective. To his disadvantage Hryntschak presented his work only during the discussion period of the congress and identified the compounds used only by code numbers. Such secretive behavior is much despised by scientific audiences. The situation was most likely forced on him by the supplying industry; apparently relevant process patent applications had not yet been filed.[81] Despite having promised to reveal in due course the identity of the chemicals he had experimented with (Hryntschak 1929), he never did so. It is reasonable to doubt that he ever was made privy to their full chemical structures, except for their classification as either iodinated or brominated organic molecules.

The situation at the Berlin congress was worsened by Hryntschak's use of stilted and excessively cautionary language and his measurement of success against the very high self-imposed goals, cited above. Would he have refrained from wanting to visualize the renal parenchyma and focused mainly on evidencing renal calyces and ureters, he could have reported more optimistically. Be that as it may, von Lichtenberg, with the aplomb of the leading authority in urography, immediately

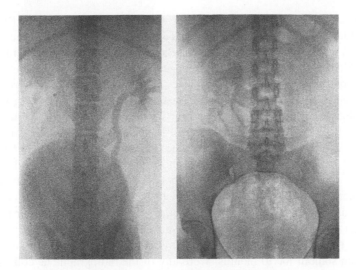

FIGURE 4.18 Promising signs badly miscommunicated.

Pioneering intravenous urographies obtained by Theodor Hryntschak before fall 1927. (Reprints of Figures 7 and 8 in Hryntschak, T. 1929. Studies of the roentgenological rendering of the renal parenchyma and the renal pelvis by the intravenous route [in German]. *Z. Urol.* 23: 893–904).

Left: Radiograph of a dog whose left ureter had been restricted to its natural diameter by a ligature near the bladder and which has been injected intravenously a 1:1 mixture of an iodinated preparation (#13) and a brominated preparation (#48) at a dose of about 1 g/kg and a concentration of 10%. The left renal pelvis, ureter, and filled bladder are nicely visible, whereas the right side is invisible in part because of the lack of ureteric constriction and in part by obscuration by the liver.

Right: Radiograph taken on February 12, 1927,[82] of a woman 1 h after having received intravenously an iodinated preparation (#13) at a dose of 6 g, probably at a concentration of 20%. The enlarged right renal pelvis is standing out among the various elements of the complete urinary system.

following Hryntschak's intervention, vented his frustration with his own attempts at excretion urography and somewhat sardonically wish (ed) "Messrs. Roseno and Hryntschak that, with perhaps more perseverance than we had and through fortunate inspiration, they would arrive at a solution. But I do not believe, for the time being, that they will succeed on this basis"[26] (von Lichtenberg [1928] 1929). Nonetheless, summarizing the communications at the congress for the *Münchener Medizinische Wochenschrift* at least the congress chairman, Ludwig Kielleuthner, critically judged both Roseno's and Hryntschak's successes "to be highly interesting" and "already permitting us to be hopeful for the near future"[26] (Kielleuthner 1928). Fact is that in 1928 the consensus on the eventual feasibility of contrast agents for excretion urography was still shaky.

This combination of circumstances can explain why the medical community, apparently Schering-Kahlbaum AG as well, failed to realize that a significant step forward had been made. The selective, incomplete, and distorted quotation that

Hryntschak's intervention received in chronicles by contemporaries (Binz, Räth, and von Lichtenberg 1930; Binz, Räth, and Junkmann 1930; von Lichtenberg 1932; Binz 1937a), and which is reiterated in the secondary historical literature (e.g., Rathert, Melchior, and Lutzeyer 1975), suggests an additional motive. His success, if accepted, partially undermined the primacy aspirations of the other participants in the race for intravenous urographic contrast agents. Although Schering-Kahlbaum AG profited from his experience, his past closeness to I. G. Farbenindustrie AG must have been a source of concern. The downplaying of his role by researchers more exclusively associated with them came in handy. This is a form of "deletion of modalities" (see Section "Technological Innovation") practiced by an industry. Also, the assertion by von Lichtenberg (1932) that Hryntschak had limited success with selectan neutral, without mentioning that he had experimented also with uroselectan sodium, qualifies as a "deletion of modalities." Whether intentional or not, it was convenient for bolstering his role in the advent of excretion urography.

By the time of the urology congress in Berlin, Hryntschak had had access to uroselectan sodium for 10 months. Were any of the radiographs he collected actually obtained with this compound or with methiodal sodium? Lamentably he never fulfilled his promise to publish the fully decoded account of his studies. In his final full publication on the subject (Hryntschak 1929), he merely revealed that two of the preparations, for which results were shown, were iodinated (#13 and #27) and that one was brominated (#48). Of these preparations, #13 was used on February 12, 1927, i.e., before the availability of uroselectan sodium to Hryntschak. The maximal tolerated dose of preparation #27 in man was very much lower than that of uroselectan sodium (Swick [1929] 1930), or methiodal sodium (Perlmann 1931), which indicates strongly that they were not the same. Binz states that Hryntschak tried uroselectan sodium only in rabbits, whereas both preparations #13 and #27 were as well used in man (Binz 1937a). At variance with Grainger (1982a–c), one is forced to conclude that none of the results publicly disclosed by Hryntschak involved uroselectan sodium. But his results with this contrast agent were evidently disclosed to Schering-Kahlbaum AG. Binz in 1930 spoke of Hryntschak having but partial success with this compound (Binz 1930). Instead in 1937 he qualified the same involvement with uroselectan sodium as completely unsuccessful (Binz 1937a).

It is incomprehensible how uroselectan sodium could have failed to attain a kidney shadow in rabbits at iodine doses similar to those that brought success with preparation #13. According to one of his collaborators, misfortune appears to have caused Hryntschak to abandon uroselectan sodium prematurely (Meuser 1953), but the nature of such an incident could no longer be reconstructed. In any event, if Hryntschak couched his conclusions on uroselectan sodium in sentences as overburdened with cautionary figures of speech as he did for other compounds at the 8th Congress of the German Society of Urology in Berlin (Hryntschak [1928] 1929), presided by von Lichtenberg, he may well have quenched some of the expectations at Schering-Kahlbaum AG. The company appears to have been disappointed with its collaborator and his results. Apparently, his progress was slow, perhaps understandable if one realizes that he had just assumed the directorship of the Urological Ward of the Wilhelminen Hospital. Binz blamed him of having caused a 2-year delay in the discovery of a suitable urographic contrast agent (Binz 1937a). Communication failed somewhere and it took clinical evaluation of uroselectan sodium by others to convince Schering-Kahlbaum AG of the product's value.

Since the summer of 1927 Lichtwitz in Altona was involved in clinical trials with iodinated products selectan and selectan neutral (Figure 4.9, **1 & 2**) for the treatment of sepsis in man (Lichtwitz 1930a; Binz 1937a), chiefly in the genitourinary tract (Binz 1937a). Lichtwitz reported to Binz and Räth and almost inevitably also to Schering-Kahlbaum AG, on his successes in streptococcal infections on January 20, 1928. He promised to quickly elucidate the renal and hepatic excretion of the compounds (Binz 1937a), and he assigned this work to his assistant, the young medical doctor Kurt Erbach.[83]

Early in 1928 the young American physician Moses Swick from New York (Figure 4.19; see *curriculum vitae*),[84] on an Emanuel Libman Fellowship, went to work for 7 months as informal post-doctoral fellow with the surgeon August Bier (in German: Hospitant) at the Friedrich-Wilhelms-University in Berlin (Rathert, Melchior, and Lutzeyer 1975; Loughlin and Hawtrey 2003), while also visiting patients with the urologists von Lichtenberg (Urology, St. Hedwig Hospital) and Joseph (Urology, Surgical University Clinic). Both urologists had a track record in radiological examination of the urinary tract. The former, together with Voelcker, had introduced retrograde pyelography with stabilized colloidal silver (Voelcker and Lichtenberg

FIGURE 4.19 *Moses Swick* in 1934. (Reprinted with permission from Dr. Susan B. Sterling (-Swick), Snoqualmie, WA, USA.)

CURRICULAM VITAE

Swick, Moses [until Apr. 17, 1924, called Goldstein, Moses or Morris] (b. February 16, 1900, New York City, New York, USA // d. August 7, 1985, New York City, New York. USA)

Townsend-Harris Hall School, New York City: High-school dipl. (1918)/ Columbia College, New York City: BS (1920–1922)/Columbia Univ. College of Physicians and Surgeons: Stud. med. (1922–1924); MD (1924)/Mount Sinai Hosp., New York City: Intern, 1st Surg. Serv. (November 1, 1925–November 1, 1927)/Emanuel Libman Fellowship awardee (1928)/Univ. Berlin, Surg. Univ. Clinic: Informal post-doctoral fellow [Hospitant] (Early 1928–September 1928); Allgemeines Krankenhaus Altona, Germany, Dept. of Internal Med.: Informal post-doctoral fellow (September 1928–April 1929)/St. Hedwig Krankenhaus, Berlin, Urol. Sect.: Informal post-doctoral fellow (April 1929–November 1929)/Mount Sinai Hosp., New York City, after 1968 called Mount Sinai School of Med., City Univ.: Volunteer, 2nd Surg. Serv. (January 14, 1930–November 1, 1931); Research Asst., Chem. Dept. (February 1, 1930–May 1, 1931), informally collaborating with the Radiol. Dept.; Volunteer Asst., Dept. Surg. Pathol. (December 8, 1931–July 1, 1932); Adjunct Surgeon (May 10, 1932–January 12, 1948); Assoc. Urol. (January 13, 1948–April 30, 1953)/Assoc. Attending Urol. (May 1, 1953–May 7, 1956)/Attending Urol. (May 8, 1956–February 28, 1960)/Consulting Urol. (March 1, 1960–July 1, 1982)/Clin. Prof. Emeritus of Urol. (1982)/Harlem Hosp., New York City: Attending Urologist (1934–1935)/ City Hosp., New York City: Assoc. Urologist (1936–1937)/Private clinical urol. practice, New York City (1935–August 1942) and (1946–1982)/Wartime military serv., last as major, Volunteer Mount Sinai Med. Div., US Army Third General Hosp., Alabama (September 1, 1942–May 1943), North Africa (May 1943–May 1944), Italy (May 1944–September 1944) and France (September 1944–August 1945)/Sinai Hosp., Baltimore, Maryland: Post-graduate position (September 1945–January 31, 1946)/Billings Gold Medal (Class I), Am. Med. Assoc. (1933)/Jacobi Medallion, Mount Sinai Alumni, New York City (1957)/ Ferdinand C. Valentine Award, New York Acad. of Med., Urol. Sect. (March 17, 1965)/200th Anniversary Medal of Columbia Univ. Coll. of Physicians (February 1967)/Silver Platter award, Dept. Radiol., Yale Univ., New Haven, Connecticut (1968)/Dr. med h.c., Freie Univ., Berlin, Germany (January 1975)/ Honorary membership, Berlin Urological Society, Germany (January 1975)/ Honorary membership of the American Urological Association (September 1976)/New York City Award for Distinguished and Exceptional Serv. (July 1978)/John K. Lattimer Medal, American Urological Association (1978).

1905, 1906) and the latter had performed the initial clinical trials on lithium iodide (Joseph 1921) for the same purpose. Given the large urological patient population in their clinic, it appears likely that Swick was familiar with retrograde pyelography, its promises and problems. At the end of September 1928, in coincidence with

Räth's departure from Binz's institute, Swick changed to the Municipal Hospital in Altona, today part of Hamburg. There he worked with the chief of internal medicine Lichtwitz (Swick 1930c), who was renowned for the breadth of his experience and interests, including colloid chemistry, and ennobled by his human qualities, not the least his humor (Heubner 1946). Within the frame of the aforementioned clinical trials of selectan sodium and selectan neutral as antiseptics (Figure 4.9, **1** & **2**), Swick joined Erbach in studies of their pharmacokinetics and excretion pathways in animals and man, studies that were ongoing already for half a year (Swick [1929] 1930, 1930a, 1978). Swick learned from Erbach, who never published his results, of the substantial contribution renal excretion made to the total elimination from the body of these compounds. Refering to selectan neutral and pointing out his awareness of the roentgenological properties of iodine, Swick has written "When I heard of its composition, I asked Prof. Lichtwitz if I could work with it" (Swick 1981). He obtained cooperation and with that, for roughly 7 months, experimented principally with selectan neutral as intravenous, oral or rectal contrast agents for excretion urography (Swick [1965] 1966). The radiologist, whom he necessarily had to call on in Altona, was never acknowledged for his collaboration and remains thus unknown. Swick obtained encouraging results in that the kidney parenchyma and urinary bladder could be visualized not infrequently after the intravenous injection of the compounds; the renal pelvis and ureter, on the other hand, were for the most part poorly seen[26] (Swick 1978). Even oral administration showed some success (Swick [1929] 1930). On November 29, 1928, Lichtwitz informed Binz that selectan neutral showed promise in excretion urography (Binz 1937a). Hence, the statement of Binz and Räth (1930a) reported above, and pertaining to their unwitting entry into the area of urology while collaborating with Lichtwitz, may as well be read in terms of urological radiology.

Since by October 1928 Räth had taken a position at I. G. Farbenindustrie in Leverkusen, it is unlikely that he participated in meetings with Swick and Lichtwitz. From this moment on Binz took the reins of the project into his own hands, at least all components that involved his institute. Five years later, when his ties to Schering-Kahlbaum AG were terminated, he even engaged on his own in the search for new contrast agents and novel syntheses for existing ones. In particular, he described and patented a new synthetic process for methiodal, dimethiodal and their triiodinated analog (Binz 1935a). Unfortunately, as contrast agent the latter compound was insufficiently stable. Nonetheless, Binz was able to sell the patent to the owner of methiodal and dimethiodal sodium, *I. G. Farbenindustrie AG* (1935).

According to Lichtwitz, Swick was a clever and energetic man, one with the deepest devotion to and passion for research work[26] (Lichtwitz 1930c). Lichtwitz credited him alone with having performed the studies that demonstrated the feasibility of excretion urography with products of the selectan group (Lichtwitz 1930a–c). But he remained curiously unclear as to who had generated the idea, declaring simply that their excretion by both kidneys and liver, combined with an absence of iodism, naturally led to the idea of testing them for the widely desired intravenous pyelography[26] (Lichtwitz 1930c) and that "on the basis of these (excretion)-experiments and our rather substantial experience with the innocuousness of the preparations on oral and intravenous injection, the application of the selectans for contrast generation appeared promising"[26] (Lichtwitz 1930b). As to selectans and selectan neutral, Swick originally stated: "It was noticed that they were excreted both by the kidney and through the bile (Dr. Erbach). This led to the question of whether selectan neutral, whose tolerability relative to its iodine content (54%) was rather good, could be used as a roentgenologic contrast agent after intravenous or oral administration"[26] (Swick 1929 [1930]). Later Swick replaced the impersonal story of the idea with the personal phrase "it occurred to me …." (Swick [1965] 1966, 1974, 1975). Lichtwitz asserted that Swick was not aware that Hryntschak had been given selectans, among which also uroselectan sodium, for testing as contrast agent for excretion urography in animals (Lichtwitz 1930c), von Lichtenberg confirmed this assertion (von Lichtenberg [1930] 1931). The affirmation is convincing, since Swick would surely have insisted on evaluating uroselectan sodium earlier, had he known of its existence.

Swick never explained his reasons for joining Lichtwitz's clinic. He never even hinted at pursuit of excretion urography as motivation. Instead he explained that upon arrival in Altona he learned from Erbach about the tolerability and renal excretion of the iodine-containing therapeutic selectans, and how this prompted him to realize that the same compounds could be useful for excretion urography. Another question likewise has no direct answer. What role did Hryntschak's results that were divulged at the 8th Congress of the German Society of Urology in Berlin at the end of September 1928 (Hryntschak [1928] 1929) play in Swick's decision to explore excretion urography? Swick had participated at the congress as evidenced by being listed as a freshly admitted member of the German Society of Urology on the membership list attached to the official record of the meeting (Deutsche Gesellschaft für Urologie 1929). Lichtwitz, who was not a member, too had the occasion to learn about what was communicated and could have read summaries of the presentations, such as that by Kielleuthner, published on November 9, 1928, which commented on Hyntschak's results optimistically (Kielleuthner 1928).

Between November 9 and Lichtwitz's report to Binz on November 29, 1928, there was plenty of time for X-ray experiments in animals, which do not require more than a few hours each. Swick's exposure to Hryntschak's results at the congress end of September must have in fact afforded him still more time to experiment before the results were reported. All these circumstances suggest that Hryntschak's results constituted a strong impetus for Swick to pursue excretion urography with the selectans in Lichtwitz's laboratory. Did Schering-Kahlbaum AG that according to Lichtwitz (1930a) supplied at least some of the material for the therapeutic trials in Altona, suggest roentgenological application potential? This question is not a devious one, given that this company had experience in the field through Hryntschak. All in all, it seems clear that if indeed Swick had the idea of using compounds of the selectan group as intravenous urographic contrast agents independently of von Schickh, Schoeller, and Hryntschak, he did so in an environment that was heavily poised for it.

The astute choice of uroselectan sodium for urographic testing at Hryntschak's clinic involved most probably Schoeller and/or von Schickh at Schering-Kahlbaum AG. Whether they used exclusively their own judgment or took advice from others within the company is not determinable. All we know is that after the acquisition of the research program by Schering-Kahlbaum AG and the discovery of the low toxicity of uroselectan sodium further toxicity data related to selectan derivatives were generated at the company in the laboratory of Karl Junkmann (Binz and Räth 1930b; Binz, Räth, and Junkmann 1930).

Swick in the meantime determined that the availability of patients for clinical experimentation in Altona was not congruent with his urographic ambitions. He believed the Urological Section of the St. Hedwig Hospital, Berlin, a very large and famous unit caring at any moment for about 200 urology patients (Spence 1990), to be a better place. This structure was at the time under the direction of the world-renowned von Lichtenberg (Figure 4.20), whom Swick knew from his earlier stage in Berlin.[85] He suggested to Lichtwitz that he and his research on urography be transferred to Berlin. Initially hesitant, but then with praiseworthy vision and altruism, Lichtwitz helped to arrange this transfer. As will become understandable, later Swick came to declare this move to be his fundamental mistake. This detail was related to me in the course of several contacts I had with Ronald G. Grainger. In 1981, he had enjoyed an interview as well as an exchange of letters and documents with Swick, which were clearly intended to serve as material support in historical efforts.

Announced by Lichtwitz, Swick paid a first visit to Binz in March 1929 (Binz [1930] 1931, 1937a; Swick 1930b,c) and disclosed his observations and wishes for improved compounds. In response, he apparently received one or two candidates (Swick 1930b,c; Swick [1965] 1966). Still in March a second meeting took place in Berlin, involving Binz,[86] Lichtwitz and Swick (Swick 1930b,c). All aspects were

FIGURE 4.20 *Alexander von Lichtenberg* around 1930. (Courtesy of Dr. Franz von Lichtenberg, Boston, MA, USA.)

CURRICULAM VITAE

von Lichtenberg, Alexander (= Sándor) (b. January 20, 1880, Budapest, Kingd.
of Hungary, Austro-Hungarian Empire // d. April 12, 1949, Mexico City, Mexico)
Königliches Katholisches Universitätsgymnasium, Budapest: High-school
dipl. [Érettségi] (1897)/Royal Hungarian Univ. of Sci. in Budapest: Stud. of
med. (1897–1903); State med. lic. (March 1903); MD (1903); Anatom. Inst.,
Asst. (September 1902–November 1903)/Univ. Heidelberg, Germany: Surgical
Univ. Clinic, Asst. (November 1903–March 1908)/Univ. Strassburg, Germany:
Surgical Clinic (April 1908–December 1913 and May 1914–June 1917); *Venia
legendi* [surg. and orthopedic surg.] (February 1910)/Surgical Policlinic,
Budapest, Hungary: Chief Surgeon (December 1913–April 1914)/Wartime mil-
itary serv., Royal Hungarian Army: Regimental Physician and Chief Surgeon,
Pecs, Hungary (August 1914–March 1916); Director, Royal Hungarian Bureau
of Invalides and Consulting Surgeon of the Military Command, Kassa,
Hungary (March 1916–November 1918)/Univ. Strassburg, Germany: Titular
prof. of surgery (June 1917–December 1918)/Bábakèpsö Hosp., Miscolc,
Hungary: Chief surgeon (1919)/Private practice in orthopedic surg. within the
private clinic of Rudolf Jahr, Berlin, Germany (1920)/Univ. Berlin, Germany:
Recognition of *Venia legendi* and conversion of designation to surg. (Septembet
1920); Assoc. Prof. [untenured] of Surgery (1922); Withdrawal of teaching permission by Nazi regime, followed by appeal (November 1933); Partial reinstatement (July 1934); Retroactive withdrawal valid for 1935 and extension of
research permission until end of 1936 (July 1936)/St Hedwig Hosp., Berlin:
Consultant urologist (1922–1924); founding chief of Urol. Dept. (December
1924–June 1936)/Univ. Pennsylvania: Honorary PhD (June 1930)/Escape to
Hungary (June 1936)/Private urol. practice at Siesta Sanatorium, Budapest
(1937–1939)/Escape to Mexico (1939); Private practice of urol. and urol. surg.
at Clinica Londres and at Sanatorio Dr. Alfonso Ortiz Tirado (1939–1944),
upon incorporation renamed Sanatorio Durango, Mexico City (April 1944–
April 1949)/Hosp. Colonia de Ferrocarriles Nacionales de México, Mexico
City: Consultant/Co-founder and co-editor, Z. f. Urolog. Chir. (1913–1936).

again covered. At one of these March meetings it was explained to Binz that Swick
would continue his work at von Lichtenberg's clinic (Swick 1930b; Binz [1930]
1931, 1937a). In April 1929, Swick transferred to Berlin and initiated clinical trials.
With the assistance of the head of the Roentgenology Department, Werner Rave,
he focused on excretion urography after oral consumption of selectan neutral (von
Lichtenberg and Swick 1929; von Lichtenberg [1930] 1931).

For the time span following the March 1929 meetings[87] until the arrival of uroselectan sodium in Swick's hand there exist contrasting chronicles not only from Binz
(1930b, 1937a) and Swick (1978) but also from von Lichtenberg (von Lichtenberg
1932), whose pronounced self-serving behavior is rendering him least credible of all.
Mostly the stories recounted by Binz and Swick shall be detailed.

Binz reports that after the March 1929 meeting with Lichtwitz and Swick, he introduced the participants, together with von Lichtenberg to Schering-Kahlbaum AG (Binz 1937a). Thus, according to him his opening interaction with von Lichtenberg occurred in this context (Binz [1930] 1931, 1937a). If Swick's claim that the two met first only in September of 1929 at the Munich meeting of the 9th Congress of the German Society of Urology (Grainger 1982b) is accepted, it must be concluded that earlier contacts between the two were restricted to letters and phone calls.

In any case, before in April Swick transferred to Berlin an agreement was reached according to which he would be supplied by Binz with candidate contrast agents for clinical trials, to be performed under von Lichtenberg's responsibility. Binz was to select candidate agents on the basis of maximal iodine content, stably bound iodine, solubility, and tolerability. Actually, von Lichtenberg informed Binz that he had already finished selecting the patients on which the contrast agents were to be tested since he would be leaving for the United States in June 1929 (Binz 1937a).

There is consensus (Binz 1937a; Swick 1978) on the fact that Swick, imagining a connection between the known toxicity of methanol and his observations with selectan neutral (Figure 4.9, **2**) in patients, at one time suggested to Binz that the compound may suffer from splitting off of the methyl radical as a cause of some adverse reactions in man—double vision, headache, and vomiting (Swick 1933a,d). Although this theory did not hold up, Swick's willingness to converse with Binz in terms of medicinal chemistry must be seen as helpful in the interaction between actors with different knowledge cultures. Binz dates Swick's suggestion to after his arrival in Berlin and describes having provided the latter with three diiodo derivatives (Figure 4.11, **4–6**) (Binz 1937a). One of these (**4**) had been known since 1887 (Pfeiffer 1887). Since another of the compounds (Figure 4.11, **5**) carried an *N*-methyl group, this account is puzzling, although it is likely that all along Binz had reservations with regard to Swick's methyl group hypothesis (Binz 1937a).

Swick has repeatedly expressed the belief that his methyl group theory had influenced the chemical design of uroselectan sodium (Swick 1930b,c, [1965] 1966, 1978; Grainger 1982b). Binz (1937a) has brushed this opinion aside noting that Swick was not in a position to appreciate the full extent of previously acquired experience on structure–activity relationships gained from more than a decade of research on this class of chemicals. In fact, uroselectan sodium with an acetic acid group replacing the methyl group that Swick blamed for toxicity had been characterized by Räth both chemically and toxicologically years earlier and covered by a patent (Räth 1927b). Schering-Kahlbaum AG had even identified this compound as being one of the more promising for excretion urography, when it singled it out for testing in Hryntschak's clinic in October 1927. Out of self-interest Binz steered clear of using the latter fact to counter Swick's contentions. The conclusion that Swick had nothing to do with the original synthesis of uroselectan sodium is inevitable.

Swick (1978) dates the discussion of the methyl group theory with Binz to the period he worked in Altona. It probably occurred during the March visit to Berlin. Swick recounts having obtained in response several (Swick 1930b) diiodo derivatives (Binz and Räth 1930a) of the selectan type but without a methyl group. Given that Räth by that time had left the university, they were prepared by Hillgruber (Binz, Räth, and von Lichtenberg 1930; Swick 1930b). One of the compounds was only soluble in strong alkali (Swick 1930b) (Figure 4.11, **4**).[88] Because of its low solubility, in Altona it was evaluated initially in dogs only after oral administration. No significant absorption was observed. In defiance of this negative result, testing was extended to a male patient after oral administration in form of gelatine capsules. The effort was in vane (Swick 1930b). Just 1 week after receipt of the diiododerivatives and before Swick had had time to work with all available substances, the second March meeting involving Binz and associates, Lichtwitz and Swick took place.

Following his move to Berlin in April, Swick continued to study in man, albeit with little success, selectan neutral after oral administration (Swick [1929] 1930, 1930b; Binz 1937a; von Lichtenberg [1930] 1931, 1932). He also again tried the insoluble compound previously tested in Altona (Figure 4.11, **4**) with the same negative results. Similarly, unsuccessful were his experiments with the diiodo analogs of both selectan neutral (Figure 4.11, **5**) and uroselectan sodium (Figure 4.11, **6**). The latter, which carried an *N*-acetic acid moiety as in uroselectan sodium, was significantly more water-soluble than the others but not sufficiently so to permit intravenous injection (Swick 1930b).

It is interesting to ponder why Binz did not include uroselectan sodium in his selection of compounds for Swick. Perhaps it was perceived as having definitively failed in the hands of Hryntschak. Another possibility is that for prestige reasons Binz was trying to discover a suitable candidate not previously singled out by Schoeller and von Schickh, and experimented with by Hryntschak. Consistent with this interpretation is the fact that he never considered diiodochelidamic acid (compound **7**), the synthesis of which had been described much earlier by another research group (Lerch 1884).

Swick found the diiododerivatives unsatisfactory, in part because of low solubility and in part because of excessive biliary excretion. As a result of Swick's reports on their clinical failure, Binz felt constrained to return to monoiododerivatives. Sometime around June 1929, when von Lichtenberg had arrived in the United States (Binz 1937a; Binz, Räth, and von Lichtenberg 1930), Binz's institute supplied Swick with uroselectan sodium, the excellent water solubility, low toxicity, and renal excretion of which Räth had established much earlier (Binz [1930] 1931, 1937a). It is difficult to say whether Schering-Kahlbaum AG was involved in the decision to give the compound a second chance, but it seems likely. Swick has offered details of the crucial events that followed in handwritten letters to Grainger (Swick 1980, 1981). In July, just 1 month into the clinical trials with uroselectan sodium, Swick informed Binz of his success with this contrast agent candidate (Figure 4.21, Left) and cabled the exciting news through his American benefactor, the bacteriologist and serologist Emanuel Libman at the Mount Sinai Hospital in New York, to von Lichtenberg on visit in Portland, Oregon. The latter proceeded right away to informally divulge the news in the United States without acknowledging Swick (Rathert, Melchior, and Lutzeyer 1975). Then, he hurriedly returned to Berlin,[89] where he set an aggressive clinical research program in motion. It is astounding that von Lichtenberg and Swick could report on the experience with 84 patients examined in a lapse of time of just about 2 months (von Lichtenberg and Swick 1929, 1930). The results formally became public at the 9th Congress of the German Society of Urology in Munich, September 26–28, 1929 and appeared in print soon thereafter (Swick [1929] 1930; von Lichtenberg and Swick 1929; von Lichtenberg [1930] 1931). On October 13,

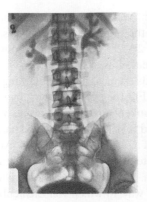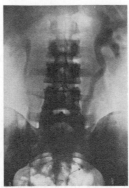

FIGURE 4.21 Two competitive and successful products on each other's heal, uroselectan sodium (UROSELECTAN™) and methiodal sodium (ABRODIL™).

Left: Excretion urogram taken about 3 h after intravenous injection of 40 g of uroselectan sodium, i.e., 17 g of iodine, in form of a 30% (w/v) solution. The clinical trial, performed by Moses Swick between July and September 1929 and presented at the 9th Congress of the German Society of Urology in Munich in September 1929, shows small dilation of renal calyces in a woman suffering from recurrent ambilateral pyelonephritis without efflux impediment. (Reprinted by permission from Springer Nature Customer Service Centre GmbH of Figure 1 in Swick M. 1930. Darstellung der Niere und Harnwege im Röntgenbild durch intravenöse Einbringung eines neuen Kontraststoffes, des Uroselectans. *Klin. Wochenschr.* 8(45): 2087–2089.)

Middle: Excretion urogram of a patient with a ureteral stone, taken 10 min after intravenous injection of 28 g of methiodal sodium, i.e., 15 g iodine, in form of a 20% (w/v) solution. Conspicuous congestion in the left ureter and renal pelvis above of the stone, indicated by the black arrow on white background. (Reproduction of Figure 2 in Bronner, H., G. Hecht, and J. Schüller: 1930. Excretion-pyelographie with "Abrodil" [in German]. *Fortschr. Geb. Rontgenstr.* 42(2): 206–218.)

Right: Excretion urogram of a woman pregnant in the 10th month with fetus in breech position taken 24 min after injection of 25 g of methiodal sodium, i.e., 13 g of iodine, in form of a 20% (w/v) solution. The strong enlargement of both renal pelvises and ureters is noticeable. (Reproduction of Figure 3 in Bronner, H., G. Hecht, and J. Schüller: 1930. Excretion-pyelography with "Abrodil" [in German]. *Fortschr. Geb. Rontgenstr.* 42(2): 206–218.)

the radiologist Rave involved in the studies divulged the same results to his peers in Dresden (Rave [1929] 1930).

Between the dates of success and its communication at the congress lies an ugly struggle between two actors in drastically different power positions. On the one hand, there was Swick, the young foreigner without any formal hospital or academic appointment anywhere, who carried the ball in the clinical trial and who perceived himself as the mastermind behind uroselectan sodium and the "uroselectan method" (Swick 1978). On the other hand, there was von Lichtenberg, the great German professor (who in reality was Hungarian) with a somewhat impulsive and choleric but otherwise jovial temperament (Del Rio 1974; Westermann 1978; Spence 1990). He saw himself as carrying the burden of responsibility for everything that happened in his clinic even while abroad and who derived from the stiffly regimented hierarchy reigning in the institutions of those times the right to communicate the results obtained therein how he pleased. Upon his precipitous return to Berlin, von Lichtenberg prepared to present, under his sole authorship, Swick's results at the impending 9th Congress of the German Society of Urology, to be held September 26–28, 1929 in Munich. Swick saw what he considered to be his discovery, under threat of usurpation.

Unbeknownst to Swick, his cause received some assistance. Through Grainger I have obtained a copy of a letter dated April 15, 1965, Flushing NY, to Swick by Otto A. Schwarz, in 1929 a physician at the St. Hedwig Krankenhaus. In it the events were recalled. According to his recollection, he had procured Swick with the second patient ever to be studied with uroselectan sodium. A beauty of a picture of a tubercular pyelo-nephrosis had been obtained. Schwarz and von Lichtenberg met at the Linkstrasse editorial office of the Julius Springer Verlag in Berlin. The former tried to convince the latter that he could not treat Swick in the intended way, to which von Lichtenberg responded with irritation, denying his counterpart any right to mingle in his affairs. Swick himself called for help from Libman, who managed to get Lichtwitz activated (Rathert, Melchior, and Lutzeyer 1975). Lichtwitz in turn alerted Victor Salle,[90] who was in charge of the medical section within Springer Verlag and senior medical editor of *Klinische Wochenschrift* (Heubner 1943; Sarkowski 1996), the journal that was to be the first to publish some records of the congress. Like Schwarz, Salle engaged in heated arguments with von Lichtenberg over the matter. The situation was eventually straightened out at a final meeting between Swick, von Lichtenberg, Salle and Alfred Renner from Breslau,[91] who, as his trusted colleague, substituted for the bed-ridden Lichtwitz (Binz 1937a; Grainger 1982a). The tenuous outcome of the meeting foresaw that Swick as sole author would orally report at the congress, ahead of his antago-nist, on his pilot experiments in Altona and Berlin, and von Lichtenberg would orally present the jointly authored clinical evaluation of uroselectan sodium. The corre-sponding publications would have to reflect that agreement. Edited by Salle, *Klinische Wochenschrift*, was the first to publish the two reports and they obviously reflected the agreement (Swick 1929 [1930]; von Lichtenberg and Swick 1929). In the subsequently published official meeting report with the same wording, von Lichtenberg appeared as sole author of the clinical communication (von Lichtenberg [1929] 1930a). Swick, Räth, Binz, Erbach, Hillgruber and Rave were acknowledged for their involvement, von Schickh and Schoeller, as well as Schering-Kahlbaum AG, were ignored. The insistence with which von Lichtenberg and clinical authors in his sphere of influence, continued to use in their later publications, even after launch of the commercial prod-uct, the product's name misspelled with k instead of c, i.e., Selektan, hints at a poorly disguised effort to provide the new contrast agent with a mark that would detract from its origin in Räth's selectan series of compounds and Swick's achievements.

At Schering-Kahlbaum AG the role Swick had played was well known and appre-ciated. Swick's drive and open enthusiasm was infective. His advocacy of the selectan class of chemicals for excretion urography constituted probably as crucial a contri-bution to the eventual attainment of the goal as his animal studies or the execution of clinical testing. He managed to convince Lichtwitz, Binz, von Lichtenberg and Schering-Kahlbaum AG of his success. Undeterred by the existing frictions among some of the actors, in the autumn of 1929 von Lichtenberg in company with Swick was sent by Schering-Kahlbaum AG on a tour arranged by Carlo Ravasini, urologist at the Hospital Regina Elena in Trieste, Italy, to some Italian sites of uroselectan sodium clinical testing (Report on Lectures 1930) and to the 8th Congress of the Italian Society of Urology on October 26, 1929, in Genoa, where they gave separate presentations (von Lichtenberg [1929] 1930b; Swick [1929] 1930). Later Schering-Kahlbaum AG rewarded the important contribution of Swick through a donation directly or indirectly through him to the Emanuel Libman Fellowship Fund that had funded his stay in Germany in 1928 and 1929.[92]

Back in the United States, Swick was invited to present his work on January 15, 1930, to the Section of Genito-Urinary Surgery of the New York Academy of Medicine (Swick 1930a). In his closing remarks, Swick thanked at least Binz, Lichtwitz and von Lichtenberg for their help and cooperation. Clearly at this point Swick was willing to keep the spirit of the priority dispute resolution.

Schering-Kahlbaum AG began selling UROSELECTAN for the visualization of the urinary tract after intravenous injection on February 1, 1930 in Germany (Binz, Räth, and von Lichtenberg 1930; Schering AG and Wlasich 1991) and the product became soon available in many other countries. In the United States Schering Corp. Bloomfield, NJ, gave it the trade name IOPAX™.[93] In Germany, the single-dose pharmaceutical presentation (Figure 4.22) was a powder contained in a ground glass

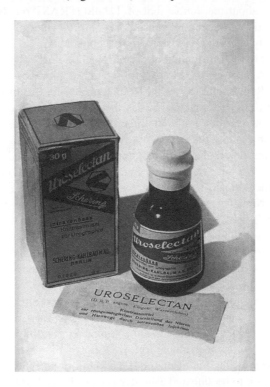

FIGURE 4.22 Pharmaceutical presentation in 1930 of uroselectan sodium (UROSELECTAN™), a breakthrough product but with some shortcomings.

Application for a German patent (D. R. P.) and the trademark registration (Warenzeichen) are disclosed, emphasizing the proprietary status of the product. (Reprinted with permission from Schering Archiv, Bayer AG, Berlin.)

mushroom-stoppered and wax-sealed conical blue glass vial and comprised sufficient powder of uroselectan sodium to prepare an excess over the maximally recommended dose of 30 g for injection, equivalent to 12 g of iodine.[94] One of the two
alternative instructions for use read that the powder (40 g) was to be dissolved in 110
mL of doubly distilled water, filtered through paper and sterilized by boiling for 10
min before use (Speck 1995). Excessive boiling led to precipitation (Sgalitzer 1930).
Taking into account an estimated 10 mL loss of water during boiling, the resulting
concentration became 1.34 mol/kg$_{Water}$. Making some educated approximations, this
concentration corresponds to one of about 1.2 M, 35% (w/v), or 150 mg(Iodine)/mL,
and an osmolality at 37 °C, never actually measured, of roughly 2,300 mosmol/kg[41].
The exceedingly complicated preparation of the solution for injection, starting with
the solid contrast agent presentation, was an important shortcoming that it shared
with the soon to appear competitive product, ABRODIL.

When considering prices, let us start with a comparison of the costs of doses for
a minor application of UROSELECTAN, namely retrograde pyelography. That is
the clinical indication for which various other contrast agents already existed. A
dose of the new contrast agent for retrograde pyelography cost only about half that
of UMBRENAL, about the double that of THOROTRAST but 10–20 times more
than sodium bromide produced by pharmacies on prescription. No wonder sodium
bromide took many years to be replaced in this application.

More pertinent to the affirmation of the novel technology was the price of a dose
of UROSELECTAN for excretion urography. Such a dose was initially sold at the
retail level for 18 Reichsmark (Schering-Kahlbaum AG 1930), corresponding to $63,
if adjusted to 2016. In packages for clinics the price was reduced to around 60%–
70% (Fränkel 1930; Sartorius and Viethen 1933). A dose for excretion urography
thus cost 10–20 times more than a dose for retrograde pyelography of the same compound, and 100 times more than a corresponding dose of sodium bromide. The large
price differences were mitigated mainly by increased safety and efficacy of the new
procedure over the competitive procedure and the reduced time of involvement of
a physician in it. Nonetheless, prices became a subject of heavy debate and the elevated one of UROSELECTAN did slow down the rate at which excretion urography
became established (Moll 1995).

On December 25, 1929, merely 3 months after the congress at which uroselectan
sodium was disclosed, and still before the product was commercially launched,
Ossenbeck, Tietze and Hecht from I. G. Farbenindustrie AG, requested a manufacturing process patent for the preparation of methiodal sodium (Figure 4.16, **9**) and
mentioned the use of its salts as contrast agents for excretion urography (Ossenbeck,
Tietze, and Hecht 1929a,b; Hecht, Ossenbeck, and Tietze 1930a). Many additional
process patents for the same chemical followed shortly thereafter.[95]

Pharmacological evaluation of methiodal sodium in animals, which lasted only
3 months, revealed it to be safe and effective (Hecht 1931).[96] Methiodal sodium was
clinically tested in excretion urography by the surgeon Hans Bronner and the dentist/
surgeon Joseph Schüller, respectively, university lecturer/assistant medical director
and assistant at the Surgical University Clinic of the Citizen's Hospital in Cologne
(Bronner and Schüller 1930; Bronner, Hecht, and Schüller 1930) (Figure 4.21,
Middle & Right). Eugen Schlief (1930) at the Marien Hospital in Osnabrück studied

it in intravenous and oral urography. Bronner and collaborators (Bronner, Hecht, and Schüller 1930) as well as Hans Köhler at the Hospital of the Jewish Community in Berlin (Köhler 1930) established its suitability for retrograde pyelography.

Methiodal sodium (ABRODIL) first arrived in the hands of the German clinician on September 1, 1930 (Bayer-Meister Lucius 1930; Hecht 1931). In the USA, it took the brand name of SKIODAN™ (Winthrop Chemical Company, Inc., New York). The rapidity with which discovery, development and commercialization of methiodal sodium followed the initial disclosure of uroselectan sodium, with a difference in commercial launch times of merely 7 months, is surprising. It becomes understandable if one realizes that the inventors worked in a company that possessed experience with manufacturing of similar halogenated chemicals (e.g., tribromoethanol) and which had apparently supplied Hryntschak earlier with them for the experiments on excretion urography described above.

Like UROSELECTAN, ABRODIL initially came as a powder in stoppered glass vials containing either 20 g for a single dose or 100 g for clinics (Bayer-Meister Lucius 1930). Before use it had to be dissolved, filtered, and sterilized by boiling to form a 40% (w/v) solution, corresponding to a concentration of 208 mg(Iodine)/mL. Subsequently also a 20% (w/v) formulation, i.e., one with 104 mg(Iodine)/mL, was recommended. With the concentration of 208 mg(Iodine)/mL, ABRODIL exceeded the contrasting ability of its competitor, UROSELECTAN, which had merely 148 mg(Iodine)/mL.

As measured decades down the road, at the concentration of 104 mg(Iodine)/mL methiodal sodium possessed an osmolality of 1,570 mosmol/kg (Holtermann 1973b). The value for the original formulation at 208 mg(Iodine)/mL was roughly 3,100 mosmol/kg[41]. This means osmolalities of ABRODIL and UROSELECTAN differed essentially in proportions to their iodine concentrations. At the same iodine concentrations, they were nearly identical. But this lack of commercially helpful distinction was not the reason contrast agent osmolality, which many years later would move into the focus of attention, was at the time not debated. Rather, the traditional focus on the chemical structure, in conjunction with blinders to solution behavior as the basis for explaining tolerability, has to be blamed.

Initially ABRODIL suffered from the same complicated preparation requirements as UROSELECTAN. Both of them were more complex to use in retrograde pyelography than ready for use lithium iodide solution (UMBRENAL), or sodium bromide and sodium iodide solutions. This fact contributed to the continued use in this modality of the older solutions for about a decade after the arrival of the newer products.

Excretion urography, which depended on the new contrast agents, encountered astonishing skepticism that only time was able to overcome (Pollack 1996). The complicated preparation of the solution for injection probably too has played a role in the slow adoption of the new modality. In any event, within a few years ABRODIL was offered as 20% and 40% (w/v) ready for injection, sterile solutions, corresponding to concentrations of 105 and 210 mg(Iodine)/mL, respectively. The product following up on UROSELECTAN, UROSELECTAN B™, to be discussed further down (discussed in Sections "From Sodium Iodide to Uroselectan Sodium and Methiodal Sodium" and "Perfecting Ionic Uro-Angiographic Contrast Agents") allowed similarly ready-to-use pharmaceutical formulations with concentrations up to 75% (w/v) or 386 mg(Iodine)/mL. With this user friendliness, adoption of the pharmaceutical presentations was dramatically improved. This quality is relevant in any phase of a specific article's or technology's life cycle, but it assumes crucial importance in the latest phases.

Beginning with the almost simultaneously entrance on the scene of uroselectan sodium and methiodal sodium, and up to the present, the great majority of successful intravascular products have been iodinated organic compounds. This fact made it possible to make comparison of prices on the basis of their iodine content, i.e., the primary determinant of radiological efficacy. Henceforth comparisons are expressed in $ (adjusted to 2016) per gram of iodine. The price of ABRODIL per gram of iodine (Sartorius and Viethen 1933) was 1.7 times higher than that of UROSELECTAN. The superior performance of the product, foremost its tolerability, helped overcome the economic hurdle and make it a success.

While the clinical trials with uroselectan sodium for excretion urography were coming to an end, other trials pursued its use in angiography. Up to that time angiography had emulated the development stages of excretion urography. Except in Germany, where strontium bromide (DOMINAL-X) continued to be sporadically used, French, Portuguese, and American clinical investigators preferred to use sodium iodide solutions. Newly uroselectan sodium was put to the test. In spite of some spectacular images of peripheral arteries (e.g., Sgalitzer et al. 1930; Demel, Kollert, and Sgalitzer 1930; Demel, Sgalitzer, and Kollert 1931; Oselladore 1930; Schmidt 1930) (Figure 4.23, Left) and veins (e.g., Ratschow 1930; Schmidt 1930) (Figure 4.23, Middle), the results were never convincing enough to induce widespread diagnostic application. Moreover, cerebral angiography had to be excluded, because uroselectan sodium caused convulsions (Demel, Sgalitzer, and Kollert 1931).

More successful was methiodal sodium. In addition to allowing excretion urography (Perlmann 1931), it demonstrated to be effective and safe for angiography (dos Santos, Lamas, and Caldas 1931a,b; Wolf and Remenovsky 1931; Loewe 1931; Sgalitzer, Kollert, and Demel 1931) (Figure 4.23, Right). As opposed to uroselectan sodium, methiodal sodium had a low neurotoxicity. It even permitted carotid angiography. But so long as the safety problems of colloidal thorium dioxide preparations were brushed aside, this product's unsurpassable angiographic performance did not allow the iodinated contrast agents to conquer general acceptance for angiographic applications. Methiodal sodium was eventually introduced for myelography as well (Arnell and Lindström 1931). Pronounced acute adverse reactions limited its utility to lower spinal canal examinations, but for this clinical indication it enjoyed a life on the European market for several decades. Other applications included for example the visualization of the milk ducts, so-called galactography (Oselladore 1937).

Notably, the successful development of contrast agents for excretion urography had yielded with methiodal sodium, as a pleasant bonus, also one for several angiographic indications.[97] This product was the first of its kind to merit the label uro-angiographic contrast agents. With this development, the second half of the vision guiding further search of uro-angiographic contrast agents was born (Table 4.1).

The advent of uroselectan sodium coincided curiously with another dramatic event in radiology. Chance would have it that just ahead of the two articles

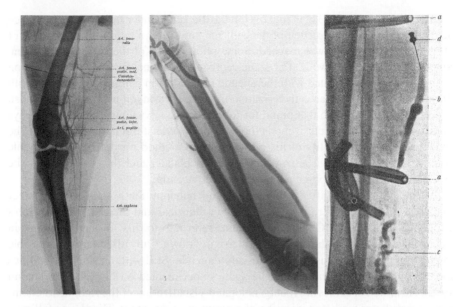

FIGURE 4.23 Angiography with uroselectan sodium (UROSELECTAN™) and methiodal sodium (ABRODIL™).

Left: Arterial angiogram of a leg. Ten mL of UROSELECTAN was injected during 6 s into the liberated femoral artery and then the artery was ligated. Around 1 s before termination of the injection the radiograph is taken, which captures the moment at which the artery is being ligated but is still patent. (Reprinted by permission from Springer Nature Customer Service Centre GmbH of Figure 3 from Schmidt, W. 1930. The Angiography with Uroselectan, a Method for the Radiographic Representation and Functional Testing of the Peripheral Circulation [in German]. *Chirurg* (Heidelberg) 2(14): 652–663.)

Middle: Venography of lower arm. In this pioneering self-experiment, the investigator injected 10 mL of UROSELECTAN into a dorsal hand vein, and upon reaching the ninth mL he took a radiograph, showing the basilic and cephalic veins with valves. (Reproduction of Figure 1 in Ratschow, M. 1930. Uroselektan in Angiography, with Special Consideration of Varicography [in German]. *Fortschr. Geb. Rontgenstr.* 42(1): 37–45.)

Right: Venography of a lower leg with the intent of mimicking with a contrast agent the spreading of a sclerotizing agent in the presence of tourniquets. ABRODIL, methiodal sodium 104 mg(Iodine)/mL, was injected (d) into a small saphenous vein (b) in the region between two applied rubber tourniquets (a). Invisible cirsoid dilatations below the outer tourniquet were revealed (c). (Reprinted by permission from Springer Nature Customer Service Centre GmbH of Fall 6, middle radiograph, from Wolf, M. and F. Remenovsky. 1931. The Practical Application of Varicography [in German]. *Wien. Klin. Wochenschr.* 44(11): 353–355.)

by Swick and von Lichtenberg (Swick [1929] 1930; von Lichtenberg and Swick 1929), which originally announced the new contrast agent, appeared the report by Werner Forssmann on his breakthrough in the catheterization of the human heart. Forssmann achieved it while working as resident in the Second Surgical Section of the Auguste Viktoria-Heim, a small-town hospital in Eberswalde near Berlin. Ignoring the ban of his superior, Richard Schneider, he had introduced on himself a 65 cm long, 4 French, metal stylet-bearing soft rubber ureteric catheter, lubricated with sterile olive oil, for 35 cm into his left basilic vein. Under fluoroscopic guidance he could continue the insertion of the catheter until after a total insertion length of 60 cm the right atrium was reached. Some photographs of the screen were taken. Since everything went well, after some reproaches, Schneider allowed application of the procedure associated with therapeutic injections on a moribund patient. With these two cases of cardiac catheterization in man, Forssmann proceeded to publish (Forssmann 1929). Up to that point, for following the advancing catheter by X-rays Forssmann had relied exclusively on the opacity of stylets. But in his publication, he speculated about the utility of his catheterization method for delivering injectable contrast agents for the visualization of the gall bladder and the renal pelvis. The awkward speculation exposed a hidden research agenda.

At the end of 1929, after a tumultuous and unhappy few weeks as a volunteer in Berlin, initially in Ferdinand Sauerbruch's Surgical and then in Otto Ringleb's Urological University Clinic, Forssmann returned to his former position in Eberswalde. Before anything else, he had to calm down. Only afterwards was he able to pick up on his previously hidden research agenda. Still without suitable facilities for experimentation with animals, he again experimented on himself (Forssmann 1972). He aimed at achieving roentgenological visualization of the right heart chamber after catheter-mediated injection of a contrast agent into it. His choice fell on a 25% (w/w) sodium iodide solution as contrast agent. He injected 25 mL of contrast agent solution using ureteric catheters bearing a tip with multiple holes. This time he employed 8 French catheters instead of the 4 French ones used previously. He was able to visualize the pulmonary artery but little more. The osmolality of the solution he used is today known to be 4,900 mosmol/kg[41]. Forssmann initially feared that the pronounced hyperosmolality could have detrimental effects (Forssmann 1931a), although he did not measure it. Pushing this fear aside he had produced initial evidence for the tolerability of the procedure. Extensive experimentation on animals for optimizing contrast agent type and administration mode, as well as for the documentation of safety became imperative.

At this point, luckily Forssmann encountered Willi Felix, a former trainee of Sauerbruch and now the Director of the Second Surgical Section of the Community Hospital Berlin-Neukölln. Felix allowed Forssmann access to radiology facilities suitable for animal experimentation but could not offer the animals. So, Forssmann bought them himself and his mother kept at least the dogs in her apartment. He began with experiments on rabbits injected with various contrast agents directly into the heart chambers through catheters placed by his procedure. The initial choice of COLLARGOL (see Chapter 3) had soon to be abandoned. Solutions of sodium iodide 20% (w/w) were more successful. Although through the aforementioned adjacency of the articles Forssmann must have become aware of uroselectan sodium, this alone did not suffice to enthuse him for it. After all, it was used only for excretion urography, where rapid intravenous injection was not of the essence. Instead, an article on venography with it (Schmidt 1930) triggered his interest (Forssmann 1931a). Upon finding rabbits unsuitable as experimental animals for intracardiac injections, he switched to dogs. He injected contrast agents rapidly into the right atrium, a place guaranteeing intensive mixing with blood. He used 20 mL of a 20% (w/w) sodium iodide solution or 20 mL of a 50% (w/w) uroselectan sodium, both approximately

200 mg(Iodine)/mL in concentration. The extra strength solution of uroselectan sodium he prepared by concentrating the regular strength solution through boiling (Forssmann 1931a).[98] In short order, he also used methiodal sodium, in the form of a solution saturated at elevated temperature (Forssmann 1931b). When his timing for taking radiographs with an exposure time of 1/20th of a second was fortunate, he obtained radiographs faintly showing the contrast agents in the right atrium, the right ventricle or the pulmonary vessels. On plain fluoroscopy, he saw fascinating swirling clouds emerging from the catheter progressing from atrium to ventricle and pulmonary vessels. Through a friend he was introduced to Viktor Gottheiner, a collaborator in sports medicine of Herbert Herxheimer, the Assistant Medical Director of the Second Medical Clinic of the Charité Hospital in Berlin, who simultaneously headed the Consulting Services in Sports Medicine (Gottheiner 1971). Gottheiner was among the earliest to succeed with cineradiology.[99] Forssmann was able to observe a dog with this new technology. On the cinestrip he could follow the contrast agents into the left atrium, the ventricle and the aorta. Forssmann was enthralled (Forssmann 1931b; Gottheiner 1971). Under the sponsorship of Felix, Forssmann was given the opportunity for an oral exposition of his results at the national surgery congress (Forssmann 1931b). Without help from Felix this would have been impossible to accomplish for a just 26-year-old gentleman from the provinces, as he was arrogantly dubbed by a luminary of the academic high-society in Berlin (Forssmann 1972). Forssmann's paper describing his studies with the heart appeared on March 20, 1931 (Forssmann 1931a). His work provided preliminary evidence for the safety of the procedure and for the cardiac tolerability of uroselectan sodium and methiodal sodium. With this he opened the door for various physiological evaluations of the heart. More importantly, he marked the beginning of a lengthy path that would eventually bring about cardioangiography and coronary angiography (Doby 1976), both procedures unthinkable without contrast agents.

Uroselectan sodium was essentially a prototype that remained on the market for very few years. Its own producer, Schering-Kahlbaum AG, replaced it merely 2 years after its introduction by the safer UROSELECTAN B™ (see Section "Perfecting Ionic Uro-Angiographic Contrast Agents"). Methiodal sodium, the second product on the market, was a success, being substantially safer than uroselectan sodium and applicable in expanded clinical indications. According to Frings (1994), methiodal sodium remained in use, at least for lumbar myelography under spinal anesthesia,[100] until 1974, after having passed as late as 1972 from Farbenfabriken Bayer AG to Schering AG. This example may be added to the many other cases that show how unfounded is the common management wisdom about the desirability of owning the pioneering product in a category, rather than a rapid follower. Others have likewise begun to realize this lesson (The Clichés 2001).

Contrasting Historical Interpretations

No sooner had UROSELECTAN been commercially launched than the simmering dispute over its invention and development erupted into the open. In von Lichtenberg's publications on excretion urography his version of history became a regular ingredient (von Lichtenberg [1930] 1931, 1931, 1932). Grainger (1982, 1994), supported by an interview with Swick and an exchange of letters (Swick 1980a, 1980b, 1981), has criticized how von Lichtenberg and others have tried to rewrite history, with varying degrees of success. Indeed, in his speeches and publications von Lichtenberg most blatantly omitted crucial facts and names and carefully crafted subtle distortions. Even in regular scientific publications he utilized "deletion of modalities" for selfish purposes.

During the Meeting of the American Urological Association in June 1930 in New York City, a symposium on excretion urography was organized by the president of the society, Joseph F. McCarthy. Swick, who by then was back in New York for several months, was not invited with the excuse that he was not a member of the society. Grainger has provided me with a copy of a handwritten letter to Moe (nickname of Moses Swick) from H. Carey Bumpus Jr., Duxbury MA, dated April 1, 1958. Therein the meeting co-president and symposium organizer, McCarthy, is quoted to have responded to the latter's endeavor about the participation of Swick at the meeting: "Bumpus, I'm not going to have any G. D. Jews and damn few Protestants!." To McCarthy, von Lichtenberg was apparently "acceptable" despite his Jewish extraction, either out of ignorance or because the latter worked at the catholic St. Hedwig Hospital in Berlin and sometimes declared himself a roman catholic.[101] One cannot be certain that von Lichtenberg and Binz were left ignorant of Swick's exclusion, but it seems likely that discussions of the program on which they were speakers allowed them to know. For their ambitions, the situation was convenient and von Lichtenberg, just like Binz, followed the invitation to present uroselectan sodium in the very hometown of Swick (von Lichtenberg [1930] 1931; Binz [1930] 1931). This occasion was not the premier exposure of uroselectan sodium in America. Swick (1930a) had already promulgated the novelty a few months earlier.

Von Lichtenberg lived up to expectations with an enthusing talk and with an overwhelming exhibit of compound excretion results and radiographs based on over 700 patients. X-ray viewing boxes with 15 by 20 cm films covered the entire circumference of a large room as well as the walls of numerous cabinets at the center of the room (Belt 1974). A number of physicians orally communicated the initial American clinical experience with uroselectan sodium (e.g., Braasch [1930] 1931). Elmer Belt was allowed to add three of his radiographs to von Lichtenberg's exhibit (Belt 1974). On occasions, von Lichtenberg attributed the synthesis of uroselectan only to Binz (von Lichtenberg [1930] 1931). To Swick he referred just twice, and called him Lichtwitz's assistant. Subtly he transmitted the impression, that Lichtwitz was the one who had conceived the idea of pursuing excretion urography (von Lichtenberg [1930] 1931, Marshall 1977), and Binz ([1930] 1931) declared that it was Lichtwitz who observed that selectan neutral had a most remarkable other effect which concerned the visualization of the urinary tract. Lichtwitz remained ambiguous on the issue (see above). In the footsteps of his master, von Lichtenberg, Wilhelm Heckenbach in his own publication cited Swick exclusively for his failure to detect the iodinated compound in human blood already 15 min after administration (Heckenbach 1930). The criticizable way of dealing with Swick's role in the advent of excretion urography should not detract excessively from the momentous contribution von Lichtenberg's in this technological breakthrough, a contribution that was to grow still more impressive in the couple of years following the described period (see Section "Perfecting Ionic Uro-Angiographic Contrast Agents").

Because of his Jewish descent, von Lichtenberg was eventually forced out of Germany by the Nazi regime[101]. In 1936, he returned to his native Budapest, from where he had to leave in 1939 for the same reasons (Spence 1990; Pérez Castro 1980). Swick told Grainger that in 1933 under similar circumstances he received Lichtwitz warmly at the debarkation in New York. Lichtwitz was rapidly licensed to practice and right away found employment in the United States, becoming Director of the Internal Medicine Ward at the Montefiore Hospital and Clinical Professor of Medicine at Columbia University in New York. In contrast, according to a letter to me from Franz von Lichtenberg, Boston MA, 1995, October 5, 6 years later his father was not offered a gratifying position in the United States. The principal reason for this situation is the fact that in the late 1930s massive immigration had saturated the American integration capacity for immigrating physicians. In addition, the large proportion of Jews among those physicians had reinforced pre-existing informal anti-Semitic quota policies at hospitals (Ludmerer 2005; Rowland 2009). But probably also von Lichtenberg's controversial behavior toward Swick played a role. Grainger (1994) mentioned to me a temporarily inaccessible letter Swick had provided him with, authored by Renner many years after the pertinent events. According to Grainger's memory Renner's letter stated that, belatedly, von Lichtenberg realized he had dealt Swick a "dirty deal" and seemed very embarrassed by it.

It has been claimed that von Lichtenberg was denied by the medical authorities a permission to practice in the USA (Pérez Castro 1980).[102] The assertion seems highly unlikely. But it helps being reminded that beginning in the late 1930s, passage of the State Board Examination was declared obligatory for a state license to practice. This requirement amounted to substantial psychological and financial hurdles (Schwarz 1972). Most likely von Lichtenberg, for one reason or another, could not face this prospect. Promises of better conditions caused him to settle in Mexico City.

Binz, who prior to the point at which primacy quarrels over the advent of uro-selectan sodium and excretion urography arose, had demonstrated himself to be quite altruistic, quickly changed his attitude under the lure of instant fame. In his publications, he began in subtle ways to portray Räth and Swick as assistants, who merely executed the plans of their superiors (Binz, Räth, and von Lichtenberg 1930; Binz 1930, [1930] 1931, 1937a), despite the evidence supporting their pivotal roles in transcending the original plans. At a time Swick's participation in the study of selectans in Altona was already well publicized, he mentioned only Lichtwitz as the one who during human antisepsis trials with selectan neutral observed a most remarkable other effect which concerned the visualization of the urinary tract (Binz [1930] 1931). He thereby incorrectly insinuated that radiography of the urinary system was part of the ordinary antisepsis study protocol. In his behavior Binz joined von Lichtenberg (von Lichtenberg [1930] 1931). One is justified in speaking of a conspiracy of the powerful.

With time the efforts of Binz to cement his perspective on the advent of excretion urography increased. His use of the first-person plural in connection with the authorship of the crucial German patent, which in reality was granted to Räth alone, and the generic designation of its applicant as technology, instead of Schering-Kahlbaum AG (Binz 1937a), is a masterpiece of obfuscation. Revealing is Binz's sentence regarding the initial clinical trials with uroselectan sodium in a blatantly self-serving publication: Only when I succeeded in 1929 to arrange clinical tests in man at the St. Hedwig hospital a sweeping success was the result[26] (Binz 1937b). Except for himself, no other instigator of clinical trials played a role. No clinical investigators were judged worth acknowledging. Addressing an audience of chemists, he dared to ignore Swick, Lichtwitz and von Lichtenberg simultaneously, not to speak of Schering-Kahlbaum AG and its representatives, Schoeller and von Schickh. Late in the game Binz, newly without Räth, began research of new iodinated contrast agents himself (Binz 1935a). His move likely was intended to strengthen his recognition in the contrast agent field. It paid off insofar Binz and Räth, in 1938 became formally nominated for the Nobel Prize in chemistry (Nobelprize.org 2014a).

Swick did not acquiesce to the situation. Following his exclusion from the urology meeting in June 1930 he promulgated multiple historical accounts that loudly proclaimed his primacy (Swick 1930c, 1965, 1974, 1975, 1978, 1981). He had the chief of the Radiology Department of his institution, Leopold Jaches, label excretion urography with uroselectan sodium as "Swick's method" (Jaches 1930) and repeatedly quoted his former mentor Lichtwitz's assessment of the situation in his favor (Swick 1974c, 1978).[103] To his life's end he pretended that the acetic acid moiety in uroselectan sodium was introduced as a consequence of his methyl group theory (Grainger 1982b), despite the fact that this claim was historically incorrect. He even managed to get this version of the story published in the American weekly news magazine, *Time*, August 24, 1931, p. 32. The professor of histology and embryology at the Jefferson Medical College in Philadelphia, Henry E. Radasch, was so impressed that he formally nominated Swick for the 1931 Nobel Prize for physiology and medicine (Nobelprize.org 2014b), but the principal evaluator did not consider the introduction of a better urographic contrast agent sufficiently meritorious (Nyman 2016). Finally, as the one that outlived all other protagonists, Swick used the advantage of having the last word. Swick's struggle was initially to no avail but succeeded well in his lifetime.

Von Lichtenberg managed for some decades to be seen as the originator of intravenous excretion urography with uroselectan sodium, with Binz sometimes described as the chemist behind the breakthrough. Räth and Swick were relegated to the role of assistants merely executing their professor's wishes. Lichtwitz was occasionally mentioned, but rightfully he did not qualify himself as a crucial player, as deducible from his generous behavior toward Swick. Schering-Kahlbaum AG, its officers and employees, Schoeller and von Schickh, were totally ignored and so was Hryntschak. In an otherwise laudable book on contrast agents published in 1970 (Barke 1970), the name of Swick and his publications are missing.

A situation began to change when in 1965 in connection with the grant to Swick of the Valentine Award of the New York Academy of Medicine (Swick [1965] 1966; Melicow 1966), an award with which the Academy honors a living scientist for an outstanding contribution to urology. For its promotion and in its justification a committee headed by Victor F. Marshall, professor of urology in surgery at the New York Hospital, Cornell Medical Center, re-examined the evidence and concluded that Swick's achievement "shines through the dust of this scramble of priorities" (Marshall 1977). During the award ceremony, the uropathologist Meyer M. Melicow (1966) read testimonial letters by academic luminaries. Therein William F. Braasch affirmed that "Not only did Doctor Swick sense the possible value of the chemical formula made by Binz and his colleague, but he altered its chemical structure so that it became less toxic." Frederick E. B. Foley declared Swick's feat as a "singlehanded, vastly important contribution to Urology." As Grainger (1982) describes it, on the occasion a group of renowned American urologists publicly apologized to Swick for 30 years of heartache and oblivion, and generously acknowledged this major innovative involvement. The German urologists followed suit, accepted the primacy of Swick's original work on excretion urography, and awarded him in 1975 honorary membership to their society (Riedel 1975). In the same year, the Free University of Berlin then bestowed onto him an Honorary Doctor of Medicine (Swick 1975; Rathert, Melchior, and Lutzeyer 1975). Senator Jacob K. Javits had the event inserted into the United States congressional record (Congressional Record—Senate 1975). In a later Citation of Distinguished and Exceptional Service by New York City, the health commissioner Reinaldo A. Ferrer hailed Swick's contribution as one of the three to five greatest achievements in medicine in this century (Mount Sinai Medical Center 1978).

Apart from the fact that both the analysis of Marshall (1977) and the declaration of some of his colleagues (Melicow 1966) pertaining to Swick's role appear inordinately superficial,[104] all chronicles to date have neglected to clarify the role that industry had played. In the 1880s, the chemical industry, especially in Germany, was the earliest industry that had built up in-house research and development capabilities. In chemical companies at the time Röntgen discovered X-rays, very limited pharmaceutical research was performed. This changed and by the 1920s this industry had clearly taken over the lead for drug discovery from university-based research.

A striking aspect in all published narratives of the advent of excretion urography is the passive role into which Schering-Kahlbaum AG is relegated. Do not the concerted actions of that company reveal an astute evaluation of the situation and a strong sense of purpose, a sharing of the vision? Considering that this industry supported the chemical research of Räth and Binz, acquired all existing and future patent rights within their research program, filed the patent application covering uroselectan sodium within just 6 months of commencing their financial research support, made compounds available to Lichtwitz, asked Binz to send uroselectan sodium to Hryntschak for clinical trials as a contrast agent for urography, met Lichtwitz, Swick and von Lichtenberg, supplied industrially prepared material for part of Swick's clinical trials before September 1929 (Swick [1929] 1930), supplied the same also for clinical trials in Italy between October and December 1929 (Gortan and Ravasini [1929] 1930; Report on Lectures 1930), in Austria before November 29, 1929 (Anonymous 1930) and in Germany before January 3, 1930 (von Lichtenberg and Swick 1929, 1930[105]; Ziegler and Köhler 1930), and in the end commercially launched uroselectan sodium (UROSELECTAN) on the market on February 1, 1930 (Binz, Räth, and von Lichtenberg 1930; Schering AG and Wlasich 1991), is it really plausible that it had not furnished some input bearing on the applications to be explored, the choice of compounds to be tested, at what time and by whom? In a more likely scenario, Schering-Kahlbaum AG participated in the decision-making but preferred to move with the discretion required of a competitive industrial environment and to avoid taking sides in academic disputes of the 1930's style.

Behavior of scientists depends strongly on their institutional environment. Academics can afford to blow their personal horn; industrial scientists must take into account consequences of their action for their company. In general, industry sees an advantage in keeping silent for a long while about products in development, their discovery, patent status, and performance in order to leave their competitors guessing. After all, the competitive game is at the core of their value system.

Regarding academic disputes, it has to be admitted that medical leaders are not only important partners for industry, primarily for clinical evaluation of candidate products but at the same time clients. Contesting them can have negative effects on the company's financial bottom line. This situation puts industrial scientists at a disadvantage in disputes over scientific authorships. Compensating for this situation there exist aspects where the industrial scientist is in a favored position. He frequently has access to a broader information base in the narrowly defined area of activity. He usually is better endowed with facilities and manpower. He operates in teams. Availability of intellectual property experts facilitates the patenting process. Taken together, these conditions diminish, in the eye of some academics, the value of the individual contribution of the industrial scientist. In effect industrial science, in general, carries less prestige than academic science. The neglect of Schering-Kahlbaum AG and its officers Schoeller and von Schickh in the story of uroselectan sodium is a consequence of the above.

A question addressed repeatedly in the literature is who really discovered or invented excretion urography with uroselectan sodium? Was it Räth because he synthesized the highly soluble low-toxicity compound and demonstrated its strong renal excretion, albeit without foreseeing a radiological use for it? Was it von Schickh or Schoeller from Schering-Kahlbaum AG, because they recognized the potential application in radiology of Räth and Binz's compounds and identified uroselectan sodium as the most likely candidate for excretion urography? Was it Hryntschak, because ahead of others he had limited success with the procedure in animals and man using soluble iodinated organic chemicals, success, which regrettably he was unable to communicate with impact to the community at large? Or was it Binz because he counseled Swick on the most suitable candidate contrast agent among those synthesized by Räth and acted as senior coordinator among various academics and industry? Was it Swick, because in his hands and with his single-minded determination, a pipe dream became reality with the clinical demonstration of uroselectan sodium's definitive suitability in man? Finally, was it von Lichtenberg after all, because he made his patients available for the crucial clinical tests and made the most massive contribution to the semeiotics of kidney diseases diagnosed with intravenous uroselectan sodium and follow-up compounds? Not surprisingly the answer is indeterminate. The problem is only aggravated, if the question is asked: Who discovered or invented excretion urography in general? Rathert, Melchior, and Lutzeyer (1975) paraphrased the inventor of endoscopy, Philipp Bozzini (1807), as follows: many have contributed to the development of the child (invention). Among the many, here we include several non-clinicians and industrial scientists. In effect, a whole network of more or less tightly communicating investigators who shared a common vision brings us much closer to an acceptable answer. At the 80th Annual Meeting of the RSNA in Chicago in 1994 the RSNA *ad hoc* committee for the centennial celebration prepared a history exhibit in which it stated "1929: Rowntree develops excretion urography. Binz, von Lichtenberg and Swick will perfect it" (Staab et al. 1994). If chauvinism is not the cause of such historical superficiality, then confusion due to an ill-posed question must be the excuse.

The present re-examination does not question the importance of Rowntree's or Swick's contribution,[106] rather it asks whether primacy for a whole technology can reasonably be assigned to any single person in this process of technology genesis, with so many steps from conception to clinical testing and so many interacting individuals and organizations. It enquires whether an interpretation of events in terms of the kind of model that is sometimes justified in describing the discovery of natural phenomena or theories, e.g., the discovery of X-rays, adequately serves description of technology genesis and shaping in the case of uro-angiographic contrast agents. It expresses serious doubts that the advent of excretion urography can be described in terms of a heroic history and stipulates that a sociological version is called for.

A notable similarity exists between the controversies surrounding the advent of excretion urography with uroselectan sodium or methiodal sodium, dealt with here and that involving the emergence of sulfa drugs at I. G. Farbenindustrie AG in Leverkusen (Lesch 1993). In both cases the chemists behind the achievements were only marginally recognized. All accolade went to pharmacologists or clinicians that conceived suitable animal models for preclinical testing and helped identify the best compound by clinical trials. The fact that the pharmacologist Gerhard Domagk received the Nobel Prize 1939 alone caused significant bitterness among his collaborating chemists, Josef Klarer and Fritz Mietsch. In 1956, prior to the time the Nobel Prize was awarded in part to Forssmann for his invention in 1929 of heart catheterization (Forssmann 1929)[107] and its application to the diagnosis of heart disease, which profited strongly from application of contrast agents (Forssmann 1931a,b), the evaluation committee also deliberated inclusion of an originator of uro-angiographic products. Wisely, no such person was honored. It could only have been a misidentification. The fact that in the case of sulfa drugs all participants were assembled in the same company, while in the present case collaboration between academic institutions and industry took place constitutes merely a minor difference. Similarities in the two cases are evident and encourage the view that technology geneses show some shared pattern. Cognizant of these, it can be hoped that in the future, futile priority disputes in situations similar to the recounted ones may be rapidly quenched.

The Dynamics of Technology Genesis

The Failure of One-Dimensional Reductionism

The intravascular uro-angiographic contrast agent uroselectan sodium and its close follower methiodal sodium, together with printed package inserts, scientific publications containing radiological illustrations, numerical data and evaluations of product performance by practitioners, positive recommendations by opinion leaders, and significant sales figures by the producers, are some of the artifacts attesting the arrival of a new technology, namely contrast agents for excretion urography. The products shared crucial features. They were water-soluble, small molecular weight, iodinated organic chemicals. They were administered as solutions for intravascular injection and eliminated from the body by renal excretion. Their tolerability allowed use in uro- and angiography. They were industrially produced. Their successors would maintain these features. Users and producers immediately perceived uroselectan sodium and methiodal sodium as a novel class of contrast agents that changed medical practice. To a large degree, they displaced a heterogeneous collection of earlier, poorly performing intravascular products, including the industrially formulated strontium bromide and sodium iodide/urea, as well as simple salt solution prepared *ad hoc*. They shifted imaging of the urinary tract from retrograde to intravenous administration of contrast agents. The magnitude of the difference in nature and performance between the new products and those available before qualified them as truly novel technology, the result of the process of technology genesis that had started with the use of sodium iodide in excretion urography. The way the new products changed the standard of care and industry's role in it marks it as a fundamental technological innovation.

While the two products shared the significance of having practically in contemporaneity established the new technology, their genesis history differed very strongly. For the clinician Hryntschak long-standing dissatisfaction with retrograde pyelography was at the origin of his pursuit of a new way. Swick instead entered the field spontaneously because results from ongoing pharmacological and therapeutic studies with compounds of the selectan family in the clinic he joined offered the opportunity to attempt excretion urography, an experimental procedure he had heard about just days before from Roseno and Hryntschak at a congress. What Hryntschak and Swick shared was the awareness of the results of Rowntree and collaborators and those of Volkmann, both of which announced the emergence of the vision of contrast agents for excretion urography. Hryntschak acted as progressive thinker and experimenter at separate historical moments for both I. G. Farbenindustrie AG and Schering-Kahlbaum AG, but the companies differed greatly in their receptivity. In coming up with uroselectan sodium Schering-Kahlbaum AG profited from prolonged and systematic pursuit of the radiological and therapeutic utility of iodinated chemicals. The way I. G. Farbenindustrie AG came to methiodal sodium combined accidental availability of the compound and related ones with opportunistic reaction to the expected economic success of uroselectan sodium, and perhaps a new awareness in the contrast agent area triggered by the company's hiring of Räth and Otto, albeit for tasks in unrelated areas. The chemists behind uroselectan worked in academia, those behind methiodal sodium in industry. Patents allowed Schering-Kahlbaum AG to develop its just then best candidate, uroselectan sodium. The precarious patent situation (*vide infra*) of the best compound known to I. G. Farbenindustrie AG at the time, dimethiodal sodium, let it give development priority to its second best, methiodal sodium. With uroselectan sodium Schering-Kahlbaum AG enriched the contrast agent market in which it was established already for several years and with multiple products. In contradistinction, with methiodal sodium, I. G. Farbenindustrie AG entered the contrast agent market for the first time. The constellation of academic and industrial actors, their motivations, and the sequences of important events in the two cases cannot be rendered by a shared scheme. In particular, the history cannot be rendered convincingly by eliminative methodological reduction to a one-dimensional sequence of scientific journal articles.[11]

The actors in a technology genesis process may be partitioned into pivotal and ancillary ones, depending on the uniqueness or ordinariness of their contributions. Let's look first at the ancillary ones. Ossenbeck and Tietze were instrumental in the synthesis of chemicals from which Hecht was able to select methiodal sodium as a successful contrast agent and they were as well the suppliers of compounds to Hryntschak. In their interaction with Hrynyschak they acted within the boundaries of their professional knowledge culture, organic chemical synthesis, neglecting to participate in guiding the project. Similar considerations apply to Rowntree's collaborators, Osborne, Sutherland and Scholl, Otto, Binz's assistant Hillgruber, Lichtwitz's assistant Erbach, Räth's various students, von Lichtenberg's assistant Heckenbach and the radiologist Rave, without mentioning the many others who were employed in the industries that developed products to marketable maturity. There is no evidence to suggest that their interests transcended the limit of furnishing chemical compounds, toxicology data or radiographs obtained by more or less routine methodologies for whatever uses others may have had for them.

Similarly, the many clinicians, other than Hryntschak, Swick and von Lichtenberg, who performed clinical trials with uroselectan sodium, or methiodal sodium, sectorially contributed the elements for the overall evaluation of the contrast agents. They acted exclusively as experts within their own knowledge culture. Their involvement may be compared to that of the many more or less anonymous team members in the Manhattan or moon flight projects. It is important to recognize that without ancillary actors the network of even the best pivotal actors would not go far.

Now let's look at the pivotal actors. The chemists Räth and Binz at the outset unconsciously and then with purpose, assumed pivotal roles. The former not only explored systematically the chemistry of a class of compounds, the selectans, but was also deeply involved in the discovery of their antimicrobial activity and their toxicological and pharmacokinetic testing. He discovered the means for rendering some of them highly water-soluble. He assured patent protection for uroselectan, a precondition for a candidate pharmaceutical product. As Grainger (1982a) pointed out, he did all this before Swick ever entered the scene. Binz acted as broker for his and Räth's chemical patrimony, making sure that commercial articles would result for both veterinary therapy and radiology, and he played an important, albeit criticizable role in the identification of the ultimately successful candidate product. The initiatives of the clinicians, Hryntschak and Swick, and the modest initial success with organic iodinated molecules they achieved, contributed in a crucial manner to the confidence that suitable contrast agents for excretion urography could be found. But it had been the industrial scientists von Schickh and Schoeller who first singled out for radiological testing by Hryntschak uroselectan sodium, the compound that in the end confirmed the feasibility of the target technology. Of relevance is that this happened long before Swick became involved. The collaboration of Swick, Binz and von Lichtenberg, sponsored by Schoeller, furnished the evidence that uroselectan sodium was the most suitable product. Although acting strictly within his medical specialty, von Lichtenberg's willingness to give Swick's pursuit of excretion urography a chance, as well as his prompt and extraordinarily massive effort of exploring the clinical utility of uroselectan, marks him as a highly influential champion of the new technology and pivotal actor.

It is therefore unacceptably reductionistic and one-dimensional when Lichtwitz (1930b), in referring to Swick, stated: "to him we owe the uroselectan method",[26] when he is stylized to "father of intravenous urography" (Loughlin and Hawtrey 2003) or worse yet, when he is elevated to "father or iodine contrast agents" (Nyman 2016). Swick was one of the pivotal actors, a champion, but not the unique father of uroselectan sodium, and even less the father of excretion urography. Considering the whole technology genesis phase, it does not seem out of place to list Rowntree, Volkmann, Roseno, Gehe & Co. AG, Swick, Räth, Binz, Hryntschak, von Schickh, Schoeller and von Lichtenberg, as fathers of the technology. One may furthermore add Hecht to this list, since it was he who identified and pharmacologically characterized methiodal sodium, and in doing so, provided the pioneering truly satisfactory and long-lasting uro-angiographic contrast agent.

There remains the question of driving forces behind the complex web of actors and actions. The vision as a source of motivation addressed sensibilities differing among actors. Association with a breakthrough product as a means of building a career was certainly a valid motive for Hryntschak, Räth, and Swick. The same association served Binz and von Lichtenberg to document their proper and their institution's continued productivity. The need for research money was a strong driving force for Räth and Binz. Although little evident in the written documentation, a tangible interest in being able to offer better care to patients as a driving force for Rowntree, Volkmann, Hryntschak, Roseno, Swick and von Lichtenberg can be presupposed. No doubt, the market provided incentives for the companies, although in this respect their precise motivations differed. Schering-Kahlbaum AG had significant activity in the field of radiology, whereas Deutsche Gold- und Silber-Scheideanstalt and Gehe & Co. AG did not. For I. G. Farbenindustrie AG the market incentives increased dramatically from the time of their initial explorations with Hryntschak to the point at which Hecht was pursuing a well-aimed industrial project. For academic actors, the market as perceived by the companies was of little relevance. The hypothesis that in the present case a principal driving force for genesis of new technology and fundamental innovation can be identified among the scientist's quest for understanding and manipulating nature, opportunities offered by an invention, medical needs, clinicians' compassion for their patients or profit promise of the market, had to be rejected.

The history of the two products of the technology genesis differed strongly with respect to the role played by motivations, triggers for action, intentions, accidents, chronologies, constellations of personalities, secrecy, and institutional environments, without being exhaustive. A multitude of pivotal actors was involved. Progress depended on many steps rather than on only a singular elementary one, no matter how uniquely enabling individual steps contributed. Technology genesis was accompanied by convergence of separate approaches, excursions into blind alleys, accidents and changes in environments, to name just some of the elements that were part of the process. It turned out to be impossible to describe the technology genesis as a linear chain of actions and progressive steps leading from identification of a technological opportunity to a technological innovation. Not even the drainage basin image with its local linearities serves well, alone considering that the breakthrough in technology took the form of two almost simultaneously commercially launched products.

The case reveals the impasse of one-dimensional reductionism in explaining the advent of uro-angiographic contrast agent technology. Hence, the analytical model, as applied to this technology, is admitted to pass its authors' own plausibility test number 1.

Technical Knowledge from Interference of Different Knowledge Cultures

The social processes of technology genesis that led to the knowledge embodied in uro-angiographic contrast agents involved, in decisive steps, actors representing a variety of professions and organizational positions. These pivotal individuals came with knowledge cultures typical of their field of professional training but nuanced by their personal experience and inclination. The knowledge culture characterizations of the analytical model proved easily applicable to the principle protagonists of the technology genesis phase. Selected instances are dissected in which interference of knowledge cultures is evident.

One set of differences in knowledge cultures of participant's concerns modes of knowledge acquisition, with regard to object, actor- and self-reference. Small differences in emphasis encountered in the behavior of the participants of the history reflect in part personality and in part differences in knowledge cultures typical of their professional background. Chemists, whether academic (Räth, Binz) or industrial (von Schickh, Ossenbeck, Tietze), acted mainly in an object-reference frame. For all those with management responsibilities (Binz, Schoeller, Lichtwitz, von Lichtenberg) society in the form of institutions (funding agencies, universities, industry, and hospitals) called for a knowledge culture with a strong actor-reference, as did students for their teachers (Räth, Binz, Lichtwitz, von Lichtenberg).

The physicians (Rowntree, Volkmann, Roseno, Hryntschak, Swick, Lichtwitz, von Lichtenberg), like the chemists, maintained a strong object-reference. Documents contain little overt evidence of seeing their patients as anything more than experimental subjects, i.e., objects of study on a par with animals and chemical compounds. In their publications, there are no prominent compassion arguments. There is not even any explicit evidence of social concerns, e.g., in the form of evaluations of medical treatment impact or of pricing of products and procedures. The price argument had been raised by other physicians many times previously and emerged again, not during but following the introduction of uroselectan sodium and methiodal sodium on the market. The fathers of the new technology did not publically contribute their own opinions regarding the price issue. Patients and society in general did not serve as important reference to them. Nonetheless, the mix of object, actor- and self-reference in knowledge acquisition by physicians contained a larger proportion of the latter two than that of the chemists. Surgeons, internists, academic chemists and industrial scientists in decreasing order claimed attention (self- and actor-reference) through primacy quarrels (von Lichtenberg, Swick, Binz, von Schickh) or showmanship (von Lichtenberg). This order may reflect a decreasing pressure for short-term performance in the public's eye.

All actors in the story showed similarly strong commitment to the acquisition of immediately usable knowledge, mostly in the form of practical products for therapy or diagnosis. Over and above that, they shared strongly manipulative-constructive modes of knowledge acquisition, i.e., synthesizing chemical entities, testing bactericidal activity, administering the compounds to animals and man. Some experimenters (Heuser, Hryntschak, Volkmann) initially acted opportunistically, using as contrast agents, pharmaceutical equivalents of the physicist's sealing wax and strings (Price 1984), provided it enabled the pursuit of their clinical research goals. Others (Rowntree and collaborators, Gehe & Co. AG and Roseno, Räth, Binz, von Schickh, Schoeller, Swick, von Lichtenberg, Hecht) got involved only when seduced by an attractive rationale or when plausible tools (pharmaceuticals) became available. On the axis of differentiation spanning observational-analytical and manipulative-constructive modes in knowledge acquisition the two groups clustered near two opposite poles.

The mix of physiological senses used in the acquisition of knowledge by chemists, surgeons, internists, and radiologists in the historical periods examined here also differed. In surgery, the sense of touch *per se* or incorporated into hand eye coordination and dexterity in general assumes a primary importance. Surgeons tend to develop an intimate relationship with their instruments that are perceived as extensions of the human hand and sense organs. Not a few of them favored retrograde pyelography over excretion urography because it called upon utilization of their preferred physiological senses, which the newer method did not. This situation slowed affirmation of the new technology down.

The relative weight of visual acquisition and conveyance of knowledge concerning contrast agents differed between those who primarily focused on radiographs (e.g., Hryntschak, von Lichtenberg) and those who concentrated on chemical properties or pharmacokinetics and toxicity (Lichtwitz, Räth, Binz). Swick's direct experimental involvement in the two types of activity documents the fact that he balanced and internalized both modes. Swick's initial radiological observations offered Lichtwitz the opportunity to expand his field of endeavor, yet he abstained from doing so. It is not farfetched to think that, as internist, he felt too little attraction to the world of images to justify a change in direction of his research. The same may be said of the surgeon Volkmann, who published his radiological studies without ever including a radiograph.

The various knowledge cultures in the history differed notably in the traditional mode of documenting and conveying their knowledge. Chemistry, toxicology, and pharmacology relied, like most natural sciences do, almost exclusively on a highly abstract symbolic representation (chemical formulas, analysis data, maximal tolerated doses, etc.). Internal medicine relied much more on documenting and conveying their knowledge in the form of demonstrations on patients, a sort of exhibition by the teacher in front of an audience (expression). In such a demonstration, the learner could complement the knowledge transmitted through not fully adequate verbal descriptions of clinical symptoms by personally verifying and repeating certain physical examinations such as inspection, palpation, auscultation, and percussion. To some degree surgery followed a similar pattern. Not uncommonly, the demonstrating surgeon, urological surgeon in particular, would add an appellative function (signal) to his theatrical performance (expression) through triumphant display of the kidney stone recovered during surgery.

In terms of his knowledge documenting and conveying behavior, the radiologist was and still is closer to the situation of the internist and surgeon than to the physical scientist. During the earliest times of radiology, live demonstrations before large audiences of the full X-ray imaging procedure and interpretation of the radiographs were the principal means of conveying one's knowledge of radiology (Kirklin 1945). Demonstrations that included the actual imaging procedure disappeared, as the equipment became too heavy to move and the dangers of radiation had become evident. What remained for the radiologist was that one picture was still worth a 1,000 words. Labeled X-ray shadow pictures then and tomographic pictures today constitute a form of raw documentation of an experiment in almost all its complexity. It is their interpretation that requires abstract concepts, such as X-ray attenuation, beam energy and beam hardening, film screen combination, motion artifacts, normal and pathological anatomy, etc., and leads to a verbalized interpretation. The radiologist relies on his acquired capacity for visual pattern recognition to make an efficient connection between the picture and its rationalization. He prefers to document his knowledge through a collection of typical pictures and to pass this knowledge on to others by a personal demonstration of visual pattern recognition in front of light boxes with radiographs. The picture-based culture of radiologists (Wackenheim 1982; Viviani 1986, 1989; Dommann 1999) is to non-radiologists only accessible with great difficulty.[108] The chemist is usually shocked in disbelief when he sees the pictorial basis on which the radiologist is capable of arriving with confidence at certain diagnoses. In the same vein the chemist and pharmacologist are often frustrated by the difficulty in conveying to radiologists the full meaning of chemical or physicochemical information about a contrast agent, information that usually takes the form of formulas, physical properties, and graphical renderings or tables of data.

Hryntschak, apart from being a highly respected surgeon (dexterity), cultivated a refined visual aesthetic. Professionally, his sensibility to the visual found its expression in his interest in radiology and his habit of transforming radiographs into handdrawn figures for publication long after this form of rendering had become outmoded. In private life, he kept an exquisite flower garden that he proudly showed off (Deutike 1952). Hryntschak's description of the pyelographic abnormalities in kidney tuberculosis as in the form of daisies (Margeritenform), which has entered the medical vocabulary, further illustrates his visual aesthetic. Others might have preferred to search for a descriptive term among Greek geometrical concepts. Hryntschak happened to be an excellent polyglot speaker, less a writer. He assiduously avoided numbers, tables, and formulas in his communications. It is not farfetched to believe that his pattern of knowledge documentation and conveyance, which was so different from that of his chemist and pharmacologist partners (Tietze, Ossenbeck, Schoeller, von Schickh, Räth, Binz) who relied essentially on chemical formulas and tables of measurements, formed a barrier that contributed to misunderstandings. It also helps explain why Hryntschak happily pursued excretion urography with compounds of which he knew, even at the end, merely code numbers. Non-radiologist readers of his congress communication were offended, and his contribution to progress was correspondingly undervalued.

Adhering to dominant practice, almost all academic physicians (Volkmann, Hryntschak, Swick, Lichtwitz and von Lichtenberg) ignored patents. One of them

(Swick) claimed novel inputs to chemistry that were in reality already patented by a chemist (Räth). Patenting did not offer occasions for constructive interference between physician's knowledge culture and that of industry. There was one exception that confirmed the rule (Roseno and Gehe & Co. AG). Entirely different was the situation with the academic chemists (Räth, Binz), who left a trail of patents whose assignments to industries (Deutsche Gold- und Silber-Scheideanstalt, Schering-Kahlbaum AG) demonstrate interference of their knowledge cultures.

The following two cases from the technology genesis phase serve as examples of interference of knowledge cultures, involving a network of actors and actions as sources. The analysis identifies actors by the conventional designation of their profession and position, and thus with the corresponding knowledge cultures, together with the personal nuances of these knowledge cultures that played a prominent role in interference. The constellations of actors and concurrences of events that led to knowledge production by interference of knowledge cultures is highlighted. The evidence adduced consists on the one hand in conceptual traces through history in the form of interpretations of chronologies of coeval documents and claims therein, as well as critical evaluation of recollections by historical protagonists in the form of printed material or interviews. On the other hand, it consists of concrete traces. These are of the documentary type (Danto 2007) and include material artifacts, including printed material, product names, employment records, etc.

CASE 1: Rowntree and his collaborators published their results with sodium iodide as contrast agent for excretion urography supported extensively by radiographs and tables of quantitative data (Osborne et al. 1923). Volkmann, unaware of the achievement of the group around Rowntree, had independently some success with the same method. He briefly presented his results, including radiographs, at congresses. However, in contradistinction to Rowntree, in print Volkmann published his findings exclusively in the format of text. He always signed as sole author, for which he prided himself openly. The price to pay for this was a weak anchoring in a network of peers. Not sufficiently taking into account the visual knowledge culture of his target audience, Volkmann revealed himself as a poor communicator, at least with the potential readers of his papers. His knowledge culture was weak in actor-reference and strong in self-reference, at least at the time of actual interest. No wonder his contribution to contrast agent progress has mostly been ignored.

Yet, one may concede to Volkmann anyhow a significant role in contrast agent technology genesis. It was he who in his publications repeatedly called for help from industry in finding suitable contrast agents, sensitized Gehe & Co. AG to the business potential of such products and thereby rendered it receptive to the proposal of Roseno. He catalyzed the entry of the company into the new field, and indirectly enticed additional ones to follow. One may see in this feat communication productive for the technology but in the end unsuccessful for Volkmann.

The mode of knowledge production and reproduction at Gehe & Co. AG stands out for strong object- and actor-references combined with scarce self-reference. The tradition at the company not to allow identification of company employees as inventors on any of its patents or as contributors to Gehes Codex,[53] may be seen as institutionalized suppression of self-reference in knowledge acquisition. In this respect, the knowledge cultures of the company differed strongly from that shared by

Volkmann and Roseno. Volkmann's playing a lone hand in publishing and Roseno's opportunistic neglect to mention the source of sodium iodide/urea in his first publication about the product's clinical performance, an illicit "deletion of modality," are indicators of pronounced self-reference in their personal knowledge cultures.

Let us for a moment construct as an alternative to reality the following hypothetical scenario. At Volkmann's suggestion Gehe & Co. AG explores as contrast agents iodinated organic molecules and is successful with a derivative of tetraiodopyrrole. As a novel compound, it is patented. Volkmann and Roseno get involved in clinical testing and report positive results. The company puts the product on the market. This scenario is well describable by simple confluence of adjacent knowledge cultures. In contradistinction to the hypothetical scenario, several elements of reality showed a more intimate interaction between industrial knowledge cultures and those of the practicing clinicians, interactions of the kind that the analytical model calls interference. In view of economic opportunities and threats the company put protection of intellectual property of a new product at the core of its attention. Through interactions among company employees a most unusual solution for how to render Roseno's suggestion patentable was found. The cumbersome pharmaceutical presentation and final preparation for injection of the commercial product, PYELOGNOST, reflected exclusively the patenting necessity and was in no way determining for its contrastographic performance. It bore this flaw as a trace of how it had come about. Against the company's practice, it now identified one of the inventors by name, i.e., Roseno. The company as primary inventor declared Roseno as co-inventor. In this case the company respected the emphasis on self-reference in the clinician's knowledge culture, overcoming its own misgivings about such emphasis. Less of such sensitivity toward Volkmann disgruntled him. A potential future client and product advocate was not only lost but even antagonized. One may ascribe the latter situation to an antiproductive interference of knowledge cultures. Roseno's report on PYELOGNOST, in coincidence with the preliminary communication of Hryntschak at a congress in 1928, contributed to Swick's decision and probably also to that of Hecht, to enter the field of endeavor and encouraged Schoeller and von Schick in their radiological pursuits.

The conceptual trace of how interference of knowledge cultures has led to sodium iodide/urea is well documented for the beginning and for the end stages by letters and publications. Except for Otto's role as assigner of a task to his researchers at Gehe & Co. AG, unfortunately no details are known about the intermediate stage. But we can be sure that pivotal actors included some with the professional knowledge cultures representing research, intellectual property protection, pharmaceutical production, and marketing. Concrete traces of the interference of knowledge cultures are the intricate and costly pharmaceutical formulation and the composition of the authorship of the patent. Obviously, the commercial product, PYELOGNOST, and published documentation of its clinical performance, as well are concrete traces, but in this case with regard to progress in the technology genesis of excretion urography in general.

CASE 2: The network of pivotal actors involved in the advent of uroselectan encompassed a rich panoply of knowledge cultures. Binz and Schoeller excelled as organizers of collaborations involving academic institutions, hospitals, and industry.

They revealed in their production of knowledge, besides the typical object-reference of applied natural scientists, a pronounced actor-reference. Swick participated in knowledge acquisition, living out quite openly his inclination for self-reference in the form of pronounced curiosity, career aspiration, and pursuit of recognition. In part thanks to these personality traits, the technology genesis gained with him an extraordinarily motivated champion for excretion urography. It may sound strange, but von Lichtenberg was a soul mate of Swick. As acrimonious as their authorship dispute eventually became, technology genesis only benefited from their interaction. The former brought to bear, apart from his clinical experience, his full organizational ability, power and showmanship (actor-reference & self-reference). In terms of self-reference, the counterparts to Swick, von Lichtenberg, and to a lesser degree to Binz, are found in Räth, Lichtwitz, Schoeller, and von Schickh, who acted very much in the background (low self-reference).

On the side of the chemists and pharmacologists involved in the technology genesis of interest here (Räth, Binz, von Schickh, Schoeller, Tietze, Ossenbeck, Hecht), patents were an important mode of knowledge documentation and conveyance. Hospital clinicians (Hryntschak, Swick, Lichtwitz, von Lichtenberg) ignored them. Practicing illicit "deletions of modality" regarding patents, Swick was able, at least in the medium term, to have his self-serving history of uroselectan accepted, but otherwise the behavioral differences between the two groups had no negative consequences. Of significance is that neither Schoeller nor von Schickh, ever published an article that included a radiograph. A poorly developed visual knowledge culture contributed to their coeval and later underrating of Hryntschak's earliest results in marking progress in excretion urography. Swick, the specialist in internal medicine, attached much importance to his proposal to Binz of desirable features of chemical compounds and the means to meet toxicological goals. The episode had the potential for interference of knowledge cultures, but given the unfoundedness and ineffectiveness of the propositions, it does not merit that classification.

Excessive self-reference in the knowledge cultures of Swick, von Lichtenberg and Binz is detectable already in their earliest scientific publications on uroselectan sodium but burdens still more their later historical tales. Only a critically interpreted aggregate of their documents can contribute to the identification of conceptual evidence for interference of knowledge cultures. One particular conceptual trace of interference is given by the mutation of veterinary antiinfective drugs into contrast agents for excretion urography. The mutation started with the independent observations of fast renal excretion of the iodinated veterinary drugs, selectan sodium, and selectan neutral, by Räth, and in the laboratory of Lichtwitz by Erbach. Schoeller and von Schickh at Schering-Kahlbaum AG let this fact interfere with their experience with iodinated contrast agents showing elevated biliary excretion. A collaboration with the urologist Hryntschak on excretion urography came about, but pursuit of this new application of selectans only took off upon Swick's entry on the scene.

The most evident concrete trace of the interference of knowledge cultures in the advent of excretion urography is the name of the first iodinated organic compound suitable as a contrast agent, i.e., uroselectan sodium (UROSELECTAN). The name stem, selectan, gives testimony of the contrast agent's origin in the eponymous veterinary antiinfective drug, whereas the first syllable obviously refers to its medical area of application. Another striking concrete trace is the conspicuous German patent number, referring to the corresponding application of Räth (1927b), on an early advertisement for UROSELECTAn™ (Schering AG and Wlasich 1991). This help for finding the pertinent patent may be explained by the fact that in it contrast agents are not even mentioned.

The interactions between Ossenbeck and Tietze from I. G. Farbenindustrie AG and Hryntschak, or that between Binz and Deutsche Gold- und Silber-Scheideanstalt, show that collaboration between academia and industry can fail to bear fruit directly. For an immediately productive outcome, constellations of actors and concurrence of events must harmonize in a way that is difficult to organize. Yet, the interactions still offered a multitude of circumscribed occasions for interference of knowledge cultures that in aggregate represent a step forward in technology genesis.

The differentiation of modes of documentation and conveyance encountered here exceeded in richness those captured by the original scheme of Dierkes, Hoffmann, and Marz (1996). Yet the proposed refinements to the scheme only reinforce the conclusion that would have been drawn on the basis of the original scheme alone. Interference of knowledge cultures did definitely play a role in some episodes of technology genesis but not in all. I suggest that in the analytical model the requirement for interference be relaxed from an essential to a favorable condition. In this modified form, the analytical model of Dierkes, Hoffmann, and Marz (1996) as applied to uro-angiographic contrast agent technology passes those authors' plausibility test number 2, in which technical knowledge is now envisioned to profit from interference of knowledge cultures.

Interference as Communication and Individuation of Knowledge

Communication channels transformed the described collection of human and institutional actors into networks within which interference of knowledge cultures at the external level could flourish. In the technology genesis phase under study the principle process components of communication, cooperation, maintenance of cooperation, and coordination of continued cooperation functioned generally at an excellent level. The exceptions that prove the rule were those between Volkmann and Gehe & Co. AG and between Hryntschak and I. G. Farbenindustrie AG. Rowntree received optimal cooperation by Osborne and colleagues and enjoyed their explicit attribution of the crucial idea to him. The network involving Räth, Binz, Schoeller, von Schickh, Lichtwitz, Swick, and von Lichtenberg worked efficiently. If the dispute over the presentation at a congress between Swick and von Lichtenberg is considered no more than an unhappy interlude, their cooperation up to the end of the technology genesis phase, i.e., the times of the commercial launches of UROSELECTAN, can be judged highly productive. Cooperation was fostered by a preference for the face-to-face meeting, which according to media richness theory constitutes the richest of the various communication agents (Daft and Lengel 1986). But when excitement and urgency needed to be conveyed, actually Swick's choice of the telegram that is considered a poor communication agent, was appropriate and achieved cooperation in the form of a change in von Lichtenberg's travel plans and acceleration of the clinical research program. Phone contacts appear in none of the chronicles offered by the protagonists.

Swick excelled as a communicator. The manner in which he succeeded in convincing others to accept or place him always at sites of maximum opportunity testifies to this capacity. His timeliness in communication is just another aspect of the same trait. Only a person with the finely tuned sense of communication that Swick possessed could have mounted such an effective defense against von Lichtenberg's attempt to gain primacy and sole authorship for communicating the data on excretion urography with compounds of the selectan family. It is fitting that Swick, through Lichtwitz, involved Salle, a medical man professionally rooted in the communication industry, in helping to resolve the dispute. Swick carried an inner mission and did not stray from it for one moment. He achieved needed cooperation with the powerful (Lichtwitz, Binz, von Lichtenberg, Schwarz). Through them he received services from less potent figures (Erbach, unnamed radiologist in Altona, Rave), although for them the cooperation offered little recognition.

The importance of communication should not be gauged by looking only at directly productive collaborations. Secretive behavior of Ossenbeck from I. G. Farbenindustrie AG in his interaction with Hryntschak hampered cooperation. Hryntschak communicated inefficiently with Räth, Binz and Schering-Kahlbaum AG. He failed to solicit constructive criticism and, left alone, fell victim to simple misfortune. As a result of his behavior, his contribution to the technology genesis, much as it may have preceded that of Swick, remained restricted.

Systematic means to reproduce cooperations are easily recognized. These means are illustrated by the following: (a) Acquisition of the full patent line of the inventors Räth and Binz by Schering-Kahlbaum AG from Deutsche Gold- und Silber-Scheideanstalt, together with the provision of a long-term horizon by signing of a plurennial research contract between the acquiring company and the university chemists, are both classical means of stabilizing cooperations. Binz and Schoeller, as orchestrators of the deal, became the primary guarantors of continued productive cooperation. (b) Swick stabilized cooperations by increasing the frequency of interactions, even to the extent of changing his domicile for reasons motivated solely by the project. (c) Swick intuitively used the tool of projecting unlimited optimism to keep collaborators motivated. This method lost its effectiveness as soon as the primary goal was reached, and centrifugal selfish interests tore the cooperation to pieces. (d) The fact that Swick and Lichtwitz weren't informed about Hryntschak's involvement with uroselectan sodium suggests that Schering-Kahlbaum AG and Binz were trying not to let a somewhat uncomfortable situation involving Hryntschak disturb the principal cooperation. (e) The successful outcome of the cooperation between Räth, Binz and Schering-Kahlbaum AG first resulting in selectan for veterinary applications, was used to reinforce trust in the future success of the cooperation toward a contrast agent for excretion urography. (f) When Räth left Berlin, Binz immediately stepped in to maintain existing cooperations and identified himself with the project to the extent of adapting the focus of his own research. (g) By organizing the mediation by Salle between Swick and von Lichtenberg, Lichtwitz reproduced their endangered cooperation.

Coordination of cooperation is part and parcel of the job description of industrial research and development managers (Schoeller, Hecht) as well as heads of academic research institutes (Binz) and hospital units (Lichtwitz, von Lichtenberg).

In general, the protagonists lived up to expectation. The grand prize for effective, and selfless, coordination of cooperation must go to Lichtwitz. He accepted Swick into his research group, allowed him to explore a research direction totally novel to the group, put him in contact with Binz, magnanimously agreed to transfer Swick to von Lichtenberg's clinic and played a crucial mediating role in the primacy quarrels between von Lichtenberg and Swick over communication of the results, all this without claiming any merits in the genesis of uro-angiographic contrast agent technology.

At the internal or individuation level of interference between knowledge cultures, personal acquisition of multicultural thinking and communication may be encountered in Swick. His proposals regarding the medicinal chemistry of selectans, irrespective of whether they were helpful to the identification of uroselectan sodium as the best candidate contrast agent, required him to slip to some extent into the knowledge culture of Binz (internalization). He had to overcome the fear of exposing himself to ridicule. His elevated self-esteem certainly helped him in this effort and in overcoming the blatant rejection of his ideas by Binz. There is no evidence that Binz made a reciprocal effort to pick up notions of urological radiology.

Von Schickh offers a second striking example of internalization. While working on the optimization of the industrial synthesis of the contrast agent iodophthalein, and probably also of the selectans, he transcended the boundaries of his knowledge culture as chemical engineer by considering the potential radiological applications of the agents. This led him to identify uroselectan sodium as a candidate for excretion urography. Schoeller, with whom von Schickh shares that merit, not surprisingly, excelled in transcultural thinking. For an industrial research director, this capacity is a fundamental requirement. It is more important than excellence in a single area.[109]

As director of a university institute, Binz was forced to defend its activity to the providers of public research funds, i.e., the university, the Ministry of Agriculture Domains and Forests and the Emergency Action Groups for German Science. To each of these entities credible rationales for the research projects had to be provided that differed necessarily in their pattern of argumentation e.g., with respect to object-relations, actor-relations, and self-relations. These rationales were not completely congruent with each other or with those of the industrial sponsors. It was part of Binz's capacity to internalize the different cultures and to articulate convincingly in the various contexts. The way the fixed subtitles of the scientific papers of Räth and Binz were cited or ignored to suit the purpose, is striking evidence of their ability to tailor the message to the audience.

Both Räth and Binz had to contend with industrial interest in patenting all innovative chemical processes. In Binz's account there is a detectable undertone of annoyance. It testifies to the exertion it cost these academic chemists to repeatedly slip into the guise of the industrially motivated inventor and illustrates that the internalization of distinct knowledge cultures leading to new technology requires continued effort. It demands ongoing work on the part of the individual himself—reproduction of internalization. Although it is difficult to find traces of this phenomenon in the available documentation, its importance can be admitted.

If Binz had to justify both his personal and his institute's activity to public funding sources and the collaborating industry, he was required to find befitting cultural terms, while simultaneously maintaining a balance of distinct rationales within himself. Traces of this difficulty are found in his somewhat contradictory reconstruction of his personal and his institute's declared goals in the early 1920s. Far from reflecting a bad memory or a lack of veracity, this phenomenon can be understood in terms of the difficulty of coordination of reproduction of internalization. Furthermore, Hryntschak's ambiguous behavior with regard to the identity of the brominated and iodinated compounds he had used in his experiments may reflect this difficulty. Such difficulty had to be faced to a greater or lesser degree by all actors operating two or more transcultural lines of communication simultaneously.

In the technology genesis phase of present concern communication and individuation of knowledge, together with their subprocesses, were readily recognized and functioned in general with elevated efficacy. They may actually be equated with the phenomenon called interference of knowledge cultures. Within the chosen theoretical framework observation and theory harmonized well. Therefore, the analytical model of Dierkes, Hoffmann, and Marz (1996) when applied to our case passes its authors' plausibility test number 3.

The Emergence of a Technological Vision

The role analogies have played in the emergence of the early vision of contrast agents in general has been amply documented in Chapter 3. In the same role analogies can lead to refinements of an existing vision. In the case of excretion urography, again, analogies were at the heart of the emergence of a revised vision. When Rowntree learned through the pharmacokinetic study of Osborne of the elevated blood iodide levels that could be achieved and thus of the inevitably elevated urinary levels, he perceived the analogy to his own pharmacokinetics experience with halogenated phenolphthaleins. The latter got concentrated in the bile and, as a consequence, iodophthalein had become a cholecystographic contrast agents. The fact that blood levels of potassium and sodium for the studies of Osborne were performed in laboratories of the department headed by Rowntree, must have favored the occurrence of these analogical arguments. The shared institution allowed him to talk to his colleague without having to overcome geographical barriers and it facilitated informal contacts. A collaborative study ensued naturally, with its economy of acquisition of initial experience.

In Volkmann's independent foray into excretion urography too, weak analogical arguments may be discerned. He solicited industry to synthesize iodine-containing organic molecules. He relied on the analogy to the iodized oil known from their use in the radiological visualization of shooting canals, in myelography and in retrograde pyelography. He called for iodinated analogs of urotropin because of this compound's elevated water solubility and renal excretion.

In the early 1920s, when Rowntree and coworkers, as well as Volkmann, were independently addressing excretion urography with encouraging results, the prospect of feasibility was added to the previously established desirability of the technology. In this way had emerged the first half of the new vision, i.e., the intravenously injectable, mostly renally excreted contrast agent for urography (Table 4.1). In this vision the earlier one of the contrast agent as a chemical substance, which absorbs X-rays but is free of detrimental pharmacological effects, as a fixed core, was implicitly incorporated. By this time, it had become a completely internalized general understanding.

Excretion urography that less than 10 years earlier had not been publicly contemplated, suddenly became deflated to an obvious thought, actually an imperative. In some publications, the impression was given that the technology had attracted attention right from the beginning of contrast experimentation. Credit to Rowntree and collaborators, as well as to Volkmann, was restricted to their timely attempts at implementation. Such is the power of a vision to create common ground of understanding that the barest of explanations suffices to communicate. In fact, no document has been found in which one actor had to explain to another the concepts behind the technology.

The discovery that the hollow spaces of the urinary tract could be radiographed after retrograde administration of various salt solutions, by analogy suggested that also hollow spaces of the cardiovascular system could be so visualized after local intravascular injection. The desirability of such visualization was self-evident. Heuser and Rowntree offered the basis for growing confidence in the feasibility. The second half of the new vision, which regarded angiographic contrast agents, had emerged. The experimental successes with sodium iodide in excretion urography and angiography fostered the expectation of finding a single compound suitable in both applications. With this expectation and the hope of applicability in demanding domains of the vascular system (cardioangiography, cerebral angiography) the vision of uro-angiographic contrast agents was completed (Table 4.1). It had come about in a stepwise fashion. Before the first uro-angiographic compounds became available (uroselectan sodium, methiodal sodium), contrast agents only adhering to the first half of the vision (sodium iodide/urea) or to the second half (strontium bromide) appeared on the market. The angiographic contrast agent colloidal thorium oxide fell even totally outside the described vision on multiple grounds, foremost because, to the limited extent it was excreted at all, it did not follow the renal route. At the cost of serious long-term health risks to the patient, it delivered angiograms in quality almost impossible to beat using water-soluble uro-angiographic products. Nonetheless, the vision of uro-angiographic contrast agents and its representative products in the end prevailed over alternatives. The technical name "uro-angiographic contrast agents" was descriptive and concise enough not to call for a metaphoric formulation of the vision.

In the cases described, the reliance of analogies resulted in the genesis of a vision that spawned progress, a positive effect. Yet, proceeding by analogy can paradoxically both foster and impede progress. Impedance of progress is exemplified by the excessive reliance on the analogy between observed X-ray shadows and bone-like calcium-containing structures that for many years caused the X-ray shadows at intramuscular injection sites of JODIPIN to be misinterpreted as calcifications, even though the X-ray absorption of iodine-containing compounds had already been established, and the correct interpretation had been reported in the literature. As a consequence, the introduction of iodized oils as contrast agents was delayed for at least a decade.

In the case of contrast agents for excretion urography and for angiography, the emergence of the technological vision could be traced. The case provided a good test for the claim of the present theory that the elementary process of interference of knowledge cultures responsible for the emergence of the vision for that technology involved the reliance on analogies. Moreover, the complete vision came about in steps involving visions with a restricted scope.

Synchronization and Interference

The fact that Rowntree with his collaborators in the USA and Volkmann in Germany independently but in near synchrony, using sodium iodide, arrived at the same partial solution to the problem of excretion urography, is best attributed to a constellation of available knowledge that they shared; the time was ripe. No interference of their knowledge cultures was involved and the reigning vision of contrast agents in general could not have synchronized the events. Rather, only the coincidental occurrences sparked the formation of the vision of contrast agents for excretion urography, and with it the first step toward the full vision of uro-angiographic contrast agents. The simultaneity in the two technologically most advanced continents reinforced the chances for a successful career of the nascent vision.

One of the institutional roles of scientific congresses is to produce, on the one hand, synchronicity of communication of knowledge among participants, and on the other hand, individuation of knowledge by individual participant's synchronous acquisition and integration of information from different sources. In addition, occasions for interference of knowledge cultures arise. The vision of excretion urography motivated the conveners of the urology congress in Berlin to organize a corresponding presentation session with subsequent discussion period. This led to the essentially synchronous scheduled presentation of Roseno and unscheduled discussion contribution of Hryntschak, both in the presence of von Lichtenberg, Swick, and Kielleuthner. The synchronicity resulted in an impact of a strength that the same communications on separate occasions and locations would not likely have had. It conditioned Swick for recognizing the potential of the selectan group of compounds as contrast agents, and prepared von Lichtenberg to eventually return to a challenge he had pursued but given up on. We can be sure that Schoeller was brought up to date and shared the new impetus. Through the new, renewed or enhanced pursuit of the vision by these actors the genesis of contrast agent technology for excretion urography was accelerated.

The nascent vision of excretion urography synchronized activities in independent networks of actors, networks that may be delineated by the company participating and coordinating activities therein, namely Schering-Kahlbaum AG and I. G. Farbenindustrie AG. The result was the commercial launch of UROSELECTAN and ABRODIL in 1930, only a few months apart, essentially in synchrony.

In the historical period analyzed examples were found for synchronization of communication and individuation, as well as of events, by a vision. But the analysis also revealed a historically relevant synchronicity that occurred without involvement of a pre-existing vision regarding contrast agents. A scientific congress was found to have played an important role in synchronization of communication and individuation. This insight appears to be generalizable to fields outside contrast agent technology. In the slightly simplified form applied by me to the first historical period of interest here, the analytical model did pass its authors' plausibility test number 4.

The Functions of the Technological Vision

The young vision of intravenously injectable, mostly renally excreted contrast agents for excretion urography oriented research interests away from the quest for improved products for retrograde pyelography. The latter effort terminated with the commercialization of lithium iodide and the prospects for excretion urography replacing retrograde pyelography. With time an increasing number of academic investigators (Rosenstein, von Lichtenberg, Hryntschak, Roseno, Swick, Binz) and industrial ones (Otto, Schoeller, von Schick, Hecht) adopted the new orientation.

Besides orientation, the vision also offered motivation to become newly active or to intensify existing efforts. This motivation function drew on expectations of procedures yielding improved delineation of renal calyces, the avoidance of painful and somewhat risky catheterization procedures, and for industries it offered a substantial market size. A few investigators and industries operated outside the vision by developing retrograde urography with colloidal thorium oxide or iodized oils. These products lost out to the first new products adhering to the vision.

The vision of the angiographic agent encountered more obstacles in imposing a new research direction than the vision of the urographic one. The main reason was the spectacular temporary success in angiography of stabilized colloidal thorium oxide (THOROTRAST). The vision instead led through strontium bromide to uroselectan sodium and methiodal sodium. Bolstered by the performance of the latter two it oriented and motivated the search for still more attractive products in this vein. All along the vision motivated a critical mass of investigators in the pursuit of a technological solution and stabilized the endeavor against reasonable doubts uttered by influential personalities (von Lichtenberg); it fostered resilience.[110]

In the case of contrast agents for excretion urography a convincing case emerged for the orientation and motivation subfunctions of the vision. It is therefore concluded that, with the modifications proposed by Kahlenborn et al. (1995), the first historical period examined provides plenty of support for the analytical model of technology genesis and shaping. The model passes its authors' plausibility test number 5.

Technology Espousal and Resistance

The uro-angiographic contrast agents based on iodinated water-soluble organic compounds constituted a suitable solution to widely perceived medical needs. Although the new technology sometimes revealed less anatomical details than the old retrograde pyelography would have, the results were satisfactory and more reproducible than with intravenous sodium iodide alone or in association with urea (PYELOGNOST). Whatever risks the procedure posed, they were smaller than those associated with either retrograde pyelography or with intravenously injected inorganic salt-based preparations. In consequence, the iodinated contrast agents for excretion urography were hailed as great accomplishments. This opinion has remained intact to date since, unlike for X-rays *per se* or contrast agents based on colloidal thorium oxide solutions, almost no late detrimental effects have turned up.[111]

One issue that slowed down espousal of excretion urography was the comparatively elevated price of the iodinated contrast agents. Arguments focused only on their costs. Physician costs were conveniently neglected and so were costs generated by adverse events due to the procedures. This illustrates the difficulty of coming up with objective economic assessment of new technology, a problem that to date has not been convincingly resolved. Due to their costs, iodinated contrast agents were initially simply reserved to those who could afford them. Unlike in the analogous case of the 1980s (*vide infra*), the debate apparently remained strictly economic. There is no evidence of ethical dilemmas being perceived or debated. Economic issues did slow down espousal of the new technology but without formation of organized resistance.

Excretion urography rendered superfluous the surgical skills retrograde pyelography required, but the two procedures remained in the same hands. This contributed to the fact that the new technology took at least 10 years to replace the old one. But replacement did take place. The creative destruction of the old ways required for the qualification as technological innovation, was met.

Already in the course of the genesis of the first uro-angiographic contrast agents some of their basic technological marks, foremost their nature as water-soluble iodinated organic molecules, became strongly imprinted upon the class of products and rendered this aspect of the technology impervious to alternative proposals. In fact, no technological solution outside the described bounds, be that metal complexes, gases, or colloidal preparations, have found lasting application, not to speak of business success.

TECHNOLOGY-SHAPING PHASE

HISTORY

Perfecting Ionic Uro-Angiographic Contrast Agents

The technology genesis phases for all major chemical classes of contrast agents in combination with their primary clinical uses reached completion with the advent of bismuth subnitrate as contrast agent for radiography of the gastrointestinal tract, iodized oil for the visualization of fistulas and body cavities as well as for myelography, iodophthalein sodium for cholecystography, colloidal thorium oxide for the visualization of the liver and the spleen and for peripheral angiography, and finally uroselectan sodium and methiodal sodium for excretion urography and angiography. Their technology-shaping phases could begin. Here such a phase is understood as a time span of substantial improvements of the technology through the selection of variations to the solution identified in the genesis phase, while operating in a highly competitive economic environment. Indeed, for all application classes of contrast agents by the end of the genesis phase market forces quickly began to shape the game with ever more companies competing in research, development, manufacturing and distribution activities. Through the choice of diagnostic procedures and their frequencies the practicing clinicians decisively influenced the market.

Uro-angiographic contrast agents, as an application class of products, entered the technology-shaping phase in temporal overlap with the final development of the prototypes, uroselectan sodium, and methiodal sodium, i.e., before the clinical and economic worthiness of the technology had been fully established. Industry took up the pursuit of follow-up products with improved performance. Intensive and imaginative clinical research in academic settings led to the extension of the applications of available contrast agents to an ever-increasing number of clinical indications. This was facilitated by the parallel development of improved X-ray equipment and accessories, with a dynamic that matched the events in the contrast agent field. Such was the spell of the technological vision (Webb 1990; Eisenberg 1992).

Beginning with their arrival on the scene in 1930 and up to the early 1980s uro-angiographic contrast agents were usually salts. They showed elevated water solubility in good measure because in aqueous solution they dissociated into anions of iodine bearing organic molecules and counter ions, most often sodium ion, protonated organic amines, or mixtures thereof. Because of this characteristic and their two-fold clinical utility, these products were classified and termed ionic uro-angiographic contrast agents. The current chapter traces the major steps in their evolution. In a conscious attempt to arrive quickly at the promised second major period of technology development to be investigated in detail, which comprises the substitution of ionic uro-angiographic contrast agents by nonionic ones, the evolution between the first and the second major period will merely be sketched to the degree necessary for the understanding of subsequent developments. Complementary descriptions of that period may be found elsewhere (Urich 1995; Rigler 1945; Hecht 1939; Hoppe 1963; Strain et al. 1964; Strain 1971, 1987, and ref. therein; Hoey et al. 1984; Felder 1986; Fischer 1987; Speck 1995).

At this point it is worth asking in what manner the successful technology genesis phase of uro-angiographic contrast agents conditioned the ways in which the vision was pursued in the technology-shaping phase. Early in the technology genesis phase scientists belonging to most variegated knowledge cultures and working in equally variegated environments, including potential end users, had searched in a large arsenal of chemical compounds and pharmaceutical formulations, available because of utility in other fields, for candidate contrast agents. Sometimes candidates were found by accident. Investigators enjoyed the benefits of interacting with colleagues from other knowledge cultures, without a typical order in which the professional disciplines came into play. This all changed with the success of the two iodinated, water-soluble organic molecules, uroselectan sodium and methiodal sodium, creations of medicinal chemists. Henceforth further progress was expected to initiate as new chemical entities in the hands of such professionals, while toxicologists, pharmacologists, pharmacists, radiologists, production chemists, and others performed the selection of useful contrast agents based on performance. The end of the technology genesis phase thus marked a change in the way new products were being sought.

The vision (Table 4.1) gave the chemists the overall goal and motivation for their endeavor. Medicinal chemists distilled from the experience with uroselectan sodium and methiodal sodium the following drug design principles, for products later called first-generation uro-angiographic contrast agents (Table 4.2).

The design principles were accompanied by selection criteria for candidate contrast agents. They included elevated water solubility, stability to heat sterilization of its solutions, lack of metabolism, rapid elimination from the body, and good tolerance at doses required for radiological effectiveness. These design principles and selection criteria shaped not only the first-generation contrast agents for uro-angiography. Partially modified, they continued to play this role also for later generation ones.

TABLE 4.2

Design Principles for First-Generation Contrast Agents

1a)   Iodine atoms serve as strong X-ray absorbers.

1b)   The iodine atoms are covalently bound to an organic chemical core that stabilizes aqueous
      solutions to heat sterilization.

1c)   The iodinated compound is rendered water-soluble by the presence of at least one negative electric
      charge, neutralized by suitable cations.

1d)   The weight percent of iodine composing the contrast agent molecule needs to be maximized.

1e)   Water solubility can be augmented by organic counter ions bearing hydroxyl groups.

In fact, the first two have remained dominant for all later generations of products. Iodine as X-ray absorber has never been successfully challenged.

In the very year in which uroselectan sodium arrived on the market, the chemist and physiologist Max Dohrn and the chemist Paul Diedrich at Schering-Kahlbaum AG asked for a patent covering a process for synthesizing its successor, iodomethamate disodium (USP), alternatively called iodoxyl sodium (B.P.) and early on coded as D40 (Figure 4.24, **11**) (Dohrn and Diedrich 1930). The compound had a chemical structure similar to that of uroselectan sodium. But the molecule possessed two iodine atoms and two carboxyl groups. Elevated solubility, rapid renal excretion and a tolerability in small animals better than that of uroselectan sodium and methiodal sodium rendered it attractive for uroradiological use.

Elevated tolerability, later preferably termed low toxicity,[112] of iodomethamate disodium was ascribed to the chemical structure in which tautomerization of the pyridone core structure was prevented, i.e., no reflections on its properties in solutions, other than solubility, were published. It became customary to specify contrast agents by the weight fraction that iodine contributed to the molecular mass, called iodine content. In these terms, iodomethamate disodium with an iodine content of 52% (w/w) was perceived as an improvement over uroselectan sodium with its 42% (w/w). This measure is expression of a chemistry knowledge culture, rooted in the practice of accompanying every synthetic compound by an elemental analysis. Biologically it is meaningless since it varies with the mass of the molecular core that carries the iodine atoms and the atomic or molecular masses of the counter ions. Meaningful instead is the number of iodine atoms per molecule, better yet, the number of iodine atoms per number of ionized particles formed in solution. More to the latter measure will be added further along our history. Notably, there is no evidence that at the relevant time at Schering-Kahlbaum AG the number of particles formed upon dissociation in solution, its influence on osmotic properties and pharmaco-toxicological consequences influenced choices of contrast agent candidates.

Given the impressive experience accumulated with the clinical assessment of uroselectan sodium in excretion urography, von Lichtenberg was the natural choice of Schering-Kahlbaum AG as collaborator in the evaluation of improved products. For excretion and urographical studies the company provided him with five new compounds that he compared with the two, by then commercial products, uroselectan sodium (UROSELECTAN) and methiodal sodium (ABRODIL) (von Lichtenberg 1931). Between June 1930 and June 1931, he added 1300 treated patients to the 700 examined earlier. He published the impressive iconography all alone, although the sheer quantity of cases absolutely required the assistance of collaborators. On one occasion he listed five names, including the urologist Heckenbach and the head of his institution's radiology department, Rave. Swick's name appeared only marginally and then just for his incorrect methyl group theory.

Initially the new candidate contrast agents were only identified by codes, but a year later the code was broken (von Lichtenberg 1932). Among the substances von Lichtenberg performed excretion urography trials with, is found compound D40, or iodomethamate disodium (von Lichtenberg 1931, 1932). The product became commercially available in Germany in 1931 as UROSELECTAN B (in USA NEO-IOPAX™, Schering Corp., Bloomfield, NJ). The concentrations offered were 50 and 75% (w/v).

In 1940 two young scientists at the Pharmacology Institute of the University of Bologna, Italy, pioneered the estimation of the osmotic properties of contrast agent solutions. Specifically, they examined iodomethamate disodium solutions by cryoscopy, using instruments whose accuracy they themselves described as problematic.[113] Anyway, the principle result was that the osmolality of iodomethamate disodium at the concentration of 75% (w/v), or 386 mg(Iodine)/mL, equaled cryoscopically that of a 23.3% (w/v) sodium chloride solution (Boriani and Boriani 1940). Today we know this to correspond to ca. 9,450 mosmol/kg.[41,114,115] Later the product was reported to match cryoscopically the osmolality of 17.3% (w/v) sodium chloride (Cotrim 1954). The observed diuretic effect of the contrast agent was ascribed to an osmotic cause.

Iodomethamate disodium rapidly replaced the original uroselectan sodium for good reasons. When given at intravenous doses achieving equal X-ray opacification in the urinary tract iodomethamate disodium produced fewer adverse reactions. Anybody who would have considered osmotic properties at all, on theoretical grounds alone, could have realized that the more highly concentrated commercial formulation of UROSELECTAN B produced an osmolality drastically higher than that of UROSELECTAN, actually about fourfold. Observation of fewer adverse reactions with the former would then have detracted attention from osmotic causes and favored others to be found in the chemical structure of the iodine bearing organic molecule. Reflecting this assessment, for decades the search for pharmaceuticals with reduced toxicity remained focused on the molecular structure of the iodine bearing moiety of the contrast agent, rather than on physical properties of the solutions for injection.

Another reason for UROSELECTAN B rapidly supplanting UROSELECTAN was its presentation as a vial containing the ready-made sterile solution for injection. This spelled enormous practical progress over its laborious precursor. No wonder, by that time ABRODIL too was offered in this improved pharmaceutical presentation. From then on, with the exception of metrizamide (AMIPAQUE™), to be discussed later, all commercial intravascular contrast agents have come as ready-made solutions for injection.

While I. G. Farbenindustrie AG was still busy with the development of methiodal sodium, work on a possible successor, dimethiodal sodium, had already gained momentum. Dimethiodal sodium was the chemical analog of methiodal sodium with two instead of one iodine atom per molecule, respectively (Figure 4.16, **10**). Again,

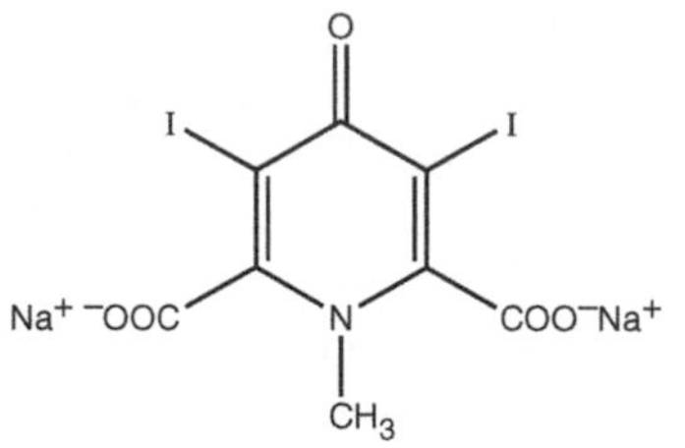

11

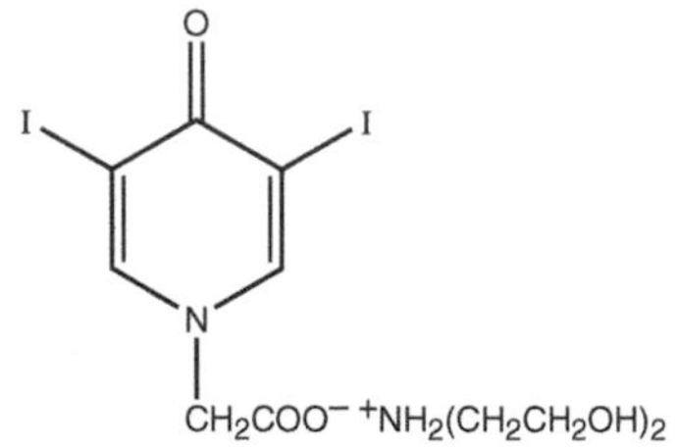

CH₂COO⁻ ⁺NH₂(CH₂CH₂OH)₂

12

FIGURE 4.24   The first-generation uro-angiographic contrast agents that, together with methiodal sodium and dimethiodal sodium, dominated the market in the 1930s and 1940s.

11: Iodomethamate disodium; iodoxyl (UROSELECTAN B™, Schering-Kahlbaum AG, Germany; NEO-IOPAX™, Schering Corp., USA). $I/P = 1.0$.
12: Iodopyracet diethanolamine = diodone diethanolamine (PER-ABRODIL™, I. G. Farbenindustrie AG, Germany). Formulated as salt of diethylamine (PELVIREN D™, I. G. Farbenindustrie AG, Germany), as mixt salt of diethylamine and diethanolamine (PER-ABRODIL FORTE™, I. G. Farbenindustrie, Germany; DIODRAST™, Winthrop Chemical Company Inc., USA), or as salt of meglumine (PER-ABRODIL M™, I. G. Farbenindustrie, Germany). $I/P = 1.0$.

the pharmacologist Hecht guided the project. In 1927 Hecht had written a chapter for the prestigious book series *Handbuch der experimentellen Pharmakologie*, entitled "The osmotic effects" (Hecht 1927). The general review of the subject, still lacked coverage of contrast agents. In 1939, after having played a crucial role in the advent of uro-angiographic products, he again wrote a chapter for the same series, this time with the title Roentgen contrast agents (Hecht 1939). Therein he very briefly remarked that contrast agent solutions then in use for retrograde pyelography were strongly hyperosmolal and thus could trigger severe mucosal inflammation.[115] He further attributed several non-severe adverse effects accompanying intravenous injection to osmotic properties. Hecht was evidently as early as 1930 sensitive to the osmolality issue.

To Hecht it must have been evident from the start of his involvement with methiodal sodium and dimethiodal sodium that the latter had the advantage of giving solutions of lower osmolality at the same X-ray opacity and that this might contribute to better tolerability. Contrast agents with two instead of just one iodine atom per molecule promised similar X-ray opacity in vivo at substantially lower molar doses. As it turned out the compound with two iodine atoms per molecule, dimethiodal sodium fulfilled its promises, while the analog with three iodine atoms was unstable.

Based on the appearance of Tietze and Ossenbeck as co-inventors in the patent covering dimethiodal sodium (Hecht, Ossenbeck, and Tietze 1930b), and the single period of their synthesis activity, i.e., 1922–1926, (Frings 1994) it can be concluded that the substance had been synthesized before 1926.[116] It is highly probable that it had been included in the aforementioned group of halogenated chemicals studied by Hryntschak. Meanwhile, an unrelated chemist had published a synthesis of dimethiodal sodium (Backer 1926), which, at the time the compound was treated as potential contrast agent at I. G. Farbenindustrie AG, precluded a composition of matter patent anywhere. Where in principle allowed, synthetic process and use patent protection could still be obtained (Hecht, Ossenbeck, and Tietze 1930b), however. But the protection of dimethiodal sodium was precarious, a circumstance that could have been a factor in the decision by I. G. Farbenindustrie AG to give methiodal sodium priority over dimethiodal sodium. It also makes plausible why the ownership of its dimethiodal sodium patents were transferred to the American company Winthrop Chemical Company, Inc., New York, of which the German company indirectly owned half (Hecht, Ossenbeck and Tietze 1930b).[79] Lastly, it has to be taken into account that the company's newest candidate contrast agent, iodopyracet salt, delivered encouraging results (*vide infra*). It appeared attractive to bet on this compound with solid intellectual property protection, rather than on dimethiodal sodium with only a weak position.

At Schering-Kahlbaum AG Hans-Georg Allardt invented a proper process for the manufacturing of dimethiodal sodium (Allardt 1931), which illustrates the intensification of the competition in the field. Yet, the existence of well-known processes distinct from the patented ones and the nonexistence of use patents outside the United States prevented the patent owners from excluding competitors from entering into manufacturing and sales of dimethiodal sodium. This fact is the likely reason none of the owners of dimethiodal sodium patents commercialized a corresponding product. In their place, smaller companies seized the opportunity to enter the radiological field with this generic compound, at least in Europe.

Well after a particular synthesis of dimethiodal sodium had been published (Backer 1926) but before in the second quarter of 1931, the patents covering other synthetic processes and the compound's use as contrast agent became known, André Guerbet at Laboratoires André Guerbet & Cie, following discussions with his father, Marcel Guerbet, had synthesized dimethiodal sodium. On July 19, 1931, he started with animal experiments and began application to man on August 7 (Bonnemain and Guerbet 1995b). In possession of preliminary evidence of clinical safety and efficacy of their compound, the two Guerbet's deposited their supposed trade secret in a so-called "enveloppe Soleau" (Bonnemain and Guerbet 1995b), an inexpensive, fast but legally weak form of documenting inventive priority in France. Given the state of the prior art in the field, it accorded them little protection. Dimethiodal sodium was manufactured anyway and introduced as TÉNÉBRYL by Laboratoires André Guerbet & Cie, starting 1932 in France (Legueu, Fey, and Truchot 1931a,b). With this action, the company expanded its LIPIODOL-based involvement in contrast agents. Beginning in 1934 in Norway, and gradually in a number of other countries, dimethiodal sodium, under the trademark UROTRAST R™ NYCO, became available. It was manufactured and sold by the pharmaceutical company Nyegaard & Co. A/S, Oslo, Norway.[117] This company, about which much more will be told, thereby made its entrance into the radiological arena (Amdam and Sogner 1994).[118] The case of dimethiodal illustrates that, at the time, for a successful contrast agent, the requirement of a strong patent protection was not absolute.

What is fascinating in the story of dimethiodal sodium is the extraordinary speed with which drug development was possible in the 1930s. The development time needed for passing from first experiments in animals to those in man in 1931 was 19 days, whereas today a reasonable estimate is 19 months. From first administration to man to product launch today easily pass 6 years of clinical trials necessary to satisfy governmental regulatory requirements and of data evaluation by the responsible agency, when at the earlier time the same was possible in 6 months.

Still in the same year in which uroselectan sodium and methiodal sodium reached the market, i.e., 1930, the chemist Joachim Reitmann at I. G. Farbenindustrie AG, Leverkusen, invented iodopyracet (USP), alternatively called diodone (INN) (Figure 4.24, **12**) (Reitmann 1930). His contribution must be seen in the context of the earlier described effort by Hecht to enter the competitive race for improved uro-angiographic contrast agents. Iodopyracet, like iodomethamate and dimethiodal, contained two atoms of iodine per molecule. It was an isomer of the compound whose sodium salt (Figure 4.11, **6**) had failed in **Swick's** hands because of insufficient solubility and excessive biliary excretion. As the sodium salt, the new compound too suffered from poor solubility but excelled in exclusively renal excretion. Experiments with various salifying ions led to progressively more water-soluble preparations (Reitmann 1931a; Hecht 1933), from the diethylamine salt (PELVIREN™ D) to the diethanolamine salt (PER-ABRODIL™), then to the salt of a mixture of the two amines (PER-ABRODIL FORTE™), and ultimately to the meglumine (= N-methyl-D-glucamine) salt (PER-ABRODIL M™). After the last development, salification with the sugar derivative, meglumine, became a widely applied means for achieving elevated solubility. Note that the experience with counter ions having progressively increasing number of hydroxyl groups prepared the ground for the much later development of nonionic contrast agents bearing multiple hydroxyl entities (see Section "Nonionic Contrast Agents").

All the versions of iodopyracet were distributed in Europe by Pharmazeutische Abteilung Bayer-Meister Lucius of I. G. Farbenindustrie AG or its licensees. PER-ABRODIL began its career in 1932 (Progress 1932). Winthrop Chemical Comp., Inc., New York, acquired the rights to the patents of Reitmann (1930, 1931b) for the United States, and iodopyracet diethanolamine salt assumed the trade name DIODRAST™. In the late 1940s patent protection of several of these articles expired, whereupon a plethora of producers of generic versions entered the fray.[119]

The members of the most recent class of contrast agents, first of all those based on iodomethamate disodium and iodopyracet diethanolamine, enabled extended clinical indications, and were safer than the pioneering uro-angiographic uroselectan sodium and methiodal sodium. Among the two new molecules, iodopyracet diethanolamine salt could claim the better safety record. At the highest concentration, i.e., 70% (w/v) or 348 mg(Iodine)/mL, it had an osmolality at 37 °C of around 3,600 mosmol/kg.[41] At the time of product conception osmotic properties were not part of the compound selection criteria, let alone were they measured. Despite its elevated osmolality, this contrast agent's tolerability at last allowed doses elevated enough to render feasible the visualization of the heart chambers, so-called angiocardiography, and of the pulmonary veins. To begin with, these procedures were explored in children with congenital abnormalities (Castellanos, Pereiras, and García 1937). Although some additional contrast agents were introduced, those based on methiodal sodium or dimethiodal sodium, and still more often based on iodomethamate disodium or iodopyracet diethanolamine salt, dominated angiographic applications for two decades (Grainger 1982a; Felder 1986; Urich 1995). They were later classified as first-generation uro-angiographic contrast agents.

Subsequently, two notable steps toward fulfillment of the vision of the ideal contrast agent were taken. The first one once more involved Swick, known from the earlier history, and again his story of what happened, often recounted in the literature, needed revision based on documents that had previously been ignored. Let's begin with the incontrovertible facts.

At the Zonite Products Corporation,[120] an important pharmaceutical company headquartered in the Chrysler Building in New York, the petrochemist Albert P. Sachs worked as consulting chemical engineer. The company's sole relationship to contrast agents was its acquisition in 1929 of the A. C. Barnes Company and with it the manufacturing of ARGYROL™, a silver proteinate-based antiseptic earlier experimented with also in pyelography (Kelly and Lewis 1913; Maluf 1956). Contrast agents were not of strategic interest. Despite this situation, Sachs, as sole inventor with Zonite Products Corporation as assignee, on July 26, 1933, applied for a United States patent on contrast agents, their radiological use and production process (Sachs 1933). Its claim 1 was "A composition for use in the radiography of internal organs, comprising a pharmaceutically pure alkali metal salt of a derivative of hippuric acid containing nonionic heavy halogen." Claim 3 specified sodium *ortho*-iodohippurate (Figure 4.25, **13**). It astounds that the claims failed to include the contrast agent with organic amines instead of alkali metal ions as counter ions, given that the water solubility of sodium *ortho*-iodohippurate was barely satisfactory (Salomon 1937) and amines were already in use to achieve elevated water solubility in numerous commercial products. Apparently, Sachs wrote his patent application without seeking sufficiently broad input from contrast agent users, Swick included. Moreover, his cultural background as chemist may have led him to aim for maximized iodine content in the contrast agent molecule and in this respect sodium ions had an advantage over amines. Since *ortho*-iodohippuric acid could not be prepared in appropriate purity by ways ordinarily used for obtaining similar compounds, a novel synthesis was introduced. The product formed sterilizable aqueous solutions of concentrations just sufficient for its intended clinical applications. *Ortho*-iodohippuric acid was known as physiological degradation product of iodoalbumin (Mosse and Neuberg 1903) and as renally excreted detoxification metabolite of *ortho*-iodobenzoic acid in dogs and rabbits (Novello, Miriam, and Sherwin 1929). In his patent Sachs underlined the latter property as favorable for its tolerability when used as contrast agent. Rabbits, dogs, guinea pigs and rats were said to have been radiographed after administration of the contrast agent without ill effects. Intravenous administration of 20 g of sodium *ortho*-iodohippurate as a 50% (w/v) solution were described to yield suitable excretion urograms in humans with various kidney diseases, again without ill effects. Demonstration of clinical application was part of the experimental support for the patent. Sachs must have collaborated at least with one radiological investigator outside the company, but his/her identity has never been revealed. The collaborator merely provided help with exemplification of the invention, which does not give a right to co-inventorship, unless in the process an additional inventive step is made. This was not the case. In 1938 the patent was granted, which means it faced no successful opposition. In a legally binding way, Sachs was the first to conceive and reduce to practice the contrast agent sodium *ortho*-iodohippurate, and some related products. Belatedly Schering-Kahlbaum AG filed competing patents for similar products, di- and triiodohippuric acid, e.g., *Schering-Kahlbaum AG* (1936).

After his return from Germany in November 1929, still supported by the Emanuel Libman Fellowship Fund, Swick pursued his research interests in excretion urography and novel contrast agents at the Mount Sinai Hospital in New York City. In the years 1930–1932 he acted primarily as a researcher, holding, sometimes simultaneously, various formal appointments, in part as a volunteer. In May 1932, he became Adjunct Surgeon and henceforth concentrated on clinical duties (Mount Sinai Hospital. n. d.; *curriculum vitae*). In the January 1933 issue of *Surgery, Gynecology and Obstetrics*, **Swick** alone published his first results on excretion urography with sodium *ortho*-iodohippurate and announced ongoing studies with related substances, such as diiodohippurate and bromoiodohippurate. In his publication Swick acknowledged as the source of the contrast agents Albert P. Sachs from the Zonite Products Corporation (Swick 1933a). Before Sachs could file his patent application and Swick could submit his manuscript for publication in a journal, considerable time for investigation was required. Swick reached his objective first. In the United States prior to 2013, patent priority was determined by the first-to-invent, rather than first-to-file rule. Thanks to this rule, Sachs had little reason for objection to Swick publishing his results with sodium *ortho*-iodohippurate (Swick 1933a) prior to patent application (Sachs 1933). In his publication Swick mentioned rabbits as only animal species experimented with, whereas the patent of Sachs speaks of three additional species. Moreover, reported clinical cases were not identical in the two documents. There is no indication for an involvement of Swick in the investigations supporting Sachs.

In his first publication Swick thanked several staff members at the Mount Sinai Hospital for kind cooperation (Swick 1933a). He mentioned as sites of his experimentation the Radiology Division directed by Jaches and the Laboratory Division of the Second Surgical Service headed by the chemist Harry H. Sobotka. The chemistry laboratory was important for renal excretion studies and perhaps for the preparation from powder of sterile contrast agent solutions for injection. Up to this stage of history Mallinckrodt Chemical Works, the later producer and vendor of the new contrast agent, was evidently not yet involved.

At the 84th Annual Session of the American Medical Association in Milwaukee in June of 1933, Swick gave an oral presentation (Swick 1933b), accompanied by a technical exhibit (Swick 1933c). The exhibit won him the Frank Billings Gold

Medal (Class I). My efforts of finding published corroboration of this event have failed, but Swick's daughter, Susan B. Sterling, has provided me with Xerox copies of the two sides of the rectangular medal. The engraved rational for the award reads: For Original Investigative Work on Intravascular and Oral Urograms Demonstrating Various Urological Conditions by Means of Sodium Ortho-iodohippurate. Swick's name alone appears on the medal. Contrary to repeated later reports (Pollack 1999; Nyman 2016), Wallingford did not share the award. Note that at the time of the award the patent of Sachs was not yet published, let alone granted.

In the early thirties Mallinckrodt Chemical Works decided to develop and commercialize sodium *ortho*-iodohippurate as contrast agent. The conclusion is hard to escape that the company had to acquire the intellectual property from Zonite Products Corporation, at the latest when the patent was granted. Development of the process for production of the new contrast agent was in the hands of Wallingford, the director of organic chemistry at Mallinckrodt Chemical Works and a crucial player in the much earlier development of the cholecystographic contrast agent IODEIKON. It further involved heavily the chemist, Paul A. Krueger (Grainger 1982c).

After the incontrovertible facts, let us come to the subjective narratives offered by protagonists of the history, starting with that of Swick. Upon his return from Germany to the Mount Sinai Hospital in New York, Swick dreamt of contrast agents even better than uroselectan sodium. The following sentences extracted verbatim and in chronologic order from Swick's publications and letters pertaining to sodium *ortho*-iodohippurate as novel contrast agent reflect his subtly evolving interpretation of how he came to work with this compound: "The compound now proposed is sodium-iodohippurate ..." (Swick 1933a); "... it was felt that sodium *ortho*-iodohippurate should meet the necessary requisites ..." (Swick 1933d); "... one of the authors has developed a new organically-bound iodide compound ..." (Jaches and Swick [1933] 1934); In 1931, "further investigation of mine led to the formulation of another compound —— Hippuran—..." (Swick [1965] 1966, 1974c, 1975, 1978); "...I came upon the idea that the iodide element could be organically bound to an aromatic nucleus, representing a product of metabolism—that of Hippuric acid" (Swick 1982); "I will be sending you reprints of another of my discoveries, (successful) Hippuran..." (Swick 1974b); "During my work at the Mount Sinai Hospital I proposed the concept of an aromatically [sic] Iodine Compound centered around a product of metabolism—the detoxification of benzoic acid to Hippuric acid with normal glycine" (Swick, 1980a); "... and Hippuran was the stimulus and forerunner of our present compounds for intravenous and angiographic compounds — all triiodo benzoates" (Swick 1980a). While in the first of his publications Swick abstained from claiming conception of the contrast agent and thanked Sachs for having provided him with the product, with one exception, in subsequent accounts mention of Sachs disappeared and Swick claimed inventorship. As the surprisingly prompt award to Swick of the Frank Billings Gold Medal already augured, Swick's version of history, wherein he depicted himself as the sole mind behind *ortho*-iodohippurate, became endorsed early by the contrast agent community and reflects general opinion up to the present (Hoppe 1959; Grainger 1982a,c).

Swick has written: "The attempts at the synthesis of hippuran was initially carried out in the chemistry laboratory of Mount Sinai Hospital under the supervision of Dr. Harry Sabatka (should be Sobotka) but shortly thereafter discontinued. I then contacted Sachs from the Zenith (should be Zonite) Corporation where small amounts of the sodium salt of iodo-hippuric acid (Iodo-Hippuran) was synthesized and which I investigated in the laboratory in rabbits and man"(Swick 1982). Swick thus arrogated for himself the initiative to seek contact with Sachs. Unfortunately, we lack documents that would reveal how Swick and Sachs had encounter each other, who might have arranged it, on what occasion it happened, what information was passed, who payed for the investigative effort and what kind of contractual arrangements were agreed upon. Swick has written further: "After my initial success with Iodo-Hippuran, I turned the compound over to the Mallinckrodt Chemical Works where the chemist Dr. Wallingford collaborated with me in its production for commercial use" (Swick 1982). Swick's description of his working relationship with Mallinckrodt Chemical Works and Wallingford is amusing with its somewhat vainglorious tone.

Grainger has given me a copy of a letter to him by Vernon H. Wallingford, Ferguson MO, January 4, 1980. Its author wrote: "In the early thirties, when I was engaged in the development of a plant process for sodium iodo-hippurate, I had the opportunity to visit Dr. Swick in his New York apartment. What was intended as a brief courtesy call turned into several hours of discussion of X-ray contrast media and their importance in radiology. I was impressed by Dr. Swick's enthusiasm and dedicated interest." Evidently Mallinckrodt Chemical Works was engaged in commercially developing the new contrast agent already before Swick first encountered Wallingford. It has never been revealed what had triggered the company to engage in this activity, the publications of Swick (1933a–d), a letter from Swick to Wallingford, a contact of Swick with company personnel other than Wallingford, a proposal by Sachs or any other representative of the patent owner, Zonite Products Corporation (Sachs 1933).

In the course of Wallingford's visit to Swick, the latter certainly claimed his authorship of uroselectan sodium and probably also that of sodium *ortho*-iodohippurate. Given that at the time the patent application of Sachs was not yet public, conceivably Wallingford was unaware of it and had no reasons to doubt Swick's story. In any case, it seems likely that Wallingford's positive impression of his host let his company realize the potential of Swick as the flagship for marketing of the upcoming new contrast agent and cultivation of the company image, functions Sachs could not have delivered. The fact is, the company endorsed Swick's version of history. Sachs and Zonite Products Corporation fell into oblivion. According to the aforementioned letter of Wallingford, some time in the 1970s Mallinckrodt Chemical Works honored Swick for his pioneering work in urography with a plaque.

Sachs never offered his view of the course of events, but the patent speaks for itself. Accordingly, he alone conceived and reduced to practice sodium *ortho*-iodohippurate as contrast agent. But his provision of Swick with contrast agent for studies, including clinical ones, testifies to a special relationship between the two. Regrettably their actions and acknowledgments do not in a straightforward way reveal the dynamics of their interaction. An account directly contradicting Swick's would have Sachs learn about excretion urography with uroselectan sodium and about its champion, Swick, for example through the article in the news magazine *Time* of August 24,

1931, p. 32. It could have triggered him to conceive sodium *ortho*-iodohippurate as contrast agent and to select Swick as one of the clinical collaborators. Further witnesses of the history, e.g., Jaches, Sobotka, Libman, have remained silent.

It is hard to reconstruct a history that reflects the incontrovertible facts without dismissing important aspects of Swick's account. I can think of only a single scenario that can reconcile the patent situation with at least most of the narrative of Swick. Therein Swick conceived the idea of an iodinated contrast agent well tolerated because of being a renally excreted product of metabolic detoxification, but he failed to indicate a concrete realization. He mentioned his aspiration to Sachs. The latter came up with alkali salts of halogenated hippuric acids and reduced the idea to practice through chemical synthesis of, among other compounds, sodium *ortho*-iodohippurate and through demonstration of tolerability and radiological efficacy in animals and man. At least for the radiological studies he relied on an unidentified service provider who made no supplemental inventive contributions and did not insist on publishing the finding in a journal. Recognizing Swick's original intellectual input, Sachs provided him with the contrast agent. After all elements for patenting of the invention were assembled under the reigning first-to-invent patenting regime in the US, Sachs allowed Swick to publish his clinical results under his sole authorship. In essential coincidence Sachs filed his patent application. Sachs and his lawyers must have been convinced that Swick had no basis for disputing the claimed sole inventor- and ownership of the technology. Swick may have even been left in the dark about the situation. Actually, to his end he never mentioned the patent. Neither did Wallingford.

On occasion of the visit to Swick, Wallingford recognized the value for his employer of having such an enthusiastic promotor of contrast agents and their applications on its side. The patent of Sachs was not published for several years. It is thus conceivable that Wallingford did not know about the patent application. It was therefore convenient, and perhaps even justified, for him and Mallinckrodt Chemical Works to join the rest of the contrast agent community in endorsing the account according to which *ortho*-iodohippurate was conceived by Swick alone.

The evidence offered here renders the choice of Swick as the sole award winner of the Frank Billings Gold Medal and the accompanying specific justification problematic, at least in retrospect. But these reservations should not detract excessively from his merits. His overall contribution to the field as the great promoter of excretion urography research and clinical application more than justified the award.

In 1936 the Mallinckrodt Chemical Works began selling its formulation for injection of sodium *ortho*-iodohippurate at a concentration of 48% (w/v) of anhydrous compound or 186 mg(Iodine)/mL. The generic name iodohippurate sodium was complemented by the trademark HIPPURAN™ (Council A. M. A. 1935). The concentration and recommended dosage was essentially that indicated by Sachs (1933). The contrast agent found its major application in excretion urography. It had a roughly estimated osmolality at 37 °C of 3,300 mosmol/kg[41] although this value was at the time neither measured nor taken into written consideration. Unfortunately, the low solubility of sodium *ortho*-iodohippurate limited its potential utility, in particular for angiography. It is not known what increase in solubility organic amines as counter ions would have produced.

HIPPURAN was to have very limited success. Its importance rested in the substitution of the iodinated pyridine by the iodinated benzene ring as the basic skeleton for uro-angiographically useful molecules[121] and the validation of the concept of metabolic detoxification products as highly tolerable contrast agents. The company's interest in contrast agents and interaction with Sachs and Swick stimulated Wallingford to examine additional compounds. He had ten iodinated benzoylaminoacids, analogs of *ortho*-iodohippuric acid, prepared, but none turned out to be useful (Vernon Howard Wallingford 1953). Parallel research in a similar vein went on at Schering-Kahlbaum AG (*Schering-Kahlbaum AG* 1936). With *ortho*-iodohippurate sodium ended the historical period that had brought forth the first-generation contrast agents and began that of the second-generation.

As illustrated on the website Weldon Spring Site Interpretive Center Online Tour: Tribute to the Mallinckrodt Uranium Workers (updated June 4, 2015), during World War II significant resources at Mallinckrodt Chemical Works became dedicated to defense relevant issues, such as manufacturing of uranium oxide, uranium metal and penicillin, detracting them from the contrast agent field. Not before the late 1940s was Wallingford able to return to it. Intriguing is the cross-fertilizing effect the company's dealing with penicillin had on contrast agent research. Wallingford recounts to have become inspired by reading an article on the excretion of penicillin, which reported a surprisingly low intravenous toxicity in mice of *p*-aminohippurate. He reasoned that amino-substituted benzoic acids, which were known to be easily iodinated, could form the basic skeleton for a family of novel low-toxicity contrast agent candidates [Wallingford, paraphrased in Grainger (1982a,c)]. He also hoped to find suitable chemical derivatives with higher iodine content than those constituting the earlier generation products. Lower clinical doses in terms of weight or moles of such molecules would be required to achieve the same X-ray contrast. Better tolerability at doses of equal total weight could be the result. He began by synthesizing a series of iodinated aminobenzoic acids bearing various substituents of the amino group and had them tested for metabolism and toxicity. Soon he chose as starting material for derivatives, 2,4,6-triiodo-3-aminobenzoic acid, a rather toxic substance known since 1897 (Kretzer 1897).

Sometime earlier it had been found that in rabbits *p*-aminobenzoic acid was detoxified by acetylation. Suspecting that the acetylated compounds of iodinated aminobenzoic acids might be analogously detoxified, Wallingford directed his efforts at such derivatives. Acetylation of the amino group of 2,4,6-triiodo-3-aminobenzoic acid yielded a highly water-soluble contrast agent with a promisingly low-toxicity, acetrizoate sodium (Figure 4.25, **14**) (Wallingford 1950; Wallingford, Decker, and Kruty 1952). It achieved low toxicity without metabolization to a hippurate derivative. Typically for drug discovery research the sequence of insights and actions reflects a logic that appears tortuous. The reason for this is its modification along the way in response to unexpected experimental results.

After the clinical evaluation of acetrizoate sodium (Nesbit and Lapides 1950; Richardson and Rose 1950; Osol 1950) Mallinckrodt Chemical Works, put the product in 1950 on the American market, calling it UROKON™. It came initially as solution for intravenous injection with a concentration of 50% (w/v) or 330 mg(Iodine)/mL. A solution with a concentration of 70% (w/v), or 460 mg(Iodine)/mL, followed soon after (Barry and Rose 1953). This formulation had a roughly estimated

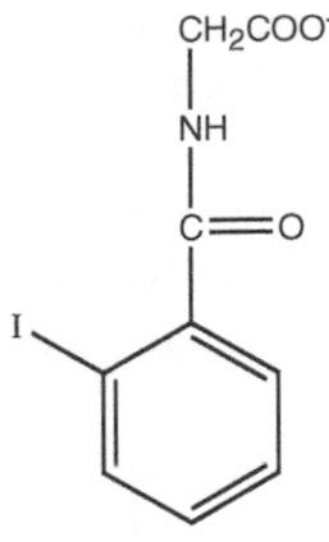

FIGURE 4.25 From prototypical product to the most successful second-generation uro-angiographic contrast agents.

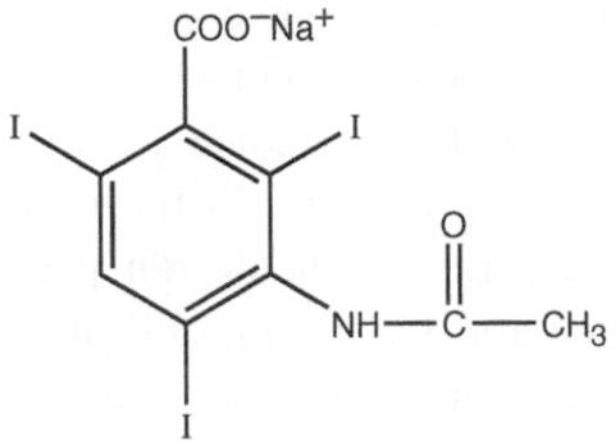

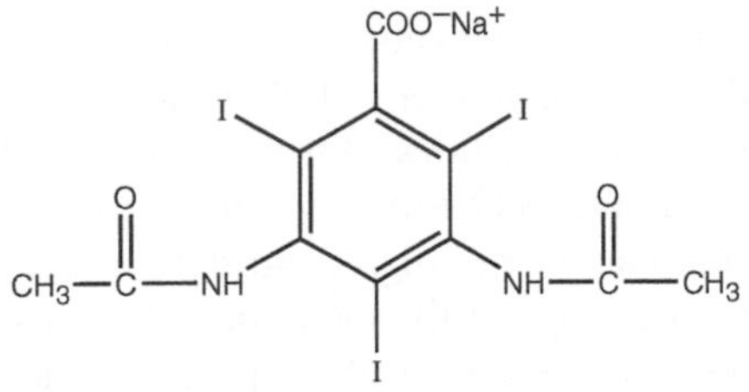

13: *Ortho*-iodohippurate sodium (HIPPURAN™, Mallinckrodt Chemical Works, USA). $I/P = 0.5$.

14: Acetrizoate sodium (UROKON™, Mallinckrodt Chemical Works, USA). $I/P = 1.5$.

15: Amidotrizoate sodium = diatrizoate sodium (HYPAQUE™, Winthrop-Stearns, Inc., USA).[79]

Amidotrizoate meglumine = diatrizoate meglumine (ANGIOGRAFIN™, Schering AG, Germany; HYPAQUE M™, Winthrop-Stearns, Inc., USA).

Amidotrizoate meglumine/sodium = diatrozoate meglumine/sodium (UROGRAFIN™, Schering AG, Germany; HYPAQUE M 60%™, Winthrop-Stearns, Inc., USA). $I/P = 1.5$.

osmolality of 2,800 mosmol/kg at 37 °C[41]. Because of its higher iodine concentration and combination of glomerular and tubular renal excretion, UROKON 70% (w/v) provided better contrast in excretion urography than the older UROSELECTAN B (= NEO-IOPAX) (Wall and Rose 1951), although such increased contrast did not necessarily mean improved diagnostic efficacy. UROKON was locally better tolerated but produced more numerous and more severe systemic adverse reactions (Tucker and Di Bagno 1956; Lance, Killen, and Scott 1959). It was a mixed blessing.

Curiously, patent protection for acetrizoate sodium was not extended to territories outside the USA. In fact, Mallinckrodt Chemical Works in the early 1950s generally did not extend its patent protection to Europe.[122] This choice allowed more than 30 competing producers and vendors outside the USA to soon sell acetrizoate-based contrast agents under their proprietary trademarks (Strain et al. 1964; Barke 1970). Acetrizoate, with its novel triiodobenzoic acid core structure and its rapid conquest of the world market, foreboded a change of eras in the field. Acetrizoate sodium was the prototype of what became known as second-generation contrast agents. In terms of performance these corresponded much more closely to the vision than the first-generation predecessors (Felder 1986).

The experience with acetrizoate sodium as the chemical structure and that with organic amines as counter ions in the case of iodopyracet-based preparations was condensed into contrast agent design principles that would guide the search for follow-up products for two decades (Table 4.3).

From the collection of the compounds satisfying the design principles, contrast agent candidates were selected by the same criteria as described for the first-generation ones. An additional important selection criterion for a good product, which is almost never aired in publications discussing contrast agent design, was the complexity and cost of industrial synthesis. Given the elevated diagnostic doses generally required and the frequency of radiological procedures with them, economical manufacturing required facilities unusually large for pharmaceuticals. Only chemicals prepared from simple and easily available starting materials and synthetic processes involving a small number of high yield steps were economically realistic. This restricted the synthetic scope of the medicinal chemists substantially. A final criterion regarded patentability of aspects of the product. The large number of competitors sharing the clientele and the ever-increasing cost of drug registration made it imperative that patent protection accorded a sufficient duration of exclusivity.

Acetrizoate sodium (UROKON) did not yet possess all the favorable attributes that the structural core, triiodobenzoic acid, could offer after further chemical substitution. A number of industrial laboratories joined the race to find compounds with better patient tolerability. Indeed, soon triiodinated acylaminobenzoic acid derivatives with improved toxicological properties (Hoppe, Larsen, and Coulston 1956) were found. Among them, the sodium salt of diatrizoic acid (U.S.P), alternatively called amidotrizoic acid (INN) (Figure 4.25, **15**), a diacetylated derivative of the long known but chemically unstable precursor, 3,5-diamino-2,4,6-triiodobenzoic acid (Lütjens 1896), revealed itself as a winner. Given the minimal difference of chemical structure between diatrizoic and acetrizoic acid it is hardly surprising that diatrizoic acid was conceived independently in two industrial laboratories.[123] In the United States Aubrey A. Larsen at the Sterling-Winthrop Research Institute in Rensselaer, NY[79], synthesized the compound. He filed a US patent application on February 19, 1954, but later he was granted a corrected priority date of March 27, 1953 (Larsen 1953a).[124] A few weeks before, on February 5, 1953, Schering AG as applicant had filed a German patent application for the same compound, the inventor, Diedrich, having chosen to remain anonymous (*Schering AG* 1953).[125]

TABLE 4.3

Design Principles for Second-Generation Contrast Agents

2a)   Substituted 2,4,6-triiodophenyl moieties form the molecular core that is responsible for X-ray absorption and stability of aqueous solutions to heat sterilization.

2b)   The iodinated compound is rendered water-soluble by the presence of at least one carboxylate group, neutralized by sodium and/or organic amines, hydroxyl-group containing ones.

As recounted in a letter to me from Larsen, Evansville IN, March 16, 1997, at the outset he encountered difficulties convincing the patent examiners of an inventive step. The conception seemed all too obvious. But then, why had Wallingford not studied diatrizoic acid earlier? Moreover, nobody could have predicted the remarkably low toxicity and elevated solubility of diatrizoic acid salts. In the end both Larsen and Schering AG got their inventions patented. Given the legal complexities in determining inventive priority in different national jurisdictions, a court battle ensued over rights in various territories of the world.[126] As a result, Schering AG was allowed to exert its rights in West Germany and Japan. In 1962 a US court instead awarded USA territorial rights to Sterling Drug, Inc.[79] In Norway and Sweden too the American company prevailed. Once territorial ownership of intellectual property was clarified in the courts, the two companies proceeded to exchange ownership of patents (Larsen 1953b), cross- and sublicenses. Through these deals diatrizoic acid-based contrast agents became available from multiple suppliers in the United States and in Europe. Purely clarifying court battles like the one around the intellectual property rights on diatrizoic acid need not destroy the ability of the parties to set up cooperations. Potentially impeding emotions in legal cases without assignment of guilt are relatively low. Nonetheless, court actions overall are a hindrance to the success of new technology and are to be avoided whenever possible.

Differences existed initially in the choice of the salifying ions in the pharmaceutical formulations of diatrizoate. In 1954 Schering AG began supplying the mixture of meglumine and sodium salts of diatrizoic acid in the weight ratio 6.6/1 and in concentrations up to 76% (w/v) or 370 mg(Iodine)/mL, calling the latter UROGRAFIN 76™. At that concentration, the osmolality at 37 °C measured 2,070 mosmol/kg (Børdalen, Wang, and Holtermann 1970). Clinical tests proved UROGRAFIN a substantial step forward for the patient in terms of safety as well as comfort, through reduction of pain during injection (May and Schiller 1954). The company soon let its original contrast agent be followed by a pure meglumine salt formulation that could be prepared with a concentration of 85% (w/v) or 400 mg(Iodine)/mL and an only slightly increased osmolality. It was given the trademark of ANGIOGRAFIN™ and was reserved for angiocardiography and aortography. Schering AG's United States licensee, E. R. Squibb and Sons Inc.,[127] quickly commercialized in the USA identical products called RENOGRAFIN™ and CARDIOGRAFIN™, respectively. Licensees of these companies in various territories followed suit and competitors formulated their own articles with analogous salt compositions. Also, lower concentrations adapted to various clinical indications were made available.

About 1 year after Schering AG had begun selling diatrizoate sodium formulations, Winthrop-Stearns, Inc., started to offer its own formulations, initially as a 50% (w/v) or 300 mg (Iodine)/mL solution, using the trade name, HYPAQUE™. This formulation had an osmolality at 37 °C of merely 1,520 mosmol/kg (Børdalen, Wang, and Holtermann 1970). To begin with, HYPAQUE had been clinically tested in excretion urography (Moore and Mayer 1955) and showed an overall incidence of adverse reactions of 10% (Speicher 1956). Other formulations of HYPAQUE followed. The one with the highest iodine concentration was HYPAQUE—M 90%™. It matched the highest one available for UROKON, i.e., 460 mg (Iodine)/mL, consisted of 30% sodium and 60% meglumine salt of diatrizoic acid (Finby, Poker, and

Evans 1956) and had an osmolality of 2,940 mosmol/kg (Fischer 1986). Because of limited solubility of the solid ingredients at room temperature, the closed vial had to be heated before use until full dissolution was attained. After sufficient cooling a supersaturated solution resulted, which was ready for injection and stable for a few hours.

The success of diatrizoic acid-based contrast agents attracted further competitors to the field. A plethora of agents very similar in chemical structure and clinical performance, sometimes deprecatingly labeled me-too products, were developed. This generation of contrast agents was eventually to include in addition to diatrizoic acid salts, metrizoic acid salt[128] [ISOPAQUE™, Nyegaard, patent priority 1961, marketed 1962], iothalamic acid meglumine salt (Figure 4.26, **16**) [CONRAY™, Mallinckrodt, patent priority 1960 and 1961, marketed 1962], iodamide, sodium salt [UROMIRO™, Bracco, patent priority 1962, marketed 1968],[129] and ioxitalamic acid meglumine or mixed meglumine/sodium salt [TÉLÉBRIX™, Guerbet, patent priority 1967, marketed 1970]. At the most widely used concentration of 300 mg(Iodine)/mL these pharmaceutical formulations had osmolalities around 1,600 mosmol/kg (Børdalen, Wang, and Holtermann 1970). They were distinguished by very modest differences in tolerability. Formulations based on these second-generation products, which beginning in the 1970s became referred to as ionic, or more precisely, as high-osmolal ionic contrast agents, dominated all major applications until the 1990s and to some extent are still used today, especially in the third world.

The second-generation contrast agents afforded substantially more safety than those of the first-generation. Still, significant nonfatal adverse reactions were common. Most predictable were sensation of heat and pain at the site of injection. Frequent too were nausea and vomiting. For the rate of fatal outcomes of angiocardiography with first-generation contrast agents, values of around 1/263 have been estimated (Dotter and Jackson 1950). For second-generation products, corresponding values are crudely estimated as one in 40,000–77,000 (Hartman et al. 1982; Cashman, McCredie, and Henry 1991). It is instructive to note that the increased safety was not so much used by clinicians to reduce the incidence of adverse events, as to increase the dose and thereby achieve improved radiological efficacy.

The large number of ionic contrast agents of quite similar chemical structure in uro-angiographic use at the end of the 1960s demonstrates that the technology had entered a process of closure, a phenomenon that characterizes certain stages of technology genesis and shaping. In a closure process a multitude of diverse technological prototypes have boiled down to a restricted variation on a single theme. Both necessary and desirable features and the meaning attributed to them have reached a consensus. Inevitably, the closure of the technology development presaged considerable difficulties in reopening the area to innovative approaches at a later time. As will be illustrated shortly, it took 20 years before that closure could be overcome. But this does not mean that for 20 years nothing happened. Rather, for 20 years a variety of different technological avenues were followed, which slowly built the basis of the next innovation cycle. Unearthed here are the most important roots of the next major innovation, to wit low osmolality,[41] and in particular nonionic uro-angiographic contrast agents.

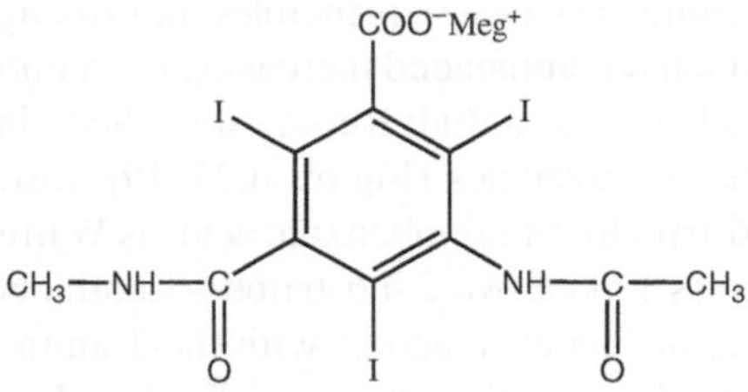

FIGURE 4.26 In pursuit of increased number of iodine atoms per molecule. Iothalamate monomer, dimer, trimer.

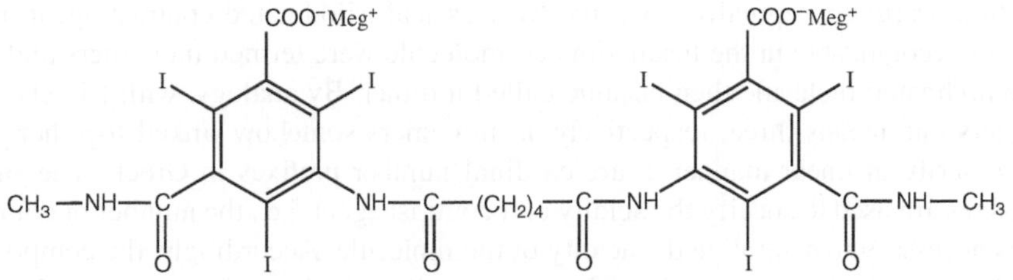

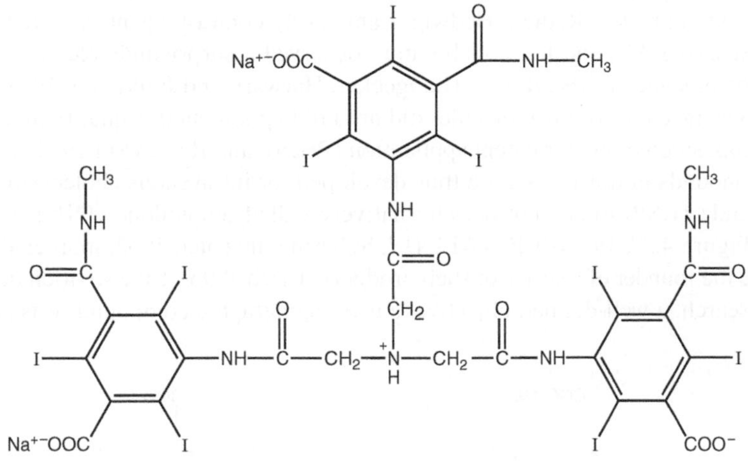

16: Iothalamate meglumine (iothalamate monomer meglumine) (CONRAY™, Mallinckrodt Chemical Works, USA). $I/P = 1.5$.

17: Iocarmate dimeglumine (iothalamate dimer dimeglumine) DIMERAY™ (Mallinckrodt Chemical Works, USA); DIMER-X™ (SA Laboratoires André Guerbet, France). $I/P = 2.0$.

18: Iothalamate trimer disodium. $I/P = 3.0$.

True to the then reigning drug design principles for uro-angiographic compounds (Tables 4.2 and 4.3), which recommended increasing the number of iodine atoms per molecule, in 1952 Hans Priewe and Rudi Rutkowski at Schering AG inaugurated the synthesis of hexaiodinated molecules (Figure 4.27, **19**). Instead of acetylating the 3-amino group of 2,4,6-triiodo-3-aminobenzoic acid, as Wallingford had done to get acetrizoic acid, they cross-linked two 2,4,6-triiodo-3-aminobenzoic acid molecules with bifunctional linker molecules reacting with the 3-amino groups. Two salified carboxyl groups assured the solubility of the molecules. As a novelty, the osmotic properties of their solutions were measured (Neudert and Röpke 1954), although the results did not become imbedded into an explicit contrast agent design theory.[130]

In a useful conceptualization, the two classical triiodinated contrast agent molecules recognizable in the hexaiodinated molecule were termed monomers and the hexaiodinated molecule then became called a dimer. By analogy, with trimers and *n*-mers one means three, respectively, *n* monomers somehow linked together, not necessarily in linear manner. *n* are cardinal number prefixes in Greek. The same prefixes are used to specify the acidity of a contrast agent, i.e., the number of counter ions necessary to neutralize the acidity of the molecule. Accordingly, the compound of Priewe and Rutkowski (1952), **22**, became a diacidic dimer. Their terminology is henceforth used.

Interestingly the dimeric molecules did not immediately promise advantages for uro-angiography. Rather, Hedwig Langecker, contrast agent research director at Schering AG, and her collaborators observed a surprisingly elevated biliary excretion of some of these dimers (Langecker, Harwart, and Junkmann 1953; Sanen 1962). The inventors of the molecules did not fail to point out this quality in the original composition of matter patent application (Priewe and Rutkowski 1952). Some of the compounds in this class were thus developed for intravenous cholecystography. Iodipamide (USP) dimeglumine, alternatively called adipiodone (INN) dimeglumine (Figure 4.27, **19**) (BILIGRAFIN™, Schering, invented 1952, marketed 1953), became the founder of a series of such products (Urich 1995). Like so often in industrial research, a well-defined objective, a uro-angiographic compound, was pursued

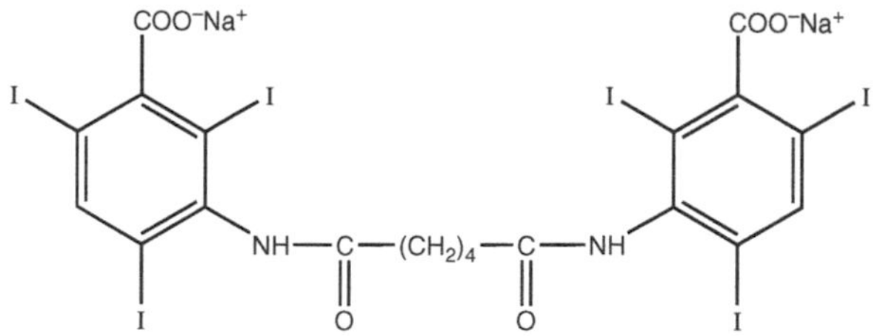

19

FIGURE 4.27   The search for uro-angiographic contrast agents that leads to the discovery of hepatobiliary agents.

19: adipiodone dimeglumine; iodipamide dimeglumine (acetrizoic acid dimer). *I/P* = 2.0. BILIGRAFIN™ (Schering AG, Germany); CHOLOGRAFIN™ (E.R Squibb & Sons Inc., USA).

and as a result a promising chemical with application elsewhere, i.e., in cholecystography, was identified.

In the mid-1960s, the design principles for uro-angiographic contrast agent molecules began to incorporate osmotic properties of their solutions for intravascular injection. The fact that these solutions at the concentrations required for sufficient radiological contrast were hyperosmolal and the possibility that this might be biologically detrimental had sporadically been cogitated since 1918 (e.g., Cameron 1918; Hryntschak 1929; Forssmann 1929; Boriani and Boriani 1940; Broman and Olsson 1949; Odén 1955). There had actually been made some measurements on contrast agent solutions by cryoscopy (Boriani and Boriani 1940; Broman and Olsson 1949; Cotrim 1954; Neudert and Röpke 1954; Hammarlund and Pedersen-Bjergaard 1958; Odén 1955; Bernstein, Reller, and Grage 1962).[113] A contribution of osmolality to toxicity was admitted but at the same time downplayed (Olsson 1954). In a scholarly review in 1953 of the history of contrast agents and the involved chemical design principles by the director of research and development of an important producer, namely Wallingford at Mallinckrodt Chemical Works (Wallingford 1953) osmotic properties received no attention. If osmotic properties had been contemplated seriously, the observation of strong detoxification by N-acetylation of 2,4,6-triiodo-3-aminobenzoate sodium in the face of expected similar or even increased osmolalities of the solutions of the acetylated compound, must have convinced Wallingford, a synthetic chemist by formation and mentality, that toxicity was determined essentially by the covalent chemical structure of the contrast agent molecules, so-called chemotoxicity, and not by the physicochemical properties of their solutions, such as osmolality and dynamic viscosity. At the time radiologists too thought primarily in those terms, emphasizing the iodine content of the molecules (Sandström 1953). As late as 1970 an otherwise solidly researched book on contrast agents by the academic radiologist Reinhard Barke still completely missed explicit reference to osmotic properties (Barke 1970). So did in 1971 a book chapter on probable future developments in angiography by Erik Boijsen (1971), chairman of the Department of Diagnostic Radiology at the Malmö General Hospital.[131] This is all the more astounding as he was the superior of Torsten Almén, who at that time had already published his design principles for low-osmolality contrast agents (Almén 1969a) (*vide infra*). Luckily, there were exceptions. Some biologists demonstrated the participation of excessive osmolality of contrast agent solutions in deleterious effects on the kidney (Boriani and Boriani 1940) and on the blood brain barrier (e.g., Broman and Olsson 1949, 1956; Olsson 1954; Odén 1955). Thereby they called attention to this physicochemical behavior of contrast agent solutions, although they often qualified its role. A bit clearer was the situation in angiography, where osmotically driven alterations in plasma volume, changes in other blood parameters and hemodynamic consequences were documented (e.g., Iseri et al. 1965; Standen et al. 1965; Brown et al. 1965; Hilal [1965] 1966, 1966; Björk 1966).

In 1952 the zoology student at Uppsala University, Jan-Erik Kihlström, in connection with using solutions of iodopyracet meglumine for the determination of the density of sperm, set out to find the osmolalities of solutions in concentrations up to 80% (w/v) (Kihlström 1952). He realized that the freezing behavior of such solutions did not meet the validity requirements for cryoscopic measurements, a fact since confirmed by others (Børdalen, Wang, and Holtermann 1970). He built and then applied a vapor pressure osmometer. This pioneering effort could have signaled a breakthrough in characterization of contrast media. Unreasonably he analyzed solutions of iodopyracet and meglumine in an ill-defined stoichiometry but clearly very different from the 1:1 stoichiometry of radiological preparations. In defiance of the fact that his results were useless for the concrete case of UMBRADIL™ (iodopyracet meglumine), he was able to publish them in a radiological journal. Such was the radiological culture in the field of osmolality.

In 1954 at Schering AG the osmolalities of solutions of acetrizoate sodium and the acetrizoic acid dimer sodium salt, iodipamide, were estimated by cryoscopy (Neudert and Röpke 1954). Remarkably low values for the dimer solutions were reported, curiously without comment concerning their relevance for tolerability. Osmolality estimates of contrast media commenced entering patents in connection with the first nonionic contrast agent candidates (Habicht and Zubiani 1962, 1963), a subject to be expounded in due course. In academia isolated studies of osmolality at low (Hammarlund and Pedersen-Bjergaard 1958) and also at elevated concentrations (Boriani and Boriani 1940; Broman and Olsson 1949; Bernstein, Reller, and Grage 1962) were performed. But it is puzzling how slowly the insight into the role of this parameter in adverse reactions to contrast media fell on fertile ground. It is still more puzzling if one realizes that the problem of excessive osmolality had become exacerbated in the case of angiocardiography. This procedure had since its beginning in 1937 become an important diagnostic procedure, requiring very high injection rates and elevated doses and concentrations.

In the 1950s osmotic properties of solutions in general were already measurable and well understood for many decades, but the community of contrast agent researchers did not feel a need to call upon the available knowledge earlier than described. Beginning only slowly in the 1960s osmotic effects of contrast agent solutions were attributed an increased potential for toxicity. But this did not yet lead to widespread and systematic study of osmolalities.[113] Drug discovery researchers kept their typical focus on the structure of contrast agent molecules. But whereas previously maximization of the iodine content in % (w/w) in a molecule was sought, slowly an osmolality-related criterion gained importance. Osmolalities reflect the concentrations of osmotic particles in solution. Water-soluble contrast agents at the time were salts that upon dissolution dissociated into ions. The maximum theoretical number of ions per triiodinated molecule formed in solution is equal to the theoretical number of osmotic particles per contrast agent molecule, P. For most ionic contrast agents, P being either two or three. Clearly desirable was a combination of a low P with a high number of iodine atoms per molecule, I.

Only much more recently, in 1980 (Almén 1980), was the ratio I/P formally introduced as a means of systematically describing the various generations of contrast agents and formulating design principles. The I/P ratio became a splendid tool for rationalizing with hindsight what had happened and prospecting what else might be worth exploring. But projecting this concept into the minds of individuals before the 1960s leads to anachronistic distortions. The same holds for osmolalities of contrast agent solutions. This needs to be remembered when I/P ratios and osmolalities are mentioned even for contrast agents synthesized at a time, when this concept and physical measurement were not yet part of the active patrimony of the relevant research community. In any case, pursuit of contrast agents with an elevated *I/P* ratio turned out to be worthwhile. The *I/P* ratio can explain 82% of the variation in the acute intravenous median lethal dose ($LD_{50, i.v.}$) in mice among all historical uro-angiographic contrast agents, sodium iodide included (Rosati and de Haën [1990] 1991).

On May 12, 1960, the chemist Geo. Brooke Hoey of Mallinckrodt Chemical Works deposited at a notary public, a formal disclosure of hexaiodinated diacidic contrast agents having as skeleton two 5-aminoisophthalic acids covalently connected through the amino group on each moiety by an appropriate molecular link. This type of molecule may be seen as a covalent dimer of the monomeric ionic contrast agent iothalamic acid, salified with sodium or meglumine (CONRAY™, Mallinckrodt Chemical Works) (Figure 4.26, **16**). As opposed to dimeglumine iodipamide, these dimers showed predominantly renal excretion. On March 6, 1963, the corresponding patent application was filed (Hoey 1963) describing among other examples, iocarmate dimeglumine (Figure 4.26, **17**). The compound was given to academic investigators for radiological evaluation (e.g., Hayes et al. 1966; Hilal [1965] 1966, 1966, 1970; Gonsette and André-Balisaux 1969). One of them, Sadek K. Hilal, who then was assistant professor of radiology at the Neurological Institute of the Columbia-Presbyterian Medical Center in New York, meritoriously discussed the pharmacological effects of the contrast agent in terms of osmolality. He called the product conray dimer (or bis-additive of iothalamate) because it was the result of chemical linkage of two molecular moieties related to iothalamic acid (CONRAY). In a couple of publications (Hilal [1965] 1966, 1970) he claimed to have been the one to have suggested to Mallickrodt Chemical Works, the synthesis of iothalamate dimer and even trimer. In personal communications with me Hoey from the company denied this, and the patenting record supports him.

Since iocarmate dimeglumine had a favorable neurotoxicity profile in small animals it was developed for intrathecal application. For uro-angiographic purposes its cost of manufacturing appears to have been a major obstacle. In 1971 under the names of DIMERAY™ (Mallinckrodt Chemical Works) or DIMER-X™ (SA Laboratoires André Guerbet) the product was introduced on the market. At a concentration of 280 mg(Iodine)/mL, it was destined for myelography rather than as uro-angiographic contrast agent. It had an osmolality of 1,040 mosmol/kg (Aspelin 1979). Because of a single over-dose fatal accident (Sovak 1984) and because it was perceived to cause epileptoid effects and arachnoiditis (Ahlgren 1973), in 1980 the United States Food and Drug Administration ordered it off the market in its territory, while it continued to be used in some European countries for additional years.

Independently of the described activities, in 1961 the chemist Hugo Holtermann (Figure 4.34; see *curriculum vitae*), director of contrast agent research at Nyegaard & Co. A/S and his company collaborators requested a patent covering similar dimers (Holtermann et al. 1961). None of the examples excelled in elevated renal excretion. Other companies followed suit, including Sterling Drug Inc., New York (Larsen 1963), Pharmacia AB, Uppsala, Sweden[132] (Björk et al. 1966b) and Bracco Industria Chimica SpA, Milan, Italy[129] (Felder and Pitrè 1968). The latter companies did have candidate compounds with the renal excretion required for uro-angiographic application, but because of low tolerability none of the diacidic dimers was successful in more than biliary imaging (for review of diacidic dimers see Hoey, Wiegert, and Rands 1971).

Around 1968, Philip E. Wiegert at Mallinckrodt Chemical Works synthesized a trimeric analog of iothalamate (Figure 4.26, **18**) and submitted a first version of a US patent application (Wiegert 1968). The molecule possessed three carboxylic groups, one of which was neutralized by internal salt formation with a tertiary ammonium cation. The trimeric molecule carrying nine iodine atoms per molecule responded convincingly to the reigning contrast agent design principle. At the same time, it introduced the idea of offsetting a negative electric charge by a positive charge internal to the molecule. This idea contributed to the next generation of contrast agent design principles, that of maximizing the *I/P* ratio. More about this later. In the trimeric molecule, this ratio assumed a value of 3.0. Like before with the dimeric molecule, iocarmate, Hilal was involved in testing biologically the trimeric relative (Hilal and Morgan 1968; Hilal 1970). Hilal's concern with the osmolality of contrast agents and previous collaboration with Mallinckrodt Chemical Works may have fostered the decision of the company to synthesize a trimeric molecule, but as already mentioned for the dimeric analog, Hoey has denied having followed a suggestion by Hilal (1970). Efforts to procure a copy of Hilal's claimed presentation of the trimer at the annual congress of the RSNA in 1968 and cited as Hilal and Morgan (1968) in Hilal (1970) have failed. In any case, Wiegert's (1968) initial patent application preceded by at least 2 days the congress. Notwithstanding presumably favorable osmolalities of its solution, the trimeric molecule was never developed further. Oligomers even higher than trimers, as well as polymers, will be dealt with later in this chapter.

In the 1970s, at SA Laboratoires André Guerbet high expectations existed for dimeric compounds. But its employees pursued an original version. They synthesized monoacidic dimeric molecules with an *I/P* ratio of 3.0 as a novel approach to low-osmolal uro-angiographic contrast agents. On May 31, 1974, a patent application covering ioxaglic acid and its salts (Figure 4.28, **20**) was filed. The application was laid open on June 12, 1975 (Tilly, Hardouin, and La[u]trou 1974). The corresponding product, ioxaglate mixed salt of meglumine and sodium, with trade name HEXABRIX™, was commercially launched in France in 1980 and has since become

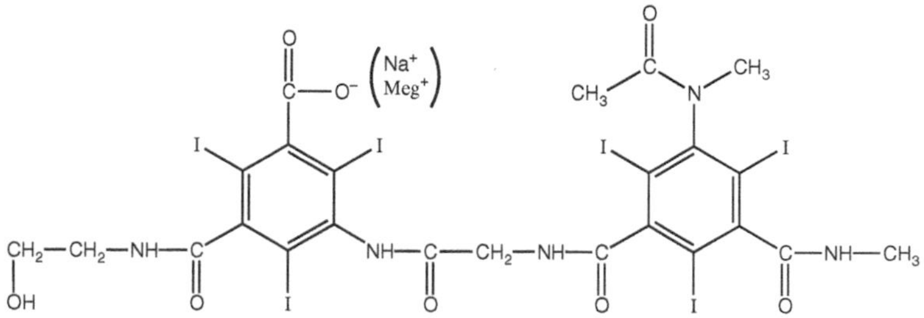

20

FIGURE 4.28 Ioxaglate meglumine/sodium: Ionic but low-osmolal contrast agent.

20: ioxaglate meglumine/sodium (HEXABRIX™, Guerbet SA, France). *I/P* = 3.0.

available in many other countries. At the concentration of 320 mg(Iodine)/mL, it has an osmolality of 580 mosmol/kg. HEXABRIX fulfilled some of the expectations long held for dimers, i.e., reasonable tolerability, which in part was explained by the low osmolality of the injectable solutions (*vide infra*). Yet it left plenty of room for further improvements in tolerability. The company complemented its offer of dimeric products in 1971 by distributing in its traditional European territories the diacidic dimeric compound iocarmate dimeglumine (DIMER-X) for lombosacral radiculography, under license from Mallinckrodt Chemical Works (see earlier in this chapter).

Beginning in 1959, the young Uno Erikson, radiologist and candidate for a doctoral degree[133] under the purely formal tutorship of Folke Knutsson at the Radiodiagnostic Clinics of the Uppsala University Hospital, investigated the angiographic visualization of the blood circulation in amputee stumps and its pharmacological manipulation (Erikson 1961, 1965; Erikson and Hulth 1962). The following account of this line of investigation is based, other than on public documents, to a large degree on information I obtained from him through letters, e-mails, phone conversations, and interviews in the period 1996–2017. Erikson initially used UROGRAFIN 60% as contrast agent. Later, in animals, he switched to THOROTRAST (Erikson and Olerud 1966). Around 1963, he was asked by the doctoral candidate in neuroanatomy Jan Ekholm for help in answering the following scientific question: When the knee joint of a cat was chilled there was an increase in blood flow in the inner part of the knee. What was the anatomical basis of this reflex? The couple of investigators tried radiography after intravenous administration of THOROTRAST, without conclusive results. But they noted that after several doses of the contrast agent the whole vascular system became contrasted. Based on this observation Erikson became interested in how to achieve arterial angiography by intravenous injection of a contrast agent. He hypothesized that an agent that stayed in the blood for about 30 min, a blood pool contrast agent, would make catheters superfluous and very primitive radiological equipment would suffice for angiography. In his pursuit of such agents and their applications he continued to enjoy the collaboration of Ekholm, who contributed experience with animal experimentation that he was still missing. Conveniently, Ekholm's dissertation tutor, the neurophysiologist Sten Skoglund, provided access to animals.

From the outset, it was clear that to achieve the arteriographic objective, the blood pool contrast agent had to be found among colloids or macromolecules. Erikson decided to test his hypothesis first with a model product not suitable for the intended use in humans. Repeated intravenous doses of THOROTRAST were administered to cats, dogs and rabbits and cineradiography was performed (Ekholm, Erikson, and Skoglund 1964). The results were encouraging. Erikson told me that the cineradiographies showed the moving heart and moving muscles.

In late 1963, Erikson felt obliged to accept as additional collaborator his superior, the assistant chief physician of cardiac diagnostics, Lars Björk. The newcomer had notable clinical and research experience in angiography and other thoracic radiological procedures requiring contrast agents. Unfortunately, he had tarnished his local research reputation by submitting around 1959 a doctoral dissertation describing experiments on the fallopian tube of rabbits which, because of errors, had to be retracted and in 1961 was replaced by a new one on another subject. The new collaboration promised help with re-establishing his credentials in animal experimentation. For a Swedish radiology meeting an abstract with Björk as first author was submitted (Björk et al. [1963] 1964), but the presentation was given by Erikson and the work that was later published without Björk's co-authorship (Ekholm, Erikson, and Skoglund 1964). In the abstract Erikson's name was misspelled, letting one wonder about the communication between the two protagonists. This incident makes understandable why it was felt that Björk had forced himself on the group. On the positive side Björk did bring to the enterprise the authority vested in his seniority and permanence of position. These could play a facilitating role both in the university hospital setting that at the time was still obsessed with hierarchy, and in relationships with industry.

Given the inadvisability of using THOROTRAST in man, Erikson looked for alternatives through collaboration with industry. Anders Grönwall, the head of the Department of Clinical Chemistry at his institution organized an invitation to Erikson by the chemist and head of research at Pharmacia AB, Björn Ingelman. The company fabricated and sold, among other products, various generic contrast agents in Sweden. Grönwall and Ingelman had gained much experience with polymers for intravenous injection and had jointly invented dextran-based blood plasma expanders, an important source of income for the company.

In early 1964, a meeting took place at Pharmacia AB. The participants included at least Ingelman and Erikson. Probably Björk was also present, but according to a letter to me from Kirsti A. Granath, September 2, 2016, later he did not usually participate in meetings. Erikson was evidently the radiological leader of the project. A letter from Erikson to me, dated September 5, 1999, included comments by Ingelman, wherein he attributed the initiative to form a collaboration group to Björk, but in Erikson's opinion this merely recognized hierarchical constraints. Ascertained is that a search for practical blood pool contrast agents was agreed upon. Except for the lack of X-ray absorption, Pharmacia's blood plasma expander, dextran, was the blood pool agent par excellence. What was conceptually more straightforward than to render it opaque to X-rays by substituting it with classical second-generation contrast agents. In terms of the theory discussed earlier, the new and old product would constitute a pair of homologous technological solutions. Such solutions are often the first coming to mind. Accordingly, Ingelman and his collaborators initially pursued that avenue. Low solubility, excessive dynamic viscosity, and low opacity to X-rays per unit weight rendered the approach impractical, although a similar one was given a second chance by researchers at Guerbet S.A. in the 1990s (Meyer and Legreneur 1991; Doucet et al. 1991).

Erikson has declared that in a second period Ingelman conceived a more saltatory innovation. One path to macromolecular contrast agents with elevated X-ray absorption was that of somehow polymerizing existing second-generation products without significantly changing the iodine content per unit weight. In the laboratories of Pharmacia AB, 2,4,6-triiodobenzoic acid and derivatives thereof, or diatrizoic acid, were copolymerized with certain bisepoxides (e.g., Figure 4.30, **21**). This resulted in various copolymers having iodine contents of around 45% and apparent weight average molecular weights between 4,000 and 41,000. Aqueous solutions containing around 220 mg(Iodine)/mL could be prepared. Some of the high molecular weight compounds gave solutions of elevated viscosity, but intraaortic injection in rabbits was still possible. They were renally eliminated at a decreasing rate with increasing molecular weight. An opening patent application covering water-soluble copolymers was filed on December 2, 1966 (Björk et al. 1966a) and an additional application followed in short order (Björk et al. 1967). Water-insoluble copolymers suitable for forming gel particles were likewise covered (Björk et al. 1966b). Besides Erikson, Björk, and Ingelman, the list of inventors included Granath and Bernt J. Lindberg from Pharmacia AB. The latter two were experts in the chemistry and physical analysis of the blood expander dextran and related polymeric substances in solution. It is noteworthy that the order of inventors on the initial patent application got changed on the finally granted Swedish and Norwegian patents, from Björk first to Granath and Ingelman, the industrial collaborators, first. The collaboration makes it clear that the concept of macromolecular contrast agents constructed around existing triiodinated molecules or closely similar building elements, and ideas about a suitable chemistry was inaugurated shortly before initial experimentation in 1964 and was followed very soon by the concept of polymeric products. The aforementioned opening patent application was filed on December 2, 1966, but became freely accessible only after July 1968 (Björk et al. 1966a). It thus followed 3 months after polymeric contrast agents were first preliminarily described (Almén 1966a). In any case after the patent's filing Erikson talked freely with colleagues about the concept of polymeric products and even tried to convince other contrast agent producers to get involved. Incidentally, as a byproduct of searching for suitable polymers, a dimeric diacidic molecule, iozomate salt (Ph DZ-59B), was patented (Björk et al. 1966b). Pharmacia AB gave development priority over polymers to the dimer. By the end of the 1960s 600 patients had been injected with the compound (Björk, Erikson, and Ingelman 1969a,b). For undisclosed reasons, the development of iozomate salt was interrupted before completion.

Contemporary studies in animals, published only much later (Björk, Erikson, and Ingelman 1976), demonstrated that in terms of radiological efficacy the copolymers fulfilled at least the radiological efficacy expectations for them, e.g., both good arterial and good venous X-ray contrast. It is not known why Pharmacia's molecular category never led to a fully developed article. It has to be suspected that among other problems, elevated viscosity, difficulties of manufacturing, and the complexities in documenting polydisperse compounds for the health authorities rendered them unpractical. In Ingelman's opinion, the top management at Pharmacia AB was more interested in completing some other interesting projects at that time than investing in the X-ray contrast agents of Ingelman and coworkers. Contrast agent research came to a halt. In any case, there were some efforts to license the technology to competitors but apparently without success. The approaching end of patent protection was an impediment as well.[134]

The opening patent application on water-soluble macromolecular contrast agents shortly mentioned Erikson's motivation for desiring such compounds, namely the expectation that they would extravasate more slowly than small molecular weight ones. On a par with this motivation, solutions with osmolalities lower than those of classical agents could be expected, in line with the contrast agent design principles emerging just then. Echoing Holtermann et al. (1961) the introduction of hydroxyl groups into contrast agent molecules in order to increase water solubility was also enunciated (Björk et al. 1966a). With this some but not all of the design principles later proposed by Almén (1969a) were anticipated.

The driver behind the search for macromolecular contrast agents was clearly Erikson. Scientifically, Björk contributed at best modestly to the collaborative effort but profited mightily from the fact that on his insistence, almost without exceptions, Erikson's publications and patents during the time period here of interest, listed authors alphabetically. After joining the group Björk automatically became first author on all jointly produced full-sized journal articles and patents (Björk et al. [1963] 1964, 1966a,b, 1967; Björk, Erikson, and Ingelman 1969a,b, 1970, 1976). In addition, he promoted himself and the common work during long stays and through numerous lectures in the USA. In some American radiologist's eyes, he became qualified as one of the spiritual fathers of low-osmolar contrast agents, a distinction not granted to others of the collaborators [Discussion contribution by Milos Sovak in Hilal (1970)].

Based on interests detectable in the bibliographies of Björk and Erikson, they pursued common objectives with distinct priorities. Clinical studies testify to Björk's interest in the role of osmolality in detrimental effects of contrast agents (Björk 1966). As collaborator in the program to reformulate ISOPAQUE described later on (Björk [1964, 1966] 1967), he was involved in the wider discourse on this argument. When in his presentations he touched on macromolecular contrast agents, he might have given osmolality arguments more attention than blood pool behavior, while for Erikson the opposite was the case. But in light of what has been presented, this does not justify the aforementioned qualification of Björk, to the detriment of Erikson and Ingelman.

In the 1960s, polling of radiological opinion leaders in academic, hospital, or private practice environments on the medical needs in the field of contrast agents revealed a lack of interest in improved uro-angiographic products (Nyman, Ekberg, and Aspelin 2016a,b). Those radiologists who had lived through the transition from acetrizoate sodium to the more advanced ionic agents were so impressed by the progress in safety that they were unable or unwilling to imagine further progress in this applicative field. Pessimism concerning possible improvements was even expressed by industrial research leaders (e.g., Wallingford 1959). Individual clinicians failed to perceive clear differences in performance among the various products. In consequence, companies encountered difficulties in profiling their merchandise. Economic incentives for further research and development therefore gradually vanished.

Deficiencies of available contrast agents were perceived instead in the areas of hepatobiliary, myelographic, bronchographic, and lymphographic imaging. Companies active in the field of contrast agent research and development adjusted their strategy correspondingly. For instance, Bayer AG exited the field. Research at Sterling-Winthrop Research Institute was dramatically cut back (Amdam and Sogner 1994). Schering AG reduced its research and development at one point in time to just one man, Heinrich Pfeiffer, and the board of directors was on the verge of deciding to abandon the field altogether. A similar situation prevailed in 1966 at Nyegaard & Co. A/S, where contrast agent research came to a halt. But in that company already in March 1967, a renewed research effort was mounted, although addressing clinical indications different from uro-angiography (Amdam and Sogner 1994). Nyegaard & Co. A/S, as well as Mallinckrodt Chemical Works, tried their luck with myelographic agents. Bracco Industria Chimica SpA dedicated itself to the development of novel biliary contrast agents and delegated remaining activities on uro-angiographic compounds to its affiliate company Eprova AG, Schaffhausen, Switzerland.[135] Solely at Laboratoires André Guerbet & Cie. SA the research focus remained on uro-angiographic agents. Although there was still room for improved articles with this clinical indication, as shown by subsequent developments, progress was impeded in good part because the sense of clinical urgency had temporarily been lost.

In some companies with persisting research on monomeric uro-angiographic contrast agents, attention shifted from novel iodinated moieties to the choice of counter ions. In the late 1950s, the availability of diatrizoic acid formulated with various cations had led to the insight that not only the anionic triiodinated portion of monomeric ionic molecules played a role in intravascular tolerability and pain sensation but also the single cation or cation mixture used for its neutralization (Massell, Greenstone, and Heringman 1957; Sjögren 1957; Fischer and Eckstein 1961). Diatrizoate sodium/ meglumine [UROGRAFIN, Schering, RENOGRAFIN, Squibb] was better tolerated and caused less pain than diatrizoate sodium [HYPAQUE, Winthrop]. In the early 1960s, this caused HYPAQUE to significantly lose market share, most strongly in the United States. Therefore, eventually Winthrop Laboratories too formulated its compound with some meglumine as counter ion, giving it the name HYPAQUE M™.

In 1962, Nyegaard & Co. A/S had introduced its first proprietary contrast agent, the second-generation product, metrizoate sodium (ISOPAQUE). In order to be competitive with diatrizoate sodium/meglumine it was immediately faced with the need to come up with a similarly improved cation composition for ISOPAQUE. This exigency soon produced the insight that small amounts of additional calcium and magnesium ions improved tolerability of metrizoate sodium in animals (Fischer and Cornell 1967; Lindgren, Dahlström, and Fondberg [1964, 1966] 1967). A first reformulated version of ISOPAQUE based on these findings became available in 1965 (Amdam and Sogner 1994). Holtermann thought it possible that additional fine-tuning of the cation composition could yield even further improvements in tolerance of ISOPAQUE (Salvesen, Nilsen, and Holtermann 1967). Such improvement was hoped at last to allow favorable clinical profiling of the product with respect to all the more widely used ones of the competitors.

Sweden was the country where ISOPAQUE had been initially introduced. Sweden at the same time provided an environment for research in radiology and vascular pharmacology that was renowned worldwide. In the 1960s, the University of Lund acquired the reputation as the Mecca for selective angiography (Boijsen 1996; Magnusson 2008). Thus, by 1965 many centers in that country participated in a major multinational research program aimed at improving the tolerance profile of ISOPAQUE through variation of its cation composition. While Holtermann provided the overall guidance of the reformulation program and cultivated personally some of the contacts with academics (Amdam and Sogner 1994), on-site coordination of the Swedish clinical activities was assigned to Lars Fondberg (Figure 4.29; see *curriculum vitae*) (Lindgren, Dahlström, and Fondberg [1964, 1966] 1967). He was Product Manager and Clinical Trial Coordinator at Erco Läkemedel AB, Stockholm, the Swedish licensee for the the products of Nyegaard & Co. A/S.

FIGURE 4.29   Leif G. Haugen presenting at a sales strategy meeting at Nyegaard & Co. A/S, with the participation of Lars Fondberg (Third person from the left). (Reproduced with permission of Gyldendal Norsk Forlag SA, of the figure on p. 125 of Amdam, R. P., and K. Sogner. 1994. *Wealth of Contrasts. Nyegaard & Co.—A Norwegian Pharmaceutical Company 1874–1985*. Oslo: Ad Notam Gyldendal).

CURRICULAM VITAE

Fondberg, Lars [Torvald] (b. February 13, 1927, Stockholm, Kingd. of Sweden // d. March 2, 1997, Stockholm, Kingd. of Sweden) [in italics: undocumented information furnished by Lars Fondberg and approved for publication by Ann Marie Fondberg]

Södra Latins Gymnasium, Stockholm: Studentexamen (~1946)/Univ. Stockholm: Studies of chem., (1945 or 1946)/1 year peace-time military serv. (~1950)/Folkskoleseminariet för Manliga Elever i Stockholm: teachers training (~1951 to ~1956)/Högre Allmänna Läroverket i Solna: Biol. and Chem. Teacher (1956–1958)/Univ. of Stockholm: matriculated in natural science (*zootomy, physiology*) (January 1960–January 1961)/Erco Läkemedel AB, Stockholm: sales and clinical trials of Searle-, Upjohn- and Nyegaard-products, last as Head of Research and Product Manager (~1958–1967)/Nyegaard & Co. AB, Stockholm: Chief of information (1967–1969)/Bayer AB, Stockholm: Product manager (clin. development and sales) (1969–1972)/Schering Nordiska AB, Stockholm: Product manager (1972–1975)/*Landstingens Inkøbscentral AB (LIC), Stockholm, managerial position (1975–1983)*/Co-founder of AB Varilab, Gothenburg, Sweden, to 51% in possesion of Hafslund Nycomed A/S, Norway (1983)/Karolinska Hosp., Stockholm, Thorax Clin., Dept. of Exp. Surg.: part-time volunteer researcher under Per I. Olsson (1983–1992)/ Diagnova HB, Stockholm: Founder, Majority Owner and Head of Research (1985–1997).

The ISOPAQUE reformulation program involved on the academic side in Sweden a rather large number of investigators (Lindgren, Dahlström, and Fondberg [1964, 1966] 1967). Only those participants who had an involvement with the principal theme of this chapter will be identified by name. These participants were, Torgny Greitz, head of cerebral diagnostics at the Karolinska Hospital in Stockholm, who in the late 1950s had catalyzed the collaboration of two other participants, namely Percy Lindgren, researcher in the Pharmacology Department of the Karolinska Institute and Gunnar Törnell, a radiologist working on his doctoral dissertation tutored by Greitz. Further participants were Boijsen, professor of radiology at the University of Lund and the aforementioned Björk.

Progress of the clinical trials on reformulated ISOPAQUE was monitored through regular meetings at the various research sites. Of present interest are those held almost monthly at the University of Lund or at its associated institution, the Malmö General Hospital in nearby Malmö. At these meetings a group of persons usually participated most regularly the coordinator Fondberg. The technical discussions were frequently followed by a common dinner sponsored by the company.

The research and development program resulted in a second pharmaceutical reformulation of metrizoic acid (ISOPAQUE) with different counter ions. Unfortunately, the reformulation did not live up to expectation and in certain animal tests even worsened the competitive position. To the disappointment of many, at a company-sponsored symposium in Copenhagen in 1964, Greitz, Telenius, and Törnell ([1964, 1966] 1967) demonstrated a dominant role for the elevated osmolality in untoward effects, where originally a major role of cation composition was suspected (Törnell 1968). Gains in some aspects of the pharmaco-toxicological profile were compensated by losses in practicality due to increases in dynamic viscosity and consequent increases in the difficulty of rapid injection. The clinical benefits where at best exceedingly modest. I learned through a letter to me from Greitz, Stockholm, 1998, April 6, that Holtermann urged his company specifically to take the lesson on the role of osmolality to heart.

At Nyegaard & Co. A/S the disappointing outcome of the clinical trials aimed at improved ionic compositions of their contrast agents contributed to the diffidence in the business future of this line of goods. Together with the internal competition for resources, it contributed to the company's decision in 1966 to reduce its research involvement in the field to almost zero. After obtaining a reversal of that decision, in 1967 Holtermann took up research again, this time looking for radically different approaches.

Apart from the shared evaluation of a modified formulation of ISOPAQUE (Lindgren, Dahlström, and Fondberg [1964, 1966] 1967), the research stimulated creative dialog about contrast agents in the aforementioned network of investigators and with outside persons who shared related interests. For example, Björk, as a participant in the ISOPAQUE reformulation program managed in Sweden by Erco Läkemedel AB, was simultaneously involved with Erikson's academic group in Uppsala and with Ingelman and his team at Pharmacia AB. In such situations cross-fertilization of thoughts among many participants in various programs is inevitable, no matter how carefully the participants try to keep their activities separated according to lines of interest of the involved companies. Indeed, initiation of various similar lines of research and development of contrast agents in Sweden can be traced back to the times of the ISOPAQUE reformulation program and the intellectual fervor it fostered.

The Controversial Prehistory of Nonionic Contrast Agents

The prehistory of nonionic contrast agents is marred by bitter controversy among actors. The factual situation and its documentation is exceedingly complex. Therefore, I have chosen to offer the history from different perspectives. To begin with it shall be recounted with an emphasis on institutional players. Personal perspectives follow.

INSTITUTIONAL PERSPECTIVE: The institutional perspective is largely based on the historical analysis of extensive archival material by Amdam and Sogner (1994). It is complemented by critically evaluated recollections that Fondberg had offered me in personal conversations. He also provided information that allowed me to fill in lacunae in his professional curriculum vitae, which otherwise was reconstructed from documents in the public domain (see *curriculum vitae*). Regrettably I failed to request before his demise documentation with details. In any case, it is clear that for some years in the late 1950s he taught biology, health and chemistry at a state secondary grammar school. In 1960, he was registered for two semesters at the University of Stockholm, for natural sciences (allegedly he took classes in zootomy and physiology). He explained the fact that he never went all the way to an academic degree in science with his impatience with basic science and his eagerness to get involved in applicative work. By the mid-1960s, Fondberg was employed by Erco Läkemedel AB, Stockholm, a company whose owner and president was the pharmacist Erik Bjäringer. This company was distributor in Sweden of various therapeutics that it had licensed in from three important international pharmaceutical companies. Since 1951, the products included those of the Norwegian pharmaceutical company and contrast agent producer Nyegaard & Co. A/S (Amdam and Sogner 1994). At the time of interest here, Fondberg had accumulated several years of industrial experience as product manager and national clinical trial coordinator. His salesmanship, science conversational agility, organizational capabilities, networking abilities, and entrepreneurial drive were solid, but his research skills, patience, experience, and trustworthiness did not match his ambitions. This posed some pitfalls along his career and explains in part why some people he had collaborated with have characterized him as dazzling personality.

Fondberg pursued in parallel with his major task of coordinating the ISOPAQUE reformulation program in Sweden, an agenda of his own. He alleged that sometimes in 1965, independently of the group of Erikson, Björk, Ingelman, and collaborators, he devised a proper synthetic scheme for a polymerized triiodinated contrast agents, but he has never disclosed more details than that partially saponified metrizoic acid was involved (Almén and Fondberg [1966] 1967). According to a memorandum from Holtermann to Ulf Blix, dated November 15, 1966, in January of 1966 Fondberg asked Holtermann some questions about the chemistry of metrizoic acid that beyond doubt indicated to the latter the intent to somehow polymerize the compound. Holtermann replied that polymerization was possible but that the polymers would have a number of drawbacks. Be that as it may, Fondberg claims to have obtained from Nyegaard & Co. A/S chemical intermediates of the syntheses of metrizoic and diatrizoic acid for his project.

In this initial phase, Fondberg had not yet posed to himself the question whether the license and distribution agreement between Nyegaard & Co. A/S and Erco Läkemedel AB allowed independent contrast agent research by the latter company or its personnel. He was satisfied with the conviction that at least Holtermann knew about his chemistry efforts in Sweden and he took the supply of chemical intermediates as a sign of approval. Lacking his own laboratories and working for Erco Läkemedel AB, which at that time did not have chemical synthesis facilities, Fondberg looked elsewhere.

Here follows Fondberg's account of how he came at a given moment to have in hand several preparations of contrast agent polymers. With his ideas, he managed to elicit the interest of the polymer chemical engineer and long-time personal acquaintance, Bo Giertz at Du Pont de Nemours Nordiska AB, Märsta, Sweden, a company specialized in thermoplastic polymers. Giertz was willing to take the outside chance of opening a new field of activity and to organize the synthesis of the proposed polymers, provided the European headquarters of the company, DuPont de Nemours International S.A. in Geneva, Switzerland, would allow it. After having clarified that the planned polymers would not yield thermoplastic materials of interest and that the company will in no way be liable for the medical use of products, a go-ahead from Geneva was obtained. The research collaboration lasted at best a few months. Given the scarce chance of ending with a product directly benefiting the company, the assistance provided to Fondberg may in part be seen as a goodwill gesture of very limited quantity and duration.

In any case, Giertz and his collaborators in Sweden synthesized polymers related to metrizoic and diatrizoic acid of various structures, mean molecular weights, and degrees of crosslinking. The availability of the chemical intermediates of contrast agent synthesis furnished by Nyegaard & Co. A/S was an important factor in the speedy realization of polymers. The product family was given the generic name polypaque. Various preparations thereof were characterized by size-exclusion chromatography and viscosimetry at the Polymer Section of the Royal Institute of Technology in Stockholm. They were pharmaco-toxicologically evaluated in dogs and rabbits. Purportedly, intravenously injected to animals the compounds tested were well tolerated and showed good biodegradability. Regrettably verification of all these activities and identification of the researchers involved was no more possible.[136] With one polypaque preparation, Fondberg even performed a pharmacokinetic and metabolism experiment, on himself, as he has told me.[137] In August 1966, a Swedish patent application was submitted.[138] Few data, but not the product, were made available to the radiologist Torsten Almén (Figure 4.31; see *curriculum vitae*). In the fall of 1966, the polypaque project became public (Almén 1966a; Almén and Fondberg [1966] 1967).

Beginning 1966, Fondberg had informed his employer, i.e., Bjäringer from Erco Läkemedel AB, about his polypaque project. The two signed a contract to split the ownership of the technology equally between them. By early 1967, Erco Läkemedel AB had some intriguing polypaque data in hand, but it knew that it had neither the desire nor a realistic possibility to develop on its own a polypaque-based contrast agent to commercial maturity. Thus, it officially informed its Norwegian licensor about the project and offered a joint development of the product. At the meeting in Oslo participated from Nyegaard & Co. A/S, besides Holtermann, Olav Bjørnson, a pharmacist and executive director, as well as probably Ulf Blix, technical director and co-owner of the company. The participants from Erco Läkemedel AB were Bjäringer and Fondberg. The latter recollected that the project was fully disclosed but did not elicit interest on the other side. The issue of the legitimacy of independent research and development at their company was touched upon, without coming to a conclusion. The situation led to a stop in the polypaque program. Fondberg's patent application was abandoned before it was ever examined, published, granted, or extended to other national territories. Therefore, it could no more be retrieved. Understandably, there could be no follow-up to the interest in polypaque that James O. Hoppe, head of the pharmacology section of the Sterling-Winthrop Research Institute in Rensselaer, New York,[79] had expressed to Fondberg at the end of 1966.

Contrasting somewhat with the description of the events just given, sources with access to the archives of Nyegaard & Co. A/S (Amdam and Sogner 1994) claim disclosure of the polypaque project by the Swedes was late and incomplete. It appears interactions with Almén and Du Pont de Nemours Nordiska AB were not disclosed. On occasion of a pharmaceutical convention in June 1966 Fondberg requested Holtermann to avoid talking to Almén about polymeric contrast agents, a request that was honored. But on June 15, 1966, in a memorandum to Ulf Blix, Holtermann pointed out the drawbacks of polymeric contrast agents, thereby testifying to his awareness of related research ongoing at Erco Läkemedel AB. Remarkably, there appears to be no evidence that anybody at Nyegaard & Co. A/S took notice of the publication of Almén's thesis in September 1966 and the subsequent radiology meeting presentation on polypaque by Almén and Fondberg. The latters claim that personnel from the Norwegian company participated at the meeting is plausible. But if true, could failure to fully realize the state of research in Sweden simply reflect a lack of attention by the management, due to the disappearance of research in the radiology area at Nyegaard & Co. A/S? To wit, by the spring of 1967, a struggle for priority between contrast agent projects and one involving a therapeutic liver factor, and another one focused on reagents, had brought contrast agent research to an almost complete halt (Amdam and Sogner 1994). This fact could also help explain the lack of interest in new compounds expressed at the intercompany meeting in Oslo.

Following the strategic decision in March 1967 to return seriously to contrast agent research, Nyegaard & Co. A/S tried but failed to get satisfactory answers to queries about chemical research at its licensee. It began suspecting disloyal behavior. The suspicion later gained support, when in September 1967 Ulf Blix communicated with the disgruntled young Swedish radiologist Almén, but more to this later.

Shortly after the aforementioned intercompany meeting in Oslo an employee of Nyegaard & Co A/S surprised Fondberg with the news that soon a tight collaboration between the company NV Koninklijke Zwanenberg-Organon, Oss, The Netherlands,[139] and Erco Läkemedel AB would begin. As his very close collaborator Fondberg confronted Bjäringer with this information. The latter categorically denied an imminent divestiture of his company and the assured Fondberg passed the denial on to the propagator of the rumor. But on July 27, 1967, despite everything, Bjäringer did divest one-third of his company, while continuing to work as an employee but in an altered function, for the enterprise of unchanged name. Fondberg, as his erstwhile confident, rightly or wrongly, felt betrayed. Bjäringer's earlier denial and the lack of advanced warning to Nyegaard & Co A/S may be understood as excusable business negotiation necessities but had the price of total deterioration of trust. At Nyegaard & Co. A/S, the divestiture caused the bucket to spill over. Immediately the license and distribution agreements were terminated. In November 1967, a proper office in Sweden was opened. With these events, Erco Läkemedel AB was thrown out of the contrast agent market. In arbitration proceedings it requested, its interpretation of the contractual obligations was partially vindicated and the ex-licensor had to pay a hefty fine (Amdam and Sogner 1994).

In these turbulent times, the anger of Nyegaard & Co. A/S was directed selectively at Bjäringer and not at Fondberg. In accordance with this situation, in November 1967 the latter was assumed by its newly founded Swedish subsidiary, where he helped, as chief of medical information, to guarantee continuity of the activities in the previously built up clinical research and sales operations. The contrast agent business boomed.

In the one and a half years during which Fondberg was employed by the recently created Swedish subsidiary of Nyegaard & Co. A/S, he sensed the company's growing mistrust in him, which he ascribed to the growing conviction by his employer that being previously so close to Bjäringer, he must have known in advance about the divestment of Erco Läkemedel AB without conveying that information to the licensor firm. Certainly, in some Swedish radiological circles Fondberg gained a reputation for lack of openness, a behavior that in part can be excused by the delicacy of the situation bearing on the ownership change at Erco Läkemedel AB. In another part, it could have been linked to misunderstandings between him and Almén regarding the polypaque project, which remain to be examined. On the one hand, the degrading climate of trust he sensed and on the other his dazzling personality, so incongruous with the intent of Nyegaard & Co. A/S to improve control of the situation in Sweden, may both have contributed to his departure from the company after barely a couple of years. Through a noncompetition clause in his employment contract he was obliged to leave the field of contrast agents for 2 years. He actually never returned to it.[140]

At Erco Läkemedel AB and its new ownership, the loss of the license to represent Nyegaard & Co. A/S led to an initiative to develop an own proprietary contrast agents. Lacking suitable facilities and personnel, a research contract with Börje Wickberg, professor for organic chemistry at the Lund Institute of Technology was organized. The search focused on diacidic dimers and triacidic trimers, obtained by chemically linking together through two or three, metrizoic acid-like molecular moieties (e.g., Figure 4.30, **22**). The concept was the same as that experimented with earlier at Pharmacia AB, except for using different linkers as building blocks. Furthermore, the compounds marked the lower limit of the degree of polymerization of metrizoic acid derivatives explored by Fondberg in his project. While polypaque was based on partially saponified metrizoic acid (Almén and Fondberg [1966] 1967), Wickberg's products were based on unsaponified material. This difference later allowed Wickberg and collaborators as inventors, with Erco Läkemedel AB as

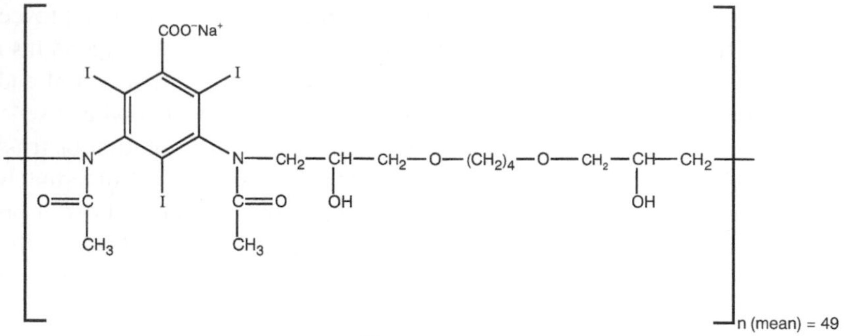

FIGURE 4.30 Contrast agents obtained by polycondensation.

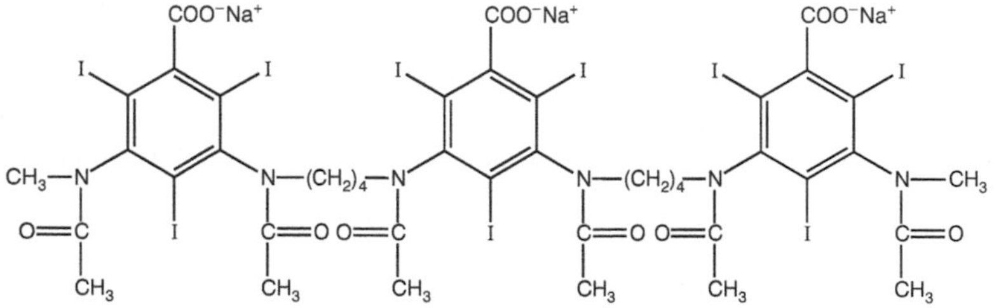

21: Pharmacia compound 602 (Björk, Erikson, and Ingelman 1976). $I/P = 2.94$.
22: Polypaque-similar trimeric compound (Ekstrand, Munksgaard, and Wickberg 1970). $I/P = 2.25$.

assignee, and without Fondberg as co-inventor, to obtain patent protection for their class of diacidic dimers and triacidic trimers (Ekstrand, Munksgaard, and Wickberg 1970).[141] Fondberg seemed to be astonished to learn from me about the described contrast agent effort organized by Erco Läkemedel AB., a fact that argues against his involvement in it. The research failed to lead to a commercial product.

Over the years, a number of university-based scientists voiced the idea of dimeric diacidic or analogous oligomeric contrast agents, sometimes even to industry and, being regrettably misinformed on the true novelty status of their ideas, subsequently claimed primacy for the concept (e.g., Almén 1966a; Hilal [1965] 1966, 1970). This pattern of behavior can only be explained by the regrettable inclination of those outside industry to ignore the patent literature, a pattern of behavior curiously tolerated up to present days by editors of scientific journals, grant review committees, and other organs of quality control in research. Obviously there exists a clash of subcultures in which there are competing ideas of what constitutes legitimate scientific literature.[142] The neglect of the patent literature corresponds to an ignorance of what constitutes primacy. The ignorance is a frequent cause of misunderstandings between the world of academia and that of industry. Such a situation was once more encountered in the following story, which for its beginning returns to the times and events surrounding polypaque. As a complement to the description given so far, which emphasized the rationales that guided the participants' organizational units, this time the contrasting personal recollections and perspectives of two protagonists, namely Almén and Fondberg, will be sequentially recounted, simultaneously commented upon and followed by further perspectives.

ALMÉN'S PERSPECTIVE: The following perspective is based heavily on two interviews I had with Almén on September 16, 1993, and June 20, 1995, as well as a letter to me from him dated November 5, 1993. At the end of the first interview and again in the follow-up letter in 1993, he asked me not to write about what he called the sorry "Erco affair," a description of which until then had never appeared in print. I indicated at least that I would take his request into account. A year later Amdam and Sogner (1994), with the help of Almén, made their perspective of the affair public. This released me to investigate and present my own opinion about it. Despite dealing mostly with failed research efforts, unproductive disputes and tainted perceptions of reality, the story is historically relevant, since it illustrates that not uncommonly, the path to technological progress is branched and presents painful excursions into blind alleys.

Around 1963, the 32-year-old physician Almén (Figure 4.31) was specializing in radiology at the Department of Diagnostic Radiology of Malmö General Hospital in Malmö.[131] He learned the established techniques of angiography the usual way, i.e., by assisting experienced clinicians at the beginning with minor and unpleasant tasks and gradually taking over more of the entire procedure. But he was eager to

FIGURE 4.31 *Torsten Almén*, 1967 in New York. (Courtesy of Torsten Almén, Falsterbo, Sweden.)

CURRICULAM VITAE

Almén, Torsten [Håkan Oscar] (b. September 2, 1931, Lund, Malmöhus Län, Kingd. of Sweden // d. January 8, 2016, Falsterbo, Skåne Län, Kingd. of Sweden)

Gymnasium Ystad, Sweden: high-school dipl. [Studentexamen], (1950)/ Peace-time military serv. 24 mo. (distributed during 1950–1976)/Univ. Lund: stud. of med. (1950–1958); Läkarexamen (1958)/Anesthesia and Women's Clinic, Malmö: temporary position (1958)/Malmö General Hosp., affil. Univ. Lund: Traineeship (1958–1960); specialization in radiol. (1960–1966); MD/ PhD (January 17, 1967); Docent of Diagn. Radiol. (1967) and October 1968– March 1970); on leave of absence (April 1970–May 1971); Docent (June 1971– December 1987); Senior Physician (December 1987–August 1996); Chairman of merged university institutes of Diagn. Radiol., Radiophysics, Clinical Physiol, Oncol., Med. Technol. (January 1995–August 1996); Prof. Emeritus (September 1996)/Temple Univ., Pennsylvania, USA: Assoc. Prof. of Physiol. and Radiol. (January 1967–September 1968)/Erco Läkemedel AB, Stockholm: Consultant (Spring 1967–August 1967)/Nyegaard & Co. A/S, Oslo, Norway: Consultant (May 1968 - ?); researcher, Dept. of Biol. (April 1970–May 1971)/ Fernstrom Great Nordic Prize (1987)/Antoine Béclère Prize, Int. Soc. Radiol. (1989)/Member of the Royal Swedish Academy of Sciences (1989–2016).

go beyond learning routine procedures and looked for a doctoral dissertation project.[133] He intended to pursue an academic career. For a dissertation project, he chose to develop a novel angiographic catheter steering device. In the period 1964/65, he applied for a corresponding patent in various countries, the one in the USA becoming granted (Almén 1965b). In 1966, he presented his dissertation on the subject (Almén 1966a, b). In parallel with his theses work he also developed another radiological device (Almén 1965a).

A couple of years after the time Almén had entered the department, Georg Theander and Lennart Wehlin published a paper entitled "Non-ultrafiltrable Contrast Agent for Renal Angiography" (Theander and Wehlin 1962). In this paper, the authors showed the improvement of the visualization of the veins in kidney angiography of the dog that could be achieved if conventional soluble and diffusible radiopaque substances were replaced by LBg 21. This contrast agent was a stabilized and sterile aqueous emulsion of ethiodol, the hydroiodized ethyl ester of the fatty acid moieties in poppy seed oil. Being actually a stabilized emulsion of the commercial product LIPIODOL-F™ (Laboratoires André Guerbet & Cᵢₑ S.A., France), it was prepared at Guerbet's Swedish licensee, Leo Läkemedel AB, Hälsingborg. It reflected an attempt to create an improved version of JODSOL NACH DEGWITZ™, in pursuit of the perfect blood pool contrast agent. This paragon was a colloidal contrast agent preparation of iodetryl, i.e., 9,10-diiodostearic acid ethyl ester. For use in arteriography it had been brought to the market as VASOSELECTAN™ by Schering AG in 1941 (Blos 1942; Schering AG ≥1941) but had been abandoned after the war.

Although radiologically efficacious in animal experiments, isotonic and presumably pain free on injection, LBg 21 produced toxic effects that prevented continued clinical development.

Almén evidently started his career in an environment strongly dedicated to progress in angiographic procedures and sensitized to the opportunities blood pool behavior and isotonicity of contrast agents presented. He never felt a need to acknowledge inspiration by the aforementioned colleagues, citing instead inspiring interactions with Japanese radiologists regarding the same subject. Almén gladly accepted the offer from his elder colleague, Göran Nylander, to collaborate on venographic (Almén and Nylander 1962) and liver angiographic research in dogs. The team contemplated the reasons why on radiographs of animals injected intrahepatoarterially with clinically used contrast agents, hepatic veins were not radiologically observable (Almén, 1966a). Except for the difference in organ the problem pursued bore close analogy to that dealt with by Theander and Wehlin. Among possible explanations Almén and Nylander identified rapid diffusion of the small molecular weight compounds into the interstitial space. They furthermore speculated that the elevated osmolality of contrast agents recruited water from the tissue leading to dilution. In order to overcome the defect of prevailing products they dreamt of macromolecular contrast agents. Whatever the merits of their assumptions were, what counts in the current context is their reasoning about osmotic phenomena and the related call for a chemical solution to their problem. Almén has told me that the main effect of the discussions with Nylander was that his curiosity on possible new contrast media was aroused.

Clinical experience as angiographer taught Almén that pain was associated with the injection of then available contrast agents but not with physiological saline (Almén [1994] 2001). Primarily through research activities ongoing in his environment he also learned that intravenous injection of stabilized colloidal thorium oxide or isoosmolal emulsified hydroiodized oily substances, such as emulsified iodetryl, or the locally studied LBg21 (Leo Läkemedel AB), did not cause pain. Apart from the compassionate desire to avoid causing pain, its elimination was desirable from a purely imaging standpoint, pain often causing patient movement during angiography and, as a result, blurred images.

Regarding the very beginning of Almén's infatuation with contrast agents, he told me of an "aha-experience," that flash of inspiration in which a not previously perceived association between distant observations is made and the satisfaction of an increased level of understanding is felt. The occasion was a striking clinical case, unfortunately undated, in which he had to fight against pain-induced patient movement that caused blurring of radiographs. Here are Almén's own words:

> I think I was carrying out a femoral arteriogram. I made the same observation made by hundreds of other angiographers that day all over the world. Each time I injected the hyperosmolal X-ray contrast medium it caused pain, and each time I injected physiological saline to wash the blood out of the catheter it did not cause pain. I suddenly saw the mechanism behind the pain. I grew up in Ystad on the most southern coast of Sweden in Skåne. As a boy, I often swam in the water there. You could see the sandy bottom, the stones on the sandy bottom and the flat fish, which were well camouflaged and almost invisible if they didn't move. One summer during the World War II, I was

with my parents at the west coast of Sweden, in Bohuslän. It was not so much fun swimming in the water there. As soon as I opened my eyes to look at the bottom my eyes started to smart. The salty (hyperosmolal) water at Bohuslän drew the fluid out of the mucous membranes of my eyes and made them sore. The brackish water around Ystad did not cause my eyes to smart when I opened them. The physiological saline solution in the femoral artery did not draw fluid from the endothelium of the vessel. This did not cause pain. The hyperosmolal contrast medium in the femoral artery was drawing fluid from the endothelium of the vessel. This did cause pain.

(Almén [1994] 2001)

It was the connection with a childhood experience that gave an intimate dimension to the abstract notion of osmotic properties learned in medical school. From this point on Almén became captivated by the association between the high osmolality of contrast agents and the pain they caused, especially in peripheral angiography. He wished to take the subject further but realized his preparation in the relevant chemistry and physical chemistry was inadequate. He therefore bought books on chemistry and colloid chemistry (cited in Almén 1966a, 1969a, [1994] 2001) and expanded his horizon in this field. About in 1963, he developed the conviction that soluble macromolecular contrast agents with iodine contents equivalent to those of the well-known small molecular weight reference compounds would not only provide solutions of low osmolality and as a consequence be less painful but also would leave the vasculature more slowly than conventional ones and thereby offer beneficial contrastographic effects for the visualization of draining veins. As opposed to colloidal particle solutions and oil in water emulsions, soluble macromolecular contrast agents promised to have the advantages of heat sterilizability of their pharmaceutical formulations and the lack of risk of embolism. Almén contemplated water-soluble blood pool contrast agents (Almén 1985), and he speculated that the reduced diffusion into tissues would result in a reduced toxicity (Almén 1966a). Interpreting some work on the relationship between contrast agent osmolality and the induced dynamic viscosity of blood he estimated that macromolecular contrast agents would have less pronounced effects. In a historical account three decades later, Almén offered a detailed list of the motives for designing, in the 1960s, contrast media isotonic to human serum (Almén 1995). This list is idealized by hindsight and reflects poorly the thought processes found in the original documents. In any case, he reasoned that in order to preserve the high iodine content of ordinary contrast agents, the macromolecules had to consist of a polymer thereof. He knew some general polymers built from benzene ring containing monomers, e.g., polystyrene. By analogy, he naively thought polymeric contrast agents containing the triiodinated benzene rings equally feasible (Almén 1966a).

It is improbable that he developed his ideas in complete independence of other investigators, since similar considerations were part of the discourse within the group of scientists and clinicians involved in the ISOPAQUE reformulation program, in whose meetings he often took part. In the aforementioned memorandum Holtermann noted that the idea of polymerization had been discussed at his company some years before 1966, and Fondberg knew of Swedish radiologists that thought along the same line.

In 1965, Almén attempted by phone to take his interest in novel contrast agents and some theoretical ruminations to two Swedish companies (Almén [1994] 2001) involved in contrast agents, namely Leo Läkemedel AB and Pharmacia AB and/ or possibly others.[143] He either received negative answers with the justification that there existed no need for additional academic contacts or was passed from person to person, the last one promising to call back without ever doing so. In an e-mail to me from Almén, March 14, 2005, he mentioned that one company did concede a discussion during which it declared itself to be already for some time involved in macromolecular contrast agents. Given that Pharmacia AB, with Ingelman, Erikson and Björk, was doing research in the field of polymeric compounds, it was probably this company. There exists no indication that Almén's proposal contained the concept of nonionic contrast agents and the concreteness of other propositions could not be ascertained.

Almén recollects that in the same vein in February 1966 he talked about his ideas to Fondberg from Erco Läkemedel AB, the Swedish distributor of the contrast agents of the Norwegian company Nyegaard & Co. A/S. On September 8, 1967, he described the interaction in writing to the Norwegian company's technical director and co-owner, Ulf Blix: "In February 1966 I presented to Lars Fondberg the idea of polymerizing the contrast agents by the method described in my doctoral thesis" [as quoted from archival documents in Amdam and Sogner (1994)]. His dissertation (Almén 1966a) did in fact report, as a research side line, on some aspects of polymeric contrast agents but provided neither molecular structure nor methods of synthesis. How this may have come about is explained later.

Almén claimed to have discussed with Fondberg at the time not only polymeric substances but nonionic ones as well. In his persistent opinion, his interlocutor did not fully understand the arguments in favor of the latter kind. By talking to Fondberg, whom he viewed more as a Swedish representative of Nyegaard & Co. A/S than as an employee of Erco Läkemedel AB, he believed to have opened a communication channel with the research and development function of the former company. He was convinced that it was now researching polymeric contrast agents and that this activity was entirely due to his input to Fondberg. He thus expected to be informed of how the project was progressing and felt it nothing less than his right to be the initial presenter of polypaque to the public. Indeed, Almén (1966a) was the first to publish on polypaque. He delivered a crude product description together with some physicochemical and pharmaco-toxicological data, as well as a pharmacokinetic study in an unspecified single human being. He did so in a curiously out of place and deficient manner, as a small chapter in his printed MD/PhD dissertation. It was entitled "Development of a Water-Soluble Contrast Medium with Low Diffusion Rate" within the overall heading "A Steering Device for Selective Angiography and Some Vascular and Enzymatic Reactions Observed in its Clinical Application." The dissertation was independently printed twice. The first time it appeared with the date of September 1, 1966, as *Supplmentum* of *Acta Radiologica*, the traditional place for dissertations (Almén 1966a). The second time it appeared as thesis in coincidence with Almén's oral dissertation defense, i.e., December 10 (Almén 1966b). The next presentation of polypaque after September 1, which is documented in the literature, was authored jointly by Almén and Fondberg and occurred at the meeting of the

Swedish Association of Medical Radiologists on December 2–3, 1966 in Stockholm (Almén and Fondberg [1966] 1967). Almén later never mentioned this presentation. Almén admitted to me, up to that time not to have contributed experimental work and not to have had any product in hand. Hence, the sole source of all information was Fondberg. Almén included in the list of references to his thesis a "personal communication" by him but cited it only in support of his own affirmation that he had presented the theoretical considerations to "Fondberg at the Erco and Nyco companies, who started the synthesis of different polymers of metrizoate and the first investigations of their properties." Nowhere the source of products and data was clearly identified. In the acknowledgments Fondberg's name is missing.

Purportedly through a contract with Erco Läkemedel AB Almén signed in the spring of 1967, he became aware that, contrary to his previous conviction, polypaque was a project of that company rather than of Nyegaard & Co. A/S. Not being adjourned on a regular basis, so far not being given access to compounds and beginning to feel misled about the company behind the project, he was happy to encounter both Holtermann and Fondberg at a pharmaceutical meeting (Amdam and Sogner 1994). Allegedly on latter's request Holtermann avoided talking to Almén about polymeric contrast agents. On August 18, 1967, the latter contacted Holtermann and was answered by Ulf Blix, the technical director. In his response to Blix of September 8, Almén bitterly complained of being kept in the dark about the polypaque project (Amdam and Sogner 1994).

When for reason given earlier, Nyegaard & Co. A/S withdrew the license and distribution agreement from its Swedish distributor, the polypaque project came to a total standstill. Almén could not believe that a scientific project he cherished as his own could be stopped by others and protested vigorously with Erco Läkemedel AB. He perceived the response letter by the lawyer of that company as an attempt to interfere with his research freedom, even as harassment. He sought legal help. An exchange of letters among lawyers settled the issue at the legal, through not at the emotional level. The polypaque episode did provide Almén with some experience in the not always easy relationship between academia and industry. He was the wiser for it when later getting involved with industry a second time.

FONDBERG'S PERSPECTIVE: In Fondberg's recollection the events had quite different dynamics, as related to me in the earlier mentioned personal contacts. According to him, at the Malmö General Hospital in the early 1960s Almén participated many times at progress meetings of the clinical investigators of the ISOPAQUE reformulation project, which Fondberg coordinated (Fondberg [1966] 1967). Fondberg recollects that Almén talked to him quite informally on the desirability of low-osmolal contrast agents, e.g., macromolecular ones, probably including nonionic ones. The ideas were quite generic and did not merit any status of secrecy. Since the ISOPAQUE reformulation project focused so much attention on the cations and their possible deleterious effects, it was natural to contemplate concomitantly their elimination altogether through a switch to nonionic iodinated molecules. As both Gunnar Törnell and Fondberg affirmed to me, unspecified number of individuals did. However, skepticism concerning feasibility, in particular regarding sufficient water solubility, reigned. Apparently, no concrete chemistry proposals were put forward for how to convey to nonionic compounds elevated water solubility. Almén did not yet offer his theory regarding the substitution of the charged carboxylate group by neutral ones with an elevated number of hydroxyl groups.

Fondberg furthermore excludes that Almén had introduced him to the idea of polymeric contrast agents. According to archival documents examined by Amdam and Sogner (1994), Almén dates the opening discussion with Fondberg to February 1966 i.e., after the January date the latter had asked Holtermann at Nyegaard & Co. A/S some questions about the chemistry of metrizoic acid that allowed the conclusion, or at least suspicion, that polymerization was being attempted (Amdam and Sogner 1994). Fondberg, who claimed some chemical training, had in precedence identified possible polymerization chemistry and did not need chemistry advice from Almén. According to the archival documents the concept of macromolecular contrast agents that would yield formulations of reduced osmolality likewise was widely discussed. In any case, in several persons' recollection the cited ideas, polymeric and nonionic substances, in those years were so pervasive that attempts to identify specific individuals or groups legitimately claiming priority for having expressed them must remain futile. Quite another issue is concrete and practical proposals of how to meet the generic goals, i.e., chemical structures, practical synthetic pathways and pharmaceutical formulations.

Fondberg appreciated Almén as a young and highly enthusiastic researcher early in his career. Allegedly he helped him to understand better the physicochemical concepts behind polymeric contrast agents, which he judged more realistic than nonionic ones. He explained how polymeric compounds would allow resolution of the osmolality and thus the pain problem. He declared himself to be involved in the preparation of polymeric candidate contrast agents but avoided explaining the touchy and never to be resolved interaction among three companies potentially called upon to cooperate, i.e., Nyegaard & Co. A/S, Erco Läkemedel AB and Du Pont de Nemours Nordiska AB.

In Almén, a promising clinical collaborator for when contrast agents from the polypaque family would become available for testing in man seemed at hand. But his pronounced zeal caused Fondberg to refrain from passing chemical structure information and actual products to him, fearing both public disclosure before application for a patent and unauthorized administration to man, e.g., self-administration. Nonetheless, to foster the relationship, only weeks after having filed in August 1966 a patent application Fondberg allowed Almén to include in his dissertation, as already described, the scarce chemical, physical and toxicological information on polypaque he had provided him with (Almén 1966a). He did not challenge Almén over the claim in his dissertation of having caused the Erco and Nyco pharmaceutical companies to investigate such products. For obvious reasons industry representatives tend to avoid confrontations with clients. Fondberg took part in Almén's oral dissertation defense and in the discussion time was asked by the audience when polypaque would become available. He answered that it was, for the moment, purely experimental. On that occasion, Almén must have clearly identified his source of information.

In the spring of 1967, sometime before the meeting between Erco-Läkemedel AB and Nyegaard & Co. A/S, Almén expressed the desire to have his relationship with Erco Läkemedel AB contractually defined. Fondberg agreed, and a more active role for him in the project could at least be envisioned. Almén's lawyer proposed a contract that after small changes was signed by both parties. The contract foresaw ownership by the company of all intellectual property emerging from the collaboration, a common arrangement for academia–industry collaborations at that time. The already described developments of the interaction between the companies brought the interaction of Almén with Erco Läkemedel AB, and thus equally with Fondberg, to an unpleasant end.

FURTHER PERSPECTIVE: A further perspective is obtained if documented events are considered that neither Almén nor Fondberg adduced in support of their versions of history. Up to the end of 1966, research of Almén covered clinical diagnostic issues, some of them with commercial contrast agents (e.g., Almén and Nylander 1962; Almén [1966] 1967, 1967) and the design of radiological devices (Almén 1965a,b, 1966a). There is no contemporary document supporting earlier interests in the conception of novel contrast agents, although through his clinical activities and the research of his department colleagues he was certainly aware of numerous problems that existing ones posed.

Almén ([1966] 1967) had reported personally on celioacographic studies with a typical ionic contrast agent of the second-generation at the meeting of the Swedish Association of Medical Radiology on September 17, 1966, in Jönköping, Sweden. With an anecdote about the tax deductibility of the travel costs, disclosed to me in a letter dated February 20, 2004, Almén affirmed his personal presence at the meeting and how well he still remembered it. He recollected that at that meeting an investigator, whose identity since escaped his memory, presented a futuristic-looking cineradiographic movie showing with extraordinary sharpness the entire vasculature of a rabbit injected intravenously with a macromolecular contrast agent. In his reminiscence, the contrast agent consisted of dextran substituted with triiodinated ionic contrast agents. In the second of the aforementioned interviews I had with Almén, he described the audience of the movie, including himself, as flabbergasted, even incredulous. Although Björk, Erikson, and collaborators had 3 years earlier presented cineradiographies of cats, dogs and rabbits intravenously injected with THOROTRAST (Björk et al. [1963] 1964), there is no indication that they were the producers of the new spectacular movie.

Regarding this movie, in the aforementioned letter to me, Greitz has written: "Törnell remembers that he saw an impressive movie, in which Fondberg injected enormous amounts of polypaque into a rabbit, which seemed to be completely unaffected of the injection." Törnell and Fondberg knew each other well. According to the collection of abstracts of the mentioned meeting in Jönköping, Fondberg indeed gave a presentation, but the official theme was counter ions in established ionic contrast agents (Fondberg [1966] 1967). One has to assume that he used his formally allotted time to present, in deviation from the abstract's content, also polypaque, or he obtained extra time to do so. The published abstract reveals no deviation from the program, and accordingly gives no name of a radiologist that could have produced the movie. The fact is that to me Fondberg, independent of all other informants, declared that he had presented polypaque for the first time at the meeting referred to. Herewith the person behind the feat is identified.

At this point, some burning questions arise. How could Almén by 2004 have forgotten that the presenter of the movie that so impressed him had been Fondberg. The unflattering fact is that after the already described fallout between the two in the second half of 1967 and the company he represented, Almén repressed his name and role in the exploration of low-osmolal, macromolecular contrast agents by any means available to him. He never cited the joint presentation. In a letter to me from Almén, Funästalen, February 20, 2004, the contrast agent used in the spectacular movie is described in the following way: "I believe it was a macromolecular contrast medium which was a three-iodinated compound linked to dextran." This description is suitable to detract attention from Fondberg and to steer any contrast agent specialist in the direction of Erikson, Björk and Ingelman, investigators who indeed had explored such compounds. There remains the uncomfortable feeling of purposeful focal weakness of memory on the part of Almén, who built himself up as the only source of the idea of low-osmolal, macromolecular contrast agents. Orally he did not shy away from insinuating on the part of Fondberg the inability of understanding fully his ideas, and even questionable conduct. His aforementioned request to me to ignore the story involving Erco Läkemedel AB fits with this interpretation.

In the course of 1966, Almén must have identified research on new contrast agents as the field in which to pursue a career. In Mary P. Wiedeman, professor of physiology, at Temple University in Philadelphia (USA), he found the desired sponsor. Having a faculty position for him approved and obtaining a visa with a working permit certainly required several months. Together with the January date of his arrival in Philadelphia, this requirement allows the possibility that Almén's choices were made under the spell of Fondberg's polypaque data and movie.

Be that as it may, the movie surprised a receptive public and fostered expectation of further contrast agent innovations. Independent of whether Fondberg originally engaged in polymeric contrast agents on his own initiative or only in response to Almén's input, 1966 marks the year the latter began his infatuation with novel contrast agents. At first research focused entirely on oligo- and polymeric products. Nonionic contrast agents were not yet in the field of view. In any case, Almén's entry into contrast agent research brought a strong impetus to the field. Intermediate steps toward a new technological vision were made (Table 4.2).

Polypaque, a particular realization of a certain chemical class of polymers, merit some critical evaluation not only as focal points for human group dynamic phenomena in technology shaping but also as molecules and contrast agents. Polypaque was described as a polymer of presaponified metrizoic acid of unrevealed exact chemical constitution (Almén and Fondberg [1966] 1967; Almén 1966a). Its degree of polymerization lay between 150 and 300 and the corresponding molecular weight between 100,000 and 200,000 (Almén 1966a). Fondberg let me know that the macromolecules were weakly cross-linked. A solution of the compound at a concentration of 440 mg (Iodine)/mL was reported to have a dynamic viscosity of 0.013 Pa·s. Almén did not question this puzzlingly small value since, as already pointed out, on erroneous theoretical grounds, he was not expecting viscosity to be a problem. Most notably, Almén described rudimentary observations on a single, unnamed human volunteer who had been injected with the compound. In the aforementioned contacts, I had with Fondberg he declared the viscosity data in error and himself as the volunteer. The polymer was reportedly degraded in the volunteer's blood and was excreted completely within 24 h. Similar observations in animals were also reported on some other members of the polypaque family (Almén and Fondberg [1966] 1967). No radiographs obtained with these contrast agent solutions were made available.

Notably, not only viscosities, but also biodegradation and excretion velocities are so baffling that full disclosure of the chemical structures and the experimental methods used would be required for a reasonable assessment of their reliability. Very disturbing is that fact that up the spring of 1967, i.e., the hole time period dealt with so far, Almén had no tangible evidence for the reality of Fondberg's products, data, collaborators, and accompanying tales. What if the whole polypaque story was invented? Almén showed lack of caution, fostered by his eagerness to see his ideas realized. The situation improved only partially, when months later, as described below, he obtained for his studies on bat wings a contrast agent preparation said to be a trimeric version of polypaque (Almén and Wiedeman 1968b).

Neither Almén's nor Fondberg's narratives as they are can be accepted as definitive. In the course of writing the present work additional questions to these individuals emerged that their death prevented from being answered. Probably other participants of the events could have offered further different perceptions. It is unlikely that a unique truth can ever be distilled. By good fortune for the current endeavor no such unique truth is required. Even in this crude state the fundamental processes of technology genesis and shaping can be recognized.

Nonionic Contrast Agents

The earliest efforts to find nonionic water-soluble uro-angiographic contrast agents and the reasons why they did result in patents but neither appeared in scientific journal articles nor materialized as commercial articles, deserve a short interlude. Since 1946 the pharmaceutical company Cilag-Chemie AG, Schaffhausen, Switzerland[144] was a significant producer and international vendor of generic contrast agents, but in the early 1950s it got into financial difficulties. Research and development activities in the radiological sector were abandoned, but when in 1958 Ernst Habicht (Figure 4.32; see *curriculum vitae*) became head of research, the sector was restituted strategic importance. Merely 1 year later Cilag-Chemie AG was acquired by Johnson and Johnson Corp., New Brunswick NJ, USA, and all discovery research was declared superfluous. In spite of this edict, Habicht, with about ten technicians, managed to pursue research undeterred until 1963, when he was forced to leave.

Just before his departure from the company Habicht and his technician Ruggero Zubiani had submitted Swiss patent applications giving a refreshing answer to the hyperosmolality problem of contrast agents. They disclosed the structure and synthesis of a class of nonionic molecules, consisting in the known contrast agent iodopyracet or a similar iodocompound, esterified with oligoethylene glycol or a homolog thereof. Some representatives possessed elevated water solubility at room temperature (e.g., Figure 4.33, **23**; Habicht and Zubiani 1962, 1963). The inventors justified their choice of nonionic molecules by the low osmolality of the corresponding solutions, an attribute that could be expected to result in reduced cellular damage. They estimated osmolalities by cryoscopy. One preparation displayed unusual solubility

FIGURE 4.32   *Ernst Habicht*. (Courtesy of Walter Habicht, Arolla, Switzerland.)

CURRICULAM VITAE

Habicht, Ernst (b. November 12, 1916, Schaffhausen, Canton of Schaffhausen, Switzerland // d. March 14, 1993, Binningen, Canton of Basel-Landschaft, Switzerland)

Kantonsschule, Schaffhausen, Switzerland: High-school dipl. [Matura] (Fall 1935)/ETH, Zurich: Stud. of pharmacy (1935–1936)/Univ. Zurich: Stud. of chem. (1936–1944); PhD (1944)/Cilag AG, Schaffhausen: Chem. (1944–1958); Head of Research (1958–1963)/Ciba AG, later Ciba-Geigy AG, Basel: Chem. (1963–1981).

properties. A 52% (w/v) solution of compound **23** had a concentration of 205 mg(iodine)/mL and with 300 mosmol/kg it was isoosmolal with blood. Surprisingly compound **23** in solution demonstrated phase separation at 37 °C. This promised to be helpful principally for lymphography and bronchography. Thus, in line with perceived medical needs at the time, Habicht focused on these applications and formulated a corresponding design theory for contrast agents that included nonionicity. Given that Habicht in the same period submitted a patent application that claimed

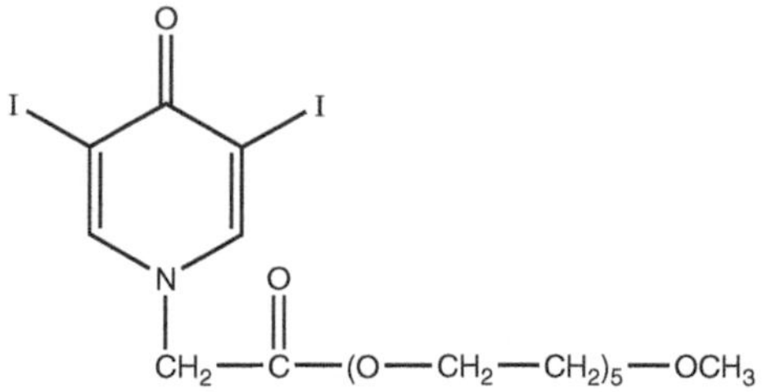

23

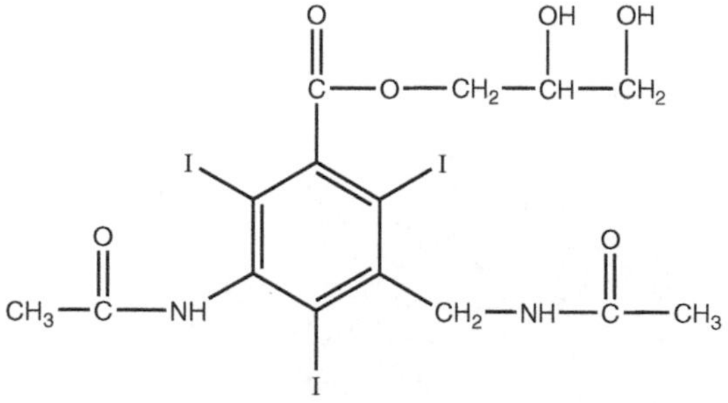

24

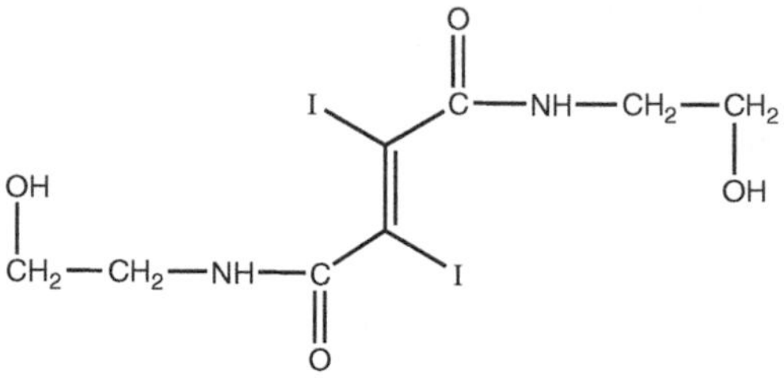

25

FIGURE 4.33   The first experimental highly water-soluble nonionic contrast agents.

23: Iodopyracet *O*-methyl triethylene glycol ester. Example from Habicht and Zubiani (1962). $I/P = 2.0$.

24: Iodamide 1-gycerol ester. Compound B-6590 from Bracco Industria Chimica SpA (Fedeli 1964). $I/P = 3.0$.

25: Example of diiodofumaric acid amides (Eprova 1968; Suter and Zutter 1975). $I/P = 2.0$.

a novel class of ionic iodinated uro-angiographic products (Habicht and Feth 1962), we can be sure he was in parallel looking for nonionic molecules for the latter purpose. After Habicht's forced departure from the company, the champion of contrast agents was gone and all related projects were abandoned. It should not surprise us that under such circumstances the results in the patent never appeared in a scientific journal, and as a result, academics in general for several years did not take notice of water-soluble nonionic contrast agents. For Habicht and his inventions, place and time were unfavorable. Good and bad luck do play roles in technological progress.

By the mid-1960s research departments of the pharmaceutical companies involved in the radiological field could be cognizant of the patents by Habicht, but it is not known how many actually were. Certainly, the patents were known at Bracco Industria Chimica SpA. After the World War II forerunners of that company and after 1958 the company with that name manufactured and distributed in Italy and some other territories pharmaceuticals of Habicht's employer, Cilag-Chemie AG. Prominent products included generic contrast agents. In the early 1950s, the latter company entered great financial turmoil. This badly affected Fulvio Bracco, investor in that company and co-owner of the Bracco group of companies. Among the commercial effects, it put his contrast agent supply line at risk. In reaction to this situation he decided to build up own research and development capacities targeting proprietary therapeutics and diagnostics. Together with Hans Suter, ex-head of research and production at Cilag-Chemie AG, in 1952 he founded Eprova AG, Schaffhausen, Switzerland. Suter became the chief executive officer. He and some other employees of the new company had left their former company together because of the precarious situation in which their employer found itself. In the late 1950s, Fulvio Bracco directed his research and development units to enhance efforts in the field of contrast agents. The corresponding activities were placed at Eprova AG, guided by Suter but under the ultimate responsibility of the director of chemical manufacturing and drug research at Bracco Industria Chimica SpA, Ernst Felder (Figure 4.37; see *curriculum vitae*).[145] Eprova AG was further charged with the role of corporate patent office, which became headed by Hans Zutter. Habicht was well known at Eprova AG, since he was an ex-colleague of Felder and Zutter, working under Suter, while Cilag-Chemie AG still employed all four at the same time. When Habicht had to leave that company, he was even offered a position at Eprova AG. For family reasons alone he chose to go elsewhere and leave the field of contrast agents. Evidently there existed networks of companies and of some of their individual employees. The latter members of the network were all Swiss citizens and native German speakers. Such networks foster mutual monitoring of activities and thereby diffusion of information. They help explain the early detection and study of Habicht's patents and the positive reception of his ideas on nonionic contrast agents at Bracco Industria Chimica SpA and Eprova AG.

Unlike today, drug discovery research in the 1960s in most pharmaceutical companies was not so much driven by medical needs, market potential, pharmacological targets and possibilities to prove pharmaco-economic benefits, as by chemists' interest in novel classes of compounds and their synthesis, as well as patenting opportunities. With such motivation Bracco Industria Chimica SpA, which by then operated as complement to its research unit at Eprova AG, also a unit in its hometown, Milan, Italy, began scrutinizing triiodobenzoic acid derivatives. Some of these were nonionic because their carboxyl group was esterified, and a hydroxyl-group-rich alcohol moiety contributed to water solubility. Larsen at Sterling-Winthrop Research Institute had earlier pursued this approach in the search for bronchographic contrast agents easily suspendable in water (Larsen 1957). In the iodamide 1-gycerol ester (Figure 4.33, **24**, B-6590), synthesized on September 7, 1964, at Bracco Industria Chimica SpA an apparently highly water-soluble, nonionic compound with potential as uro-angiographic contrast agent was identified (Fedeli 1964). The glycerol moiety was intended to contribute to water solubility through two hydroxyl groups. By bad luck, the more this candidate contrast agent was purified, the higher was its tendency to crystallize out from aqueous solution. Crystallization was the dominant justification for eventually abandoning iodamide 1-glycerol ester as nonionic uro-angiographic candidate product.

In analogy to the esterification of the carboxyl group in known ionic triiodinated contrast agents, amidation of the kind that was to become the basis for breakthrough nonionic compounds was contemplated. The industrially practical preparation of the acid chloride intermediates bearing acylated aniline functions posed an unexpected obstacle. These difficulties constituted a hurdle significant enough to impede for a while continued exploration in that direction. It is noteworthy that the value of non-ionic molecules as uro-angiographic contrast agents was not fully recognized until much later and thus at this time the motivation for overcoming the synthetic difficulties was low. Since the economic circumstances at the time were not favorable for the industrial development of uro-angiographic novelties (see Section "Perfecting Ionic Uro-Angiographic Contrast Agents"), pursuit of nonionicity did not acquire importance for many more years.

At least up to the time of Almén's dissertation defense on December 10, 1966, the efforts of Erikson, Ingelman, Björk, Fondberg, and Almén were directed primarily at contrast agents with blood pool behavior and low diffusion rates. Angiographic applications were in the focus, urographic ones being tacitly assumed to emerge in parallel. Macromolecular soluble contrast agents were seen as a possible answer. Almén's rational for wishing macromolecular contrast agents also was based on low diffusion rates (Almén 1966a). He wrote: "The use of a macromolecular contrast medium might produce only little pain because its irritating effect is reduced on account of its low rate of diffusion into and through the vascular wall." He postulated lowered osmolality to be beneficial because of expected reduction of contrast agent-induced blood viscosity increases and avoidance of microembolism. Contrary to what he tried to convey in his later narrative (Almén [1994] 2001), reduction of osmolality through nonionicity was nowhere yet on his radar screen. No abolition of ionizing groups was contemplated, given that carboxylate groups continued to convey elevated water solubility. No consideration concerning the role of hydroxyl groups in providing water solubility and reducing toxicity were offered. Given the anxiousness with which Almén elaborated theories on contrast agents extraneous to the main theme of his dissertation[146] and the eagerness with which he incorporated into his dissertation every crumb of information about polypaque he could obtain from Fondberg, it is likely that he would have mentioned in it a theory for how to obtain soluble nonionic compounds, if by then he had thought about them. Moreover, when Ulf Blix, the technical director and co-owner of Nyegaard & Co. A/S in September 1967 contacted Almén about ongoing projects with Erco Läkemedel AB (*vide infra*), the latter complained merely of being kept in the dark by the company concerning polymerized contrast agents (Amdam and Sogner 1994). In an interview with me, Almén declared that the reason Fondberg and Erco Läkemedel AB did not study nonionic contrast agents was that they had only understood part of what he had told them. There exists no corroborating evidence.

The efforts of Erikson, Ingelman, Björk, Fondberg and Almén were expression of the stepwise emergence of a new vision (Table 4.2). The first step envisioned uro-angiographic contrast agents that due to their macromolecularity showed blood pool behavior and thus sharper delineation of injected vasculature yet were still efficiently excreted. As a side-benefit they promised reduced pain at the site of injection owing to their low osmolality. Low osmolality independent of macromolecularity had found some marginal industrial attention (Habicht, Felder) but was not yet really shaping the new vision.

From January 1967 to September 1968, Almén joined the physiology laboratory of Mary P. Wiedeman at Temple University in Philadelphia. According to the university's Faculty and Student Records, he obtained a full-time joint appointment as Associate Professor of Radiology and Associate Professor of Physiology (Radiology). Wiedeman had a record of publishing on the subject of deleterious effects of contrast agents on the vasculature in animals, in particular the microvasculature of bat wings. Only in a single one of her related papers before 1967 are osmotic effects alluded to, and even then, only casually (Wiedeman 1964). This changed for the duration in which Almén joined her laboratory. In the bat wing model, he studied first the effects of commercial uro-angiographic contrast agents and solutions of mannitol as a function of their cryoscopically measured osmolalities (Almén and Wiedeman 1968a). Then, he extended the studies to the deleterious effects of solutions of macromolecules with measured osmolalities. He first compared glucose with the linear glucose polymer, dextran (Almén and Wiedeman 1968b), and later also with the branched glucose polymer, glycogen (Almén 1969a), the latter two being non-contrast agent, nonionic macromolecules. He further compared these compounds with one said to be a polymeric metrizoic acid of undisclosed exact structure and an average degree of polymerization of 3. He acknowledged to have obtained it from Fondberg (Almén and Wiedeman 1968b).[147] The studies documented the deleterious effects of contrast medium hyperosmolality and resulted in a call for products with lower values. The publications clearly document that only while working in Wiedeman's laboratory Almén began to put the osmolality concerns ahead of those involving diffusibility, viscosity, and blood pool behavior. But chemically linking existing contrast agent molecules to form polymers were the sole means for how to avoid elevated osmolality of contrast agents mentioned in his publications jointly authored with Wiedeman (Almén and Wiedeman 1968b).

While Almén worked in Philadelphia, through mutual friends he met Hilal in New York on several occasions. Hilal had published a very weighty paper about the role of osmolality in pharmacological effects of contrast agents. Alas, he did not measure osmolalities of contrast agent solutions. Insouciant of making a circular argument, he determined something he believed to reflect them, namely the sodium chloride concentration that elicited in a bioassay the same effect as the contrasting solution (Hilal [1965] 1966). Nonetheless, his activities in the field procured him the chance to evaluate iocarmate dimeglumine, a dimeric molecule, and a trimeric version (Figure 4.26, **17**, **18**), compounds that could be expected to form solutions of relatively low osmolality. These compounds had been provided to him by Hoey from Mallinckrodt Chemical Works (see Section "Perfecting Ionic Uro-Angiographic Contrast Agents") (Hilal [1965] 1966, 1966, 1970; Hilal and Morgan [1968] 1970).[148]

The trimeric molecule was intriguing, since it possessed in addition to the three carboxylate groups with their negative charges, a tertiary ammonium group contributing a positive charge. This illustrated osmolality reduction by internal salification and was related to the idea, later likewise uttered by Almén (1969a), of salification of classical, negatively charged contrast agents with similar iodinated molecules that instead of negative charges, carry positive ones. The intellectually stimulating environment Almén was working in fostered maturation of concepts. The repertory of ideas on how to achieve improved contrast agents was enriched by the one involving intervention on the electric charges.

Here he must have begun to concretize his ideas about how to obtain low-osmolal uro-angiographic contrast agents through nonionic molecules. For him it was implicit that the X-ray contrast-generating core of the molecules had to be close analogs of the 2,4,6-triiodobenzoic acid at the origin of the ionic products then dominating clinical practice, since this allowed an elevated iodine content. He then integrated his ideas on solubility, osmolality, and dynamic viscosity into a theory of chemical contrast agent design. Somewhat naively extrapolating the theory of viscosity of colloidal and macromolecular solutions to solutions of small solutes, he reasoned that desirable molecules would need to be as spherical as possible to allow solutions of low viscosities.[149] In order to address the solubility problem, he hypothesized hydrophilic substituents, at first without exemplifying them.

Almén has told me, that while still working in Philadelphia, as sole author he had submitted a partly experimental paper on contrast agent design, sequentially to various radiological journals. But he only garnered rejections. This process certainly required several months. Reluctantly, he decided to try his luck with a non-radiological journal. On August 26, 1968, he submitted the manuscript for publication a first time to the *Journal of Theoretical Biology*. This allows dating his initial publication attempt at the latest to early 1968. The *Journal of Theoretical Biology* on April 29, 1969, accepted a revised version of the manuscript and published it finally in August (Almén 1969a). If for the moment we make the reasonable assumption that the finally published paper reflects what was already present in the earliest versions of the manuscript, we can gauge the state of development of Almén's ideas at the time of his first publication attempts. The title clearly announced what the text confirmed, the objective was "Contrast Agent Design. Some Aspects on the Synthesis of Water-Soluble Contrast Agents of Low Osmolality." It theorized three principle avenues leading to low-osmolal contrast media. In the first part, viscosity and osmolality measurements on carbohydrate polymers of different shapes together with matching theories were adduced to propose oligomers and polymers of existing monomeric ionic contrast agents of the second-generation. For viscosity reasons, the molecules were required to be small and as close to spherical as possible. In the second part, ionic contrast agents in which both the anion and the cation are iodinated molecules were briefly mentioned. Lowering of osmolality by internal salt formation (Wiegert 1968; Hilal 1970) found no entry into Alméns list of options. In the third part, monomers, oligomers, and polymers were theorized which would be nonionic (in his terms, nonelectrolytic) and water-soluble because the carboxylate groups in the ionic contrast agents have been replaced by not further specified nonionic hydrophilic substituents. A mere 5% of the manuscript space was dedicated to this category. I cannot avoid the impression that at least the third part of Almén's paper in the *Journal of Theoretical Biology*, which regarded nonionicity, was grafted onto the first and second one late in the process of manuscript preparation. Maybe this part was not yet included in the earlier manuscript versions. Be that as it may, in the period 1967 to early 1968 Almén began to shift his attention from macromolecularity to nonionicity of contrast agents. This period coincides also with the time in which through his own experimentation he gained insights into role of osmolality of commercial contrast agents in deleterious effects on the bat wing microcirculation and the time of his learning about the trimeric contrast agent in Hilal's laboratory. There exists no documentation contemporary with the events contradicting this assessment. As discussed later, Almén recollected otherwise.

In the course of the first of the aforementioned interview I had with him Almén has declared that prior to joining Nyegaard & Co. A/S in 1970, he did not follow the patent literature. He behaved that way despite the fact that he had himself already deposited a patent, one covering a medical device (Almén 1965b). He admitted further to having heard at an unspecified but early time, rumors about Czech patents describing water-soluble nonionic contrast agents. He must have referred to Swiss ones (Habicht and Zubiani 1962, 1963), which with their numbers preceded by the country code CH misled him about the national jurisdiction. He did not follow up on the hint at the time. He similarly ignored the patent applications on polymeric contrast agents that by then were publically accessible (Björk et al. 1966a,b). Actually, he continued to ignore patents on these contrast agents even in a historical account (Almén 1995) written long after he had been provided with detailed information about them. He practiced a "deletion of modalities" regarding patents widely sanctioned by time-honored academic tradition but for that no less criticizable.

Almén, as academic in 1968, was shielded from gloomy economic forecasts for the contrast agent business. Hence, despite frustration experienced in connection with his earlier European collaboration with industry, once he arrived in the United States, he tried anew to get industry interested in his ideas. Contacts with Mallinckrodt Chemical Works, and Winthrop Laboratories, Division of Sterling Drug, Inc., were initially unsuccessful. What, beginning in December of 1967, appeared to be an interest and a potential job offer by the latter company, by the end of February of the following year, had still not materialized. Almén had still not been given a chance to reveal his ideas (Amdam and Sogner 1994). Happily, renewed contacts with Nyegaard & Co. A/S turned out to be more fruitful.

Up to 1965 Nyegaard & Co. A/S had supported significant research in the area of contrast agents. Disillusionment with initial sales of the resulting proprietary product, metrizoate sodium (ISOPAQUE), and some disappointing results with the version reformulated with different counter ions (see Section "Perfecting Ionic Uro-Angiographic Contrast Agents"), in combination with a company-internal struggle for research priorities, led in 1966 to a reduction of the staff assigned to this area to Johan L. Haavaldsen as single chemist (Figure 4.34; see *curriculum vitae*), apart from the research director, Holtermann. But as champion of the contrast agent field in the company the latter was stubborn. He succeeded in reverting earlier decisions and the next year he was allowed to mount a renewed research effort. He

FIGURE 4.34   Torsten Almén's industrial counterparts at Nyegaard & Co. A/S.

From left to right: Hugo Holtermann, research director at Nyegaard & Co., with Torsten Almén's co-inventors of metrizoic acid-derived nonionic X-ray contrast agents, Johan Haavaldsen, and Vegard Nordal, photographed in 1961, 7 years before the beginning of the collaboration with Torsten Almén. (Reproduced with permission of Gyldendal Norsk Forlag SA, of cut-out of the figure on p. 78 of Amdam, R. P., and K. Sogner. 1994. *Wealth of Contrasts. Nyegaard & Co.—A Norwegian Pharmaceutical Company 1874–1985*. Oslo: Ad Notam Gyldendal.)

CURRICULAM VITAE

Holtermann, Hugo (b. December 22, 1916, Bergen, Fylke of Hordaland, Norway// d. February 5, 2003, Oslo, Fylke of Oslo, Norway)

Bergen Katedralskole, Bergen, Norway: High-school dipl. [Examen artium] (1936)/Technische Hochschule Karlsruhe, Germany: Stud. of chem. (1936–1939)/Norwegian Inst. of Technol., Trondheim: Stud. of Chem. (1939–1940 & 1945–1946); Dipl. in chem. engin. (1946)/Norwegian resistence fighter (1940–1941); German prisoner in Akershus, Norway, and Sachsenhausen, Germany (1941–1945)/Nyegaard & Co. A/S, Oslo, after 1988 called Nycomed A/S[117]: Partially funded graduate student detached to Oriel College, Univ. Oxford, UK (1947–1952); PhD [Chem.] (1952); Chem. (1952–1954); Head of Research Dept. (1954–December 1976); Researcher (January 1977–1983)/Honorary prize of the Royal Norwegian Council for Scientific and Industrial Research (1969)/H. M. The King's Order of St. Olav (1995).

Haavaldsen, Johan [Lyder] (b. November 18, 1928, Oslo, Norway //)

Vestheim Skole, Oslo: High-school dipl. [Examen artium] (1948)/Gøteborgs Tekniske Inst., Gothenburg, Sweden: Stud. of chem. (1951–1952); undergraduate engineering degree (1952)/Univ. Oslo, Norway: Stud. of chem. (1953–1959); cand. real. [MSc degree] (1959)/Nyegaard & Co. A/S, Oslo, after 1988 called Nycomed A/S[117]: Chem. (1960–1978); Head of Chem. Dept. (1978–1981); Res. Admin. (1981–1982); Safety and Security Manager (1983); Safety and Environment Manager (1984–1995).

Nordal, Vegard (b. September 9, 1929, Østre Gausdal, Fylke of Oppland, Norway; Norwegian nationality // d. November 6, 1986, Oslo, Norway)

Realexamen (General School Certificate), Norway/Stockholms Tekniska Inst., Stockholm, Sweden: Stud. of chem. (December 1950–April 1954); Undergraduate engineering degree (1954)/Nyegaard & Co. A/S, Oslo, Norway: Chem. (1956–1986)

decided to focus research on niche area products, including myelographic contrast agents that would be usable for the spinal canal and fluid-filled brain cavities. He especially cultivated contacts with Swedish radiologists, including Greitz and Boijsen, because of the elevated esteem that Swedish radiology commanded. He also remembered his earlier contacts with Almén. On the occasion of a business trip to Winthrop Laboratories, a Division of Sterling Drug, Inc., on February 2, 1968, he met with a wary Almén in a New York Hotel. Probably on that occasion Almén did not yet reveal chemical details of his ideas. Anyway, following the meeting Holtermann offered Almén the prospect of an intense collaboration with Nyegaard & Co. A/S (Amdam and Sogner 1994). A letter of agreement to Almén, Temple University Medical School, Philadelphia, from Ulf Blix, Nyegaard & Co. A/S, Oslo, dated April 29, 1968, formulated the contractual obligations and delineated Almén's participation through meetings and correspondence in a renewed research effort at the company. Almén remembered the period he had to wait for that document as agonizingly long, but then he took time until May 17 to countersign it. Maybe the fact has played a role that during the waiting period he had at last received the expected job contract proposal from the Winthrop Laboratories, Division of Sterling Drug, Inc.

While Almén's job future was in limbo, he attended the 2nd International Congress of Lymphography in Miami, Florida, March 15–20, 1968. The shop sign of a notary public caught his eyes. Cautioned by his earlier experience with industry contacts, and not having the data and means to request a patent, he decided to get some minimal protection for his ideas by notarization. He joted down ten pages, comprising one with chemical formulas (Figure 4.35). The next day he presented his document at the notary public's office to a sheriff-like figure with large hat, star on the shirt, and legs with gigantic cowboy boots placed on the table. A hilarious exchange with this assistant of the notary public followed. In a broad Texan accent

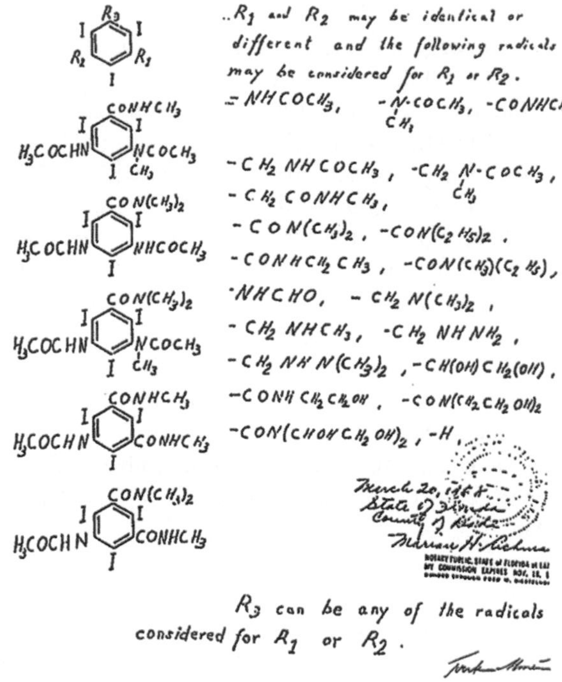
..R_1 and R_2 may be identical or different and the following radicals may be considered for R_1 or R_2.
$-NHCOCH_3$, $-N(CH_3)-COCH_3$, $-CONHCH_3$,
$-CH_2NHCOCH_3$, $-CH_2N(CH_3)-COCH_3$,
$-CH_2CONHCH_3$,
$-CON(CH_3)_2$, $-CON(C_2H_5)_2$.
$-CONHCH_2CH_3$, $-CON(CH_3)(C_2H_5)$,
$-NHCHO$, $-CH_2N(CH_3)_2$,
$-CH_2NHCH_3$, $-CH_2NHNH_2$,
$-CH_2NHN(CH_3)_2$, $-CH(OH)CH_2(OH)$,
$-CONHCH_2CH_2OH$, $-CON(CH_2CH_2OH)_2$
$-CON(CHOHCH_2OH)_2$, $-H$,
March 20, 1968, State of Florida, County of Dade
R_3 can be any of the radicals considered for R_1 or R_2.

FIGURE 4.35 Instructive page from the ten page affidavit by Torsten Almén signed on March 20, 1968, before a notary public of the County of Dade, State of Florida, USA.

Reproduction of the restored original version of the crucial page. (Reproduced with permission of Karin Almén, Falsterbo, Sweden.)

Almén was told: "Well, young man, my lack of education does not allow me to appreciate your work", upon which the addressee answered: "Don't apologize, educated people do not appreciate it." After being told to return when the notary public would be back, Almén was admonished: "You seem to be a smart young man, so pray to God and stay out of prison and there might be success with this" (Almén [1994] 2001). - Practical knowledge culture! On March 20, 1968, the affidavit that documented ideas on specific molecules was notarized and thus ready to be fully revealed to a collaborating company (Figure 4.35). Around the time of the deposition of the affidavit Almén and Wiedeman submitted for publication two manuscripts (Almén and Wiedeman 1968a,b). None of the chemical details in the affidavit entered the manuscripts. This avoided potential conflict over intellectual property. The affidavit and the secrecy around it, in combination with the restrained message in his *Journal of Theoretical Biology* paper, show that Almén had by then learned to care about his intellectual property.

Almén has told me that the complete affidavit comprised ten pages. It has never been made public in full. One page from it with chemical formulas has appeared in print a number of times, with various degrees of later alterations. Figure 4.35 shows a copy in which the primitive state has been restored.[150]

A puzzling fact is the perfectly identical benzene rings and their perfectly vertical alignment. It is hard to believe that Almén had the necessary drawing tools taken with him to Florida. This observation lets one doubt about the claimed spontaneity in the generation of the affidavit. Rather, we have to do with a blatant case of deletion of modalities. What in the affidavit page immediately becomes evident is Almén's turning away from macromolecules. The focus shifted from blood pool properties to pain avoidance. This marks the beginning of the second step in the formation of a new vision of contrast agents (Table 4.2).

The formulas in the affidavit are not straightforward to interpret. Some errors in chemistry, which Almén ascribed to the hurry in which the document had been compiled (Almén 1985), may be forgiven. Inexperience with the patent-specific Markush formulas he used may have tripped him up. Without knowledge of the explanations in the missing nine pages of the affidavit or later clarifications offered by its author, the only obvious commonality among the proposed molecules is the lack of a free carboxyl group. They are not negatively charged ionic in nature. But some substituents contain free amino or hydrazino groups that at neutral pH being protonated, create a positively charged ionic contrast agents. Possibly these positively charged ones were intended as radiopaque counter ions to classical, negatively charged contrast agents. Some proposed compounds are nonionic and bear hydroxyl-group rich substituents, while numerous others are nonionic but bear no hydroxyl groups. The latter are for sure poorly water-soluble. Completely on its own the single page from the affidavit reveals neither the intention to abolish ionicity nor that to achieve elevated water solubility mediated by numerous hydroxyl groups. Yet in historical papers Almén has always described his affidavit as doing just that, and documented it with the mentioned page, supplemented by emphasizing marks (Almén 1985, [1994] 2001). Only the missing nine pages of the affidavit could clarify the issue. In any case, for the chemist Holtermann it was trivial to filter the information on the affidavit page for relevance to Almén's orally expressed desire for low osmolality and nonionic contrast agents. The whole became effectively condensed into new design principles. These maintained some from earlier generations and modified others. They integrated what was proposed in the *Journal of Theoretical Biology* with what was found in the affidavit, while also giving them a narrower focus. The products that obeyed the design principles eventually became known as third-generation contrast agents (Table 4.4).

The design principles were accompanied by selection criteria which too had remained unchanged from earlier generations of contrast agents. But new criteria

---

TABLE 4.4

Design Principles for Third-Generation Contrast Agents

3a) One or two substituted 2,4,6-triiodophenyl moieties form the molecular core that is responsible for stability of aqueous solutions to heat sterilization.

3b) The iodinated compound bears no ionizable group and is rendered water-soluble by non-ionizable hydrophilic substituents on the 2,4,6-triiodophenyl moiety, especially substituents rich in hydroxyl groups.

3c) The I/P ratio is at least three.

had been added. The new ones included low osmolality and low viscosity of contrast agent solutions. After the recognition that solutions of compounds with the same I/P ratio and the same iodine concentration could have widely different osmolalities, this property assumed a central role in profiling of commercial contrast agents and so did nonionicity. The properties of the solutions assumed more prominent roles than those of the solid contrast agent.

Whereas in the cases of first and second-generation contrast agents, identification of design principles retrospectively generalized the experience accumulated while hunting for a prototype molecule of a new class, in the case of third-generation products the design principles were formulated ahead of identification of a valid prototype. They captured the ideas of a strongly motivated radiologist but possessed only weak anchorage in experience. It came as no surprise then, that the great majority of substances having chemical structures in accordance with the new design principles were unsatisfactory in exactly those aspects the principles were expected to resolve, i.e., solubility in water, viscosity of solutions at needed concentrations, or toxicity. What the principles really did accomplish is to stake out a field within which suitable compounds could be found by screening. In addition, it was found that the osmolality of solutions could be substantially lower than predicted by the combination of concentration and I/P ratio. As a consequence, suitable candidate contrast agents had to be selected by trial and error for low measured osmolalities in solution, besides the classical ones of low toxicity.

In the first of the aforementioned interviews I have conducted with Almén, he recollected to have developed the concept of nonionic contrast agents in the same time frame as polymeric ones, i.e., in 1965. In the interviews Fondberg conceded that he, although not alone, talked about the subject around that time. Almén affirmed to his end to have had already then been inspired by the elevated water solubility of sugar and alcohol to imagine nonionic contrast agents rendered highly water by substituents bearing numerous hydroxyl groups. There exists no objective corroboration for this but, as earlier discussed, some reasons for doubting it. Yet even if his recollection was endorsed, it can be concluded that only at a much later time did he learn to arm his proposals with sufficient concreteness to make efficacious communication of his ideas possible. Independent of the exact timing of his ideas his crucial role in the emergence of what was to become the third-generation contrast agents is undeniable.

Following the May 17, 1968, collaboration agreement of Almén with Nyegaard & Co. A/S he was invited for a meeting to the company's headquarters and research and development facilities in Oslo. In his meeting with the company's research staff in June, he revealed the ideas contained in his affidavit. Almén reports having disclosed to Holtermann and collaborators molecules from his affidavit with eight aliphatic hydroxyl groups (Almén [1994] 2001). All meeting participants received Almén's expectation of sufficient water solubility with skepticism. Despite this situation, Holtermann decided to explore in the proposed direction, but he kept the choice between angiographic and myelographic application open.

Vague as Almén's proposal was, it exceeded in focus and concreteness by far what was formulated as contrast agent design theory in the manuscript he was trying to publish and which more than a year after the meeting finally appeared in print (Almén 1969a). That design theory addressed mostly the benefit for osmolality of oligo- and polymerization of available monomeric ionic contrast agents and how the associated problem with viscosity could be minimized. Almost as an afterthought only, it considered achievement of low osmolality by mixing positively and negatively charged contrast agent molecules or by replacement of the carboxyl group in existing contrast agents with some nonionic hydrophilic one. Noteworthy is that there was no mention of hydroxyl groups.

The hydroxyl-group-bearing, nonionic hydrophilic substituents Almén proposed in his affidavit as mediators of elevated water solubility were essentially the same as those previously incorporated into more than 70 ionic compounds (Hoey, Wiegert, and Rands 1971), among which also commercial ionic contrast agents. Before Almén ever took up research, Holtermann and his collaborators had patented contrast agent molecules containing multiple N-β-hydroxyethylated acylanilinic functions, comprising some diacidic hexaiodinated dimeric molecules (Holtermann et al. 1961). Of note, the inventors stated that such hydroxyl bearing groups are expected to have water-solubilizing and detoxifying effects. Björk, Erikson and Ingelman had included similar observations in a published patent application (Björk et al. 1966a). In iodamide 1-gycerol ester (Figure 4.33, **24**) hydroxyl groups were even utilized with the intention to convey elevated water solubility to a nonionic contrast agent (Fedeli 1964; Felder 1969), although the latter effort was never made public. We know today that no matter how many hydroxyl groups are introduced into monomeric nonionic iodinated molecules, the vast majority of compounds are poorly soluble. As the mentioned case of iodomethanesulfonic acid serinolamide and isoserinolamide illustrates, the same number of hydroxyl groups in very similar configuration can lead to drastic differences in water solubility. Thus, no generally valid recipe for achieving elevated water solubility had in fact been identified. Nothing wrong with this! In research, again and again, following an intuition that eventually turns out to be totally or partially incorrect, has nonetheless led to progress. Suffices to admit that among the type of compounds envisioned by Almén, a number could be pharmaceutically formulated at the desired elevated concentrations.

In connection with the launch of the first water-soluble nonionic contrast agent, metrizamide, Holtermann wrote: "An original approach, intended to overcome the disadvantages of high concentrations of ions as well as high osmolalities, was suggested to us by Almén, who proposed the synthesis of nonionic monomeric iodinated compounds, …", and "Almén proposed that a proper choice of polar functional groups would confer water solubility on the nonionic molecule …", and "The research staff of Nyegaard & Co. selected as a basis for their investigation one of the polar groups suggested by Almén, the hydroxyalkylamido group" (Holtermann 1973b). So much reference to Almén in view of an input that in originality was modest indeed raises the question of motive. The most straightforward answer, namely that to Holtermann's research group Almén's input was truly a revelation, is unconvincing. At the time, the group was not missing the knowledge Almén conveyed, but it was low on confidence that sufficient water solubility could be achieved. In general, the group had to regain the trust of top management regarding its ability to create new products after a period of scarcity. Demonstration of openness to ideas from outside the company is a widely practiced management response to this type of challenge. Grouping activities into an initiative with a label is another. Almén was already known to Ulf Blix. As academic radiologist at a prestigious Swedish hospital, specialist in angiography, person with working experience in America, long-time and ardent promoter of the search for better contrast agents and newly proponent of chemical compounds, Almén could simultaneously be a valuable scientific collaborator from the outside and serve as a visible spearhead for the new initiative. This interpretation offers a plausible explanation for Holtermann's cited heavy reference to Almén. Later developments reinforced the interpretation. Almén became a highly visible champion of nonionic contrast agents and was in the end talked up as their unique father (Nyman, Ekberg, and Aspelin 2016a,b; Tweedle 2016). He enjoyed that standing, and Nyegaard & Co. A/S and its successor companies could profit from him not just in terms of technology but additionally for promotional purposes. Not least, by co-opting Almén the company avoided the risk of a potentially detrimental priority dispute of the kind that earlier had involved uroselectan sodium and Swick, Binz, von Lichtenberg and von Schickh. The story follows a frequently encountered pattern.

The affidavit is a singular document from the hands of a radiologist, and enables an assessment of how far Almén on his own had converted his infatuation with the physicochemical origin of undesired effects of contrast agents and his deductions on the physicochemical parameters of desired molecules into concrete chemical proposals. Yet, we must conclude that he needed more than a pair of hands in chemical synthesis to turn his concepts into practice. At Nyegaard & Co., A/S he found the necessary competent collaborators and their early endeavor was accompanied by some luck. Based on an analogical argument the chemist Vegard Nordal (Figure 4.34; see *curriculum vitae*) proposed that the passage from the meglumine salt of acetrizoic acid, a salt variant of the well-known contrast agent UROKON (Figure 4.36, **14bis**), to the covalent neutral molecule, acetrizoate meglumine amide, (Figure 4.36, **26**) would be a way of rapidly testing the ideas at hand (Almén 1985).

Within 2 or 3 months of the June 1968 meeting between Almén and the research staff at Nyegaard & Co. A/S the nonionic compound proposed by Nordal was synthesized. Luckily it had a very elevated water solubility. Thereby it gave the investigators confidence in the feasibility of the proposed technology. Unfortunately, its solutions were too viscous and its acute toxicities not favorable (Almén 1985). Many analogous products of condensation between available ionic contrast agents and various aminosugars and hydroxyalkylated amines were synthesized in rapid succession. A large fraction of them showed insufficient solubility. Water solubility of a substance, besides depending on its hydrophilicity, depends on how energetically favorable is its aggregation into a crystal or other separate phase, and this property escapes predictability. In any case, luck had it that within a few months several examples with good water solubility and unexpectedly good neurotolerability were found (Almén 1980, [1994] 2001; Holtermann 1973b).

On June 27, 1969, a patent covering many of these compounds was applied for. It demonstrated the crucial reduction to practice of one of Almén's ideas, namely that the replacement of carboxyl groups in ionic contrast agents with substituents

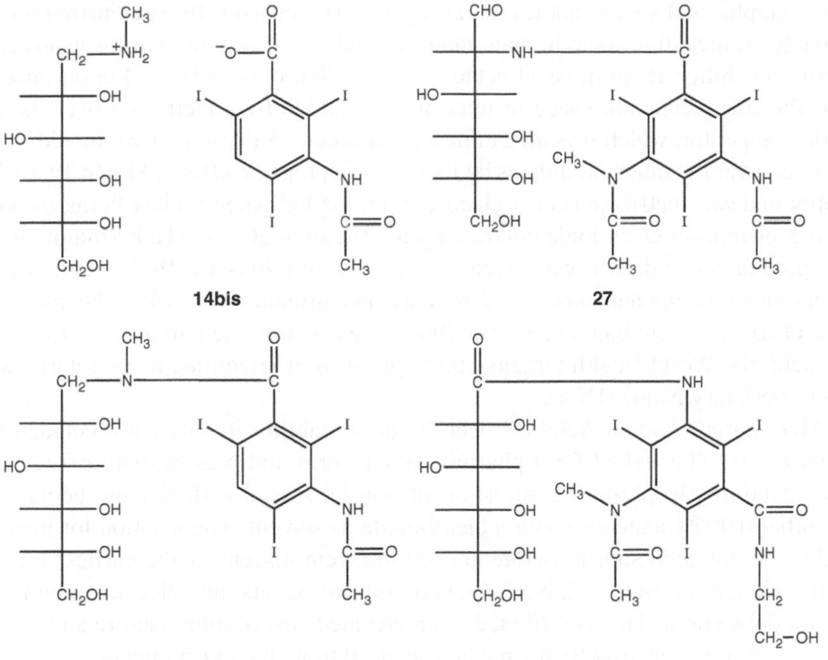

FIGURE 4.36 The transformation of Torsten Almén's concepts of nonionic contrast agents into reality by the chemist Vegard Nordal.

14bis: Acetrizoic acid *N*-methyl-D-glucamine (meglumine) salt variant of UROKON™ (14). *I/P* = 1.5.
26: Acetrizoic acid *N*-methyl-D-glucamide. *I/P* = 3.0.
27: Metrizoic acid D-glucosamide = metrizamide (AMIPAQUE™, Nyegaard & Co. A/S, Norway). *I/P* = 3.0.
28: Ioglumide; Guerbet compound P297 (Tilly, Hardouin, and Lautrou 1973). *I/P* = 3.0.

rich in hydroxyl groups can yield examples with elevated solubility and solutions of low osmolality with favorable toxicities. Almén and the company chemists Nordal and Haavaldsen figured as inventors and Nyegaard & Co. A/S as assignee (Almén, Nordal and Haavaldsen 1969). Nonionicity, low osmolality, and low toxicity of contrast agents *per se* were not the patentable features, chemicals with specified structures possessing these properties were.

The overall positive experience with the suggestion by Almén to achieve elevated water solubility of nonionic contrast agents by replacing carboxylate groups of classical ionic molecules with hydroxyl-group-rich nonionic substituents delivered the third step in the emergence of a new technological vision (Table 4.2). For a full-fledged vision that could bring on board the business arm of companies as well, demonstration of general uro-angiographic products was still outstanding.

Numerous compounds falling under the patent protection were synthesized und toxicologically screened. Five of them made it to the shortlist to become myelographic and angiographic contrast agents (Holtermann 1973b). Further preliminary tests, including some in man, made a particular one stand out for its excellent neurotolerability. Its intrathecal acute median lethal dose $(LD_{50,\ i.th.})$ upon injection into the subarachnoidal space in mice was favorable by a factor 4.3 over the best ionic competitor, which was meglumine iothalamate (Salvesen 1973). In addition, it excelled over the other candidates by its low epileptogenic effect (Almén 1980). The compound was metrizoic acid-D-glucosamide, the iodinated moiety being the company's own proprietary ionic contrast agent (Figure 4.36, **27**) (Holtermann 1973b). For preclinical studies, it was given the number 16 (Gonsette 1973a), and because of its sugar component and good toxicity performance it earned the nickname "sweet sixteen" (Amdam and Sogner 1994).[151] Reflecting the parentage with metrizoic acid, the World Health Organization approved metrizamide as its International Nonproprietary Name (INN).

Metrizamide had an Achilles' heel. Its poor stability in solution precluded heat sterilization. This called for a pharmaceutical form and presentation as two vials, one containing lyophilized contrast agent powder mixed with calcium edetate and the other 0.005% aqueous sodium bicarbonate as solvent. The solution for injection had to be prepared shortly before application, reminiscent of the earliest versions of the pioneering water-soluble iodinated contrast agents, uroselectan sodium, and methiodal sodium. This contributed to an elevated cost of manufacture and made it less practical for the user than what he was used to with older products.

Market surveys had suggested a lack of medical need for less toxic uro-angiographic contrast agents but one for myelography. Based on these, Holtermann predicted that metrizamide had the potential to monopolize the myelography market. He also guessed that the new contrast agent could command a price elevated enough to justify the expected manufacturing costs, all the more myelographic doses being small, production was technically less demanding than for large angiographic doses. Thus, at the end of 1969 Nyegaard & Co. A/S decided to develop metrizamide, to begin with for myelographic application, in line with the niche market strategy already in place. The angiographic indication could follow.

Holtermann's primary role was managerial, but he contributed also scientifically, not least as crucial mentor of the research activities and collaborations. Very important was his decision to prepare the company for measuring true osmolalities of concentrated contrast agent solutions, something the previous method, cryoscopy, could not deliver. In preparation, he commissioned an in-depth study of osmolalities of ionic contrast agents by the Central Institute for Industrial Research, Blindern, Oslo (Børdalen, Wang, and Holtermann 1970). In 1961 the Mechrolab Model 301A osmometer (Mechrolab Inc., Mountain View, CA, USA), the first commercial thermoelectric vapor pressure osmometer, revolutionized osmometry. The new technique was used in the study and allowed measurements at the physiological temperature. At last, accurate and physiologically meaningful osmolalities became easily accessible. Regrettably it had taken a decade for this easy and reliable method to begin arriving in the contrast agent research laboratories. It took several more years, before reliable osmolality measurements appeared regularly in patent applications and scientific papers on contrast media. Only gradually did researchers realize that contrast media behaved as highly nonideal solutions, meaning that real osmolalities at the high concentrations of commercial formulations could not even crudely be estimated just from molalities and I/P ratios. At the same time, they began to realize that precisely measured values could help select improved substances and give existing ones a notable clinical and commercial profile.

The problematic reception history of osmolality measurements and interpretations involving the results is best illustrated with the situation at Nyegaard & Co A/S. In preparation for the commercialization of metrizamide, an impressive effort in toxicology and safety pharmacology, both in and outside the company, was mounted. Its dimension may be gauged looking at the more than 40 articles on metrizamide in the special journal issue that helped prepare its commercial launch (Holtermann 1973a). In the introductory paper, Holtermann for the first time reports the osmolalities of metrizamide solutions (Holtermann 1973b). At the concentration 280 mg(Iodine)/mL the osmolality was only 460 mosmol/kg, as opposed to 1,460 mosmol/kg of metrizoate meglumine at the same iodine concentration. At a concentration of 170 mg(Iodine)/mL it was isotonic with blood. Neither the method of measurement nor the temperature were specified, but the similarity of the value for the ionic product with that in the commissioned study (Børdalen, Wang, and Holtermann 1970) allows the conclusion that vapor pressure osmometry at 37° was performed. Only in 3 of the rest of the 40 papers terms related to osmotic properties appear. Only in one a control substance at the same osmolality as the metrizamide solution was included in the study. The strongest statement regarding the role of osmolality in biological effects by an author other than Almén regarded suboccipital toxicity in cats and read: "The difference in tonicity between Conray Meglumine and metrizamide must therefore, at least partially, be responsible for the less toxic effect of the latter" (Oftedal 1973).

In October of 1968, Almén had returned from the USA to his *alma mater* accepting a radiology position at Malmö General Hospital. He traveled frequently to Oslo to visit Nyegaard & Co. A/S. Two years gone he requested a leave of absence from his hospital in order to fill for a limited time a gap in the company's staff and in this function to carry out studies of safety pharmacology on metrizamide. His hospital superior, Sölve Welin, threatened to fire him but retracted the threat when he realized that Almén would follow his inner call anyway. Almén thought he could always return to radiology after a stint in industry. So, from 1970 to 1971, the angiographer Almén worked as pharmacologist at the company on the preclinical development of metrizamide. For the first time, he gained access to trustworthy osmolality measurements, but very little was made of it. Having terminated this stint in industry he returned to Malmö General Hospital, where he continued work on the pharmacology of metrizamide in animals and later in man. Ever since Almén had participated in the development of this trendsetting contrast agent he continued to research the field and interact with the company and its successors. Numerous joint publications and patents resulted.[152]

Nyegaard & Co. A/S had early on involved the neurologist Richard E. Gonsette from the Catholic University of Louvin, Belgium, in some assessments of the neural tolerability of nonionic contrast agents (Gonsette 1973a). Initial clinical trials of metrizamide for myelography began in April 1972. The investigators included Gonsette (Gonsette 1973b), as well as Per Amundsen, chief physician at the Ullevål Hospital in Oslo with his collaborators (Amundsen et al. 1973). Many other radiologists participated in testing of the product's clinical utility. Already in February and March 1974 the applications for approval by the Swedish State Pharmaceutical Laboratory had been submitted. By the end of the year, the new product was approved for myelography at a concentration in the range from 170 to 190 mg(Iodine)/mL and an osmolality barely above that of blood. Metrizamide (AMIPAQUE), to begin with commercially launched in Sweden in 1974, was not only the first commercial water-soluble nonionic contrast agent but also the first water-soluble compound that could safely be injected into the entire subarachnoid space. Yet, because of its residual neurotoxicity, it was still far from fulfilling the potential of this class of nonionic compounds.

In terms of number of contrast agent doses, the myelography market was small. Only a price dramatically higher than that of available products promised a viable economic return on the investment. However, myelographic doses were small in comparison to uro-angiographic ones. This allowed contrast agents whose costs of production were substantially higher than those of the classical ones. And it worked: metrizamide became the myelographic contrast agent of choice and earned its owner a decent return (Amdam and Sogner 1994). It helped the company improve its worldwide direct or indirect sales presence.

Among the expected benefits of nonionic contrast agents for uro-angiographic applications were a decreased diuresis and consequent elevated urinary iodine concentrations, a lower vasodilation, a reduced increase in pulmonary arterial pressure and lesser pain on intraarterial injection (Almén 1980), a performance that, taken alone, predestined metrizamide for uro-angiography. In July 1977, it was decided to develop metrizamide additionally for vascular use (Andrew et al. 1981; Amdam and Sogner 1994). In order to remain compatible with the case of the myelographic dose, the price for the angiographic dose had to be 20–30 times that of conventional ionic competitors. This was simply too expensive for general uro-angiographic use. Thus, in 1978 metrizamide at a concentration of 370 mg(Iodine)/mL was recommended only for selected vascular examinations, such as coronary angiography (Amdam and Sogner 1994). It had only very limited success. This overall result was a far cry from the hopes Almén had set for the new product.

In August 1969, Almén had gotten his design theory published in the *Journal of Theoretical Biology* (Almén 1969a). Its title was "Contrast Agent Design. Some Aspects on the Synthesis of Water-Soluble Contrast Agents of Low Osmolality." Although this journal is quite respectable, it was not usually monitored and read in the community of contrast agent specialists. Nonetheless, through various abstracting services, both academics and industry people could follow developments even in such journals. Indeed, in January of 1970 Zutter at Eprova AG spotted in *Chemical Abstracts* the vacuous summary of Almén's article (Almén 1969b). Felder at Bracco Industria Chimica SpA was alerted and he procured a copy of the article that had been published half a year before. Citing it in June of 1970 he testified to his awareness of the article (Felder 1970). It is likely he also learned that Almén had ties to Nyegaard & Co. A/S. Since the article reported scientifically nothing new, the main message received was that an academic radiologist with his call for low-osmolal and nonionic contrast media might influence the research program of a competitor company. In any case, we can be sure that at least the two mentioned companies, and probably Laboratoires André Guerbet & Cie too, profited from the journal article's publication a year and a half before the earliest time at which the pertinent patent application (Almén, Nordal, and Haavaldsen 1969) became available, i.e., on

January 7, 1971. In contrast, it appears that researchers at Schering AG remained in the dark until after April 1972, since a dissertation completed in their laboratories at that time, which cites extensively any major idea in the course of the contrast agent history, does not refer to Almén (Fischer 1972). Probably this fact reflects the status of uncertainty regarding any further commitment to the field in that company.[153]

FIGURE 4.37 *Ernst Felder.* (Courtesy of Ernst Felder.)

CURRICULAM VITAE

Felder, Ernst [Hermann] (b. January 19, 1920, Zurich, Canton of Zurich, Switzerland // d. April 6, 2018, Oberrieden, Canton of Zurich, Switzerland)

Kantonsschule Zurich: High-school dipl. [Matura] (1938)/Univ. Zurich: Stud. of chem. (1938–1944); Auxiliary Asst. (1941–1943); Private Asst. of Prof. G. Schwarzenbach (1943–1944); PhD (1944)/Cilag AG, Schaffhausen, Switzerland: Head of Process Development Lab. (1944–1950); Deputy Director of Research Lab. (1947–1950)/Cilag Italiana SpA, Milano: Head of Factory (1950–1955); Director of Research (1953–1955)/Industria Chimica Dr. Fulvio Bracco SpA [ex-Cilag Italiana], Milano, becoming in 1958 Bracco Industria Chimica SpA, in 1992 Bracco SpA and in 2001 Bracco Imaging SpA: Manager of Research and Production, (1955–1960); Technical Director (1960–1969); Research Director (1969–1986); Director, R&D Div. (1986–1996); Bracco International SA, Lugano, Switzerland, Sole Administrator (1991–1996); Consultant (1996–); Various Bracco daughter companies; Member of the Board of Directors of (1988–).

On March 3, 1969, i.e., before the appearance of Almén's theoretical paper (Almén 1969a) Felder at Bracco Industria Chimica SpA had written a report for company use only, in which he alluded to the company's own water-soluble nonionic compound, iodamide 1-glycerol ester (Compound **24**). The following perspective followed a paragraph dedicated to the possibility of improving ionic contrast agents by altering their cationic composition:

> In the more remote future, the study of a contrast agent that does not generate ions seems to us to be a worthwhile goal. From such a product, we would expect a less marked disturbance of the electrolyte equilibrium of the vital organs. For intravenous pyelography, there is actually no such product in view, but it could mark an important development in angiocardiography and cerebral angiography. Some early chemistry studies have been completed towards this goal

(Felder 1969)[26]

It appears that when contemplating nonionic molecules in 1969, electrolyte balance between their solution and tissue, and not the osmolality issue raised by Habicht (Habicht and Zubiani 1962) and finally placed at the core of interest in the course of the interaction of Almén with researchers at Nyegaard & Co. A/S (Almén, Nordal, and Haavaldsen 1969), was foremost on Felder's mind. But his perception soon changed. In June 1970 Felder gave a talk at the Italo-Soviet Symposium on New Developments in the Field of Contrast Agents in Milan. Based on the published record (Felder 1970) by that time he must have seen Almén's theoretical paper. The passage of interest here went as follows:

> Another factor that is important for all angiographic contrast agents is the osmolarity of their solutions. Various authors considered the strong hyperosmolarity with respect to human plasma of currently used angiographic agents the cause of undesired vascular reactions and agglutination of red blood cells. The diminution of osmolarity is therefore one of the objectives in the search of novel angiographic contrast agents. There exist various ways to diminish the ratio of osmolarity/iodine content of the solution, which we would like to discuss briefly. The osmolarity of current angiographic contrast agents is determined to the extent of 50% by the iodinated anion and to 50% by the cation, sodium or meglumine, which have only a solubilizing function without contributing to opacification. If the cation could be eliminated there would be a reduction of the osmolarity to one half, without influence on the opacification potency. For the chemist, this opportunity means to introduce into the iodinated molecule non-dissociative solubilizing groups. As an example, we cite the glycerol ester of iodamide, B 6590 (Fig. I), which forms supersaturated solutions, but is not yet of practical interest.
>
> Another possibility consists in the synthesis of ampholytic opacifiers, i.e. molecules containing internal salts (Fig. II). Also, such compounds have at the same level of opacification half the osmolarity of the traditional salts.
>
> A third way to reduce the ratio osmolarity/iodine content is given by the synthesis of dimers, oligomers, or polymers of iodinated molecules till arriving at colloidal solutions or suspensions, respectively emulsions of iodinated substances. Recently Björk, Erikson and Ingelman have described clinical trials with a dimer in angiocardiography and in thoracic aortography. In our laboratory, we have experimented with the dimer of iodamide B 9620 (Fig. III)

(Felder 1970)[154,155]

Now the osmolality issue took center stage, but nonionic molecules were just one way among a number of others that addressed the issue. Almén's article was initially perceived as just another systematic listing of options known to be very difficult to realize. The document of Felder shows not even minimal signs indicative of an expert describing a field on the verge of a revolution.

In following the research assignment given to him by Felder from Bracco Industria Chimica SpA, Suter with the help of his collaborator Zutter at Eprova AG, addressed the problem of innovative uro-angiographic contrast agents. Being informed about the nonionic option explored by their former colleagues at Cilag-Chemie AG, Habicht and Zubiani (1962), but not yet able to know Almén's ideas, in 1968 they began searching for nonionic molecules of their own. This early activity is documented by a bill dated March 1, 1968, to Bracco Industria Chimica SpA, Milan, for execution of experiments regarding the synthesis of amides of diiodofumaric acid (Figure 4.33, **25**), and was also recollected later (Suter and Zutter 1975). The compounds turned out not to be promising because of low solubility and poor tolerability. In early 1970, N-methyl-D-glucamides of acetrizoic acid and of a substantial number of aminotriiodobenzoic acids, bearing on the amino groups various substituents, were synthesized, exactly paralleling the then still undisclosed research at Nyegaard & Co. A/S. By bad luck, all compounds turned out to be insufficiently soluble in water. It is interesting to contemplate what would have happened if instead of stumbling with their first synthesized compound upon a highly water-soluble one, the group of Almén, Holtermann, Nordal, and Haavaldsen had had the bad fortune of Suter and Zutter.

As a sideline, amides of methiodal sodium and dimethiodal sodium with hydroxyl-group-bearing amines were prepared. Zutter has told me that it was Felder who suggested exploration of amides based on serinol. We can be sure that in his thoughts, and perhaps even in his communications, the constitutional isomer isoserinol was included, all the more that the corresponding amide moiety was analogous in structure to the 1-gycerol ester moiety investigated in his own institution (Figure 4.33, **24**). Since isoserinol was more accessible than serinol, it received attention first. Given the low water solubility of the isoserinol derivative of methiodal (Figure 4.38, **29**), the excellent solubility of the corresponding serinol derivative (Figure 4.38, **30**) came as a pleasant surprise.

Based on their low-iodine content, methiodal derivatives were never expected to provide contrast agents of noteworthy low osmolality at iodine concentrations required for uro-angiography and were as a result never intended for that clinical indication. Methiodal sodium was at that time mostly used in myelography. Thus, by analogy it was thought that the nonionic methiodal serinol amide might serve the same purpose but hopefully with an improved tolerability. Some methiodal derivatives with excellent neuropharmaco-toxicological properties were indeed obtained. Unfortunately, methiodal serinol amide (Figure 4.38, **30**) lacked satisfactory chemical stability and was in consequence abandoned.

On November 17, 1971, a patent application disclosing the various nonionic contrast agents conceived and synthesized at Eprova AG was filed (Suter, Zutter, and Müller 1971), with Suter, Zutter, and the young chemist Hans Rudolf Müller as inventors. It is hard to imagine that the persons involved had failed to apprehend

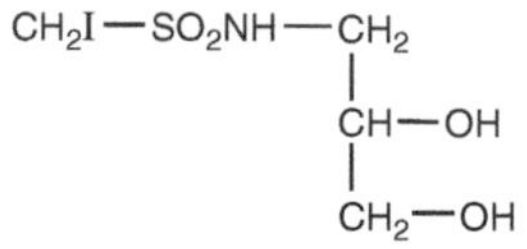

$$CH_2I-SO_2NH-CH_2$$
$$|$$
$$CH-OH$$
$$|$$
$$CH_2-OH$$

29

$$CH_2I-SO_2NH-CH \overset{CH_2-OH}{\underset{CH_2-OH}{<}}$$

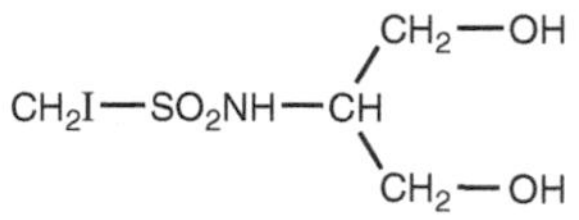

30

FIGURE 4.38 Another transformation of Torsten Almén's concepts of nonionic X-ray contrast agents to chemical reality. Amides of methiodal.

29: Iodomethanesulfonic acid isoserinol amide (Suter, Zutter, and Müller 1971). $I/P = 1.0$.
30: Iodomethanesulfonic acid serinol amide (Suter, Zutter, and Müller 1971). $I/P = 1.0$.

in time the earliest publication of the patent application of Almén, Nordal, and Haavaldsen, namely the Deutsche Offenlegungsschrift of January 7, 1971 (Almén, Nordal, and Haavaldsen 1969; Almén, Haavaldsen, and Nordal 1971). Nonetheless, they requested patent coverage for some molecules, although without exemplification, which appear in the application of Almén and collaborators. Such a difficult intellectual property situation alone did not necessarily have to be the end of the project; there existed always the possibility of acquiring a license for a compound that is worth it. More importantly, the derivatives of dimethiodal were not covered by the patent application of Almén and collaborators. Soon after Eprova AG had filed its application radiological discovery research ended, a prelude to the change of company ownership that took place in 1973.[135] For Bracco Industria Chimica SpA there remained as an important lesson for the future the positive experience with serinol as molecular appendage for achieving water solubility of nonionic contrast agents. Moreover, the fact that serinol was not widely available on the market offered the chance of opening a field of industrial activity with potential for exclusivity.

In 1970, Bracco Industria Chimica SpA was economically in a tight spot.[156] As documented by an undated, handwritten internal memorandum addressed primarily to the directors of research laboratories, Felder, and of medical affairs, Franco Bonati (Figure 4.41; see *curriculum vitae*), around that time the company's president, Fulvio Bracco, made the strategic decision to abolish ongoing discovery research in therapeutic drugs at his company and to concentrate all research and development resources in the contrast agents sector, except where ongoing contractual obligations did not permit immediate implementation. This decision was based on the realizations that (a) with its small size Bracco Industria Chimica SpA needed to focus on niche products, which it could manufacture for a worldwide supply, and (b) the dwindling number of competitors in the field engendered the possibility to be a major player in a small group. While the first of the two realizations led to the reorientation of research and development, the second would lead in early 1972 to separate license- and research collaboration agreements with E. R. Squibb & Sons Inc.

Thereby that company gained distribution rights in an important portion of the world for an existing uro-angiographic Bracco product, the ionic iodamide, and access to patented liver-directed diagnostics in development (iodoxamic and iopronic acid), as well as to all future products resulting from co-financed research at Bracco Industria Chimica SpA. The research support would eventually last 6 years.

In a company-internal talk, Felder gave on September 29, 1988, he described the beginning of the effort that eventually lead to success. With the help of literature surveillance tools Zutter at Eprova AG came in January 1971 to know of the German patent application for nonionic contrast agents of Almén, Nordal, and Haavaldsen from Nyegaard & Co. A/S (Almén, Nordal, and Haavaldsen 1969).[157] Felder at Bracco Industria Chimica SpA was alerted. Publication of this application, not the earlier article in the *Journal of Theoretical Biology*, hit Felder and his colleagues at Eprova AG, like a bomb. Here success of an approach was described that in their own hands had been deluding. The description of numerous iodinated compounds bearing hydroxyl-group-rich substituents and forming concentrated solutions with low osmolalities, triggered a profound rethinking of research goals. Suddenly nonionic contrast agents forming low-osmolal solutions potentially suitable even for angiography seemed not only achievable but were possibly already in industrial development by the competition. The insight sank in that the approach was more promising than that of ionic dimers and oligomers. At first adequate manpower to pursue the new direction was not available because development for the whole world of the biliary contrast agents, iodoxamic, and iopronic acid, activities connected to the research agreement with E. R. Squibb & Sons Inc. had absolute precedence.[158] The most that could be done was to resuscitate the nonionic candidate contrast agent, iodamide 1-gycerol ester (Figure 4.33, **24**) discovered in 1964. Due to its tendency to crystallize, renewed testing was limited to suspensions of it, mainly as a potential lymphographic contrast agent. Nothing came of it.

Triggered by the news from Nyegaard & Co. A/S, SA Laboratoires André Guerbet pursued a two-pronged search for low-osmolal contrast agents. On the one hand, the goal of suitable nonionic molecules was addressed by making chemical variations on the theme of metrizamide, i.e., the sugar glucuronic acid was linked to one of the anilinic nitrogen atoms of a substituted 2,4,6-triiodo-3,5-diaminobenzoic acid (Figure 4.36, e.g., ioglumide, **28**). Guy Tilly, Jean-Charles Hardouin and Jean Lautrou asked for a patent on December 7, 1973 (Tilly, Hardouin, and Lautrou 1973). The compound was chemically not very stable. It suffered from some of the same defects as metrizamide. In spite of favorable toxicity, it was never fully developed as a product. On the other hand, the same individuals studied monoacidic dimeric molecules. Although being ionic molecules, they have an *I/P* ratio of 3. Giving preference to this approach over nonionic compounds, the investigators pursued a path all on their own towards low osmolality. With the monoacidic dimeric ioxaglate mixed salt of meglumine and sodium (HEXABRIX) (Figure 4.28, **20**) the company had arrived at a valid exemplar of this class that in due course resulted in a substantial clinical and commercial success. At the same time, it missed the opportunity of being among the companies offering nonionic products, a category that defined the up to date standard. For a long time, the company was forced to play a most problematic catch up game.

Also among a few academics, there existed awareness of the nonionic contrast agents in the making. Sovak, in 1974 associate professor of radiology at the University of California in San Diego, began exploring nonionic molecules containing sugar moieties. He focused on molecules in which sugars were linked through glycosidic bonds to an iodinated aromatic nucleus (e.g., Sovak et al. 1975; Weitl, Sovak, and Ohno 1976; Weitl et al. 1976). The approach did not yield practical compounds.

Toward mid-1974 at last Bracco Industria Chimica SpA could liberate capacities for new research initiatives. Felder fully recognized in nonionic contrast agents, beyond the probable medical benefit for angiography identified by Almén, the enormous industrial potential created by replacement of the old-generation uro-angiographic products by better ones. Such potential was just what he needed to fulfill the wishes of his employer. The coincidence of the company's strategic decision and the prospect of a novel class of products created the optimal base for an intensive pursuit of the new ideas. Publications concerning metrizamide and its launch in 1974 injected an additional sense of urgency into the efforts.

The great breakthrough occurred, when Felder and Davide Pitrè (Figure 4.39; see *curriculum vitae*), respectively, director and vice director of the research laboratories at Bracco Industria Chimica SpA, had a series of related compounds synthesized, among which figured as the very first, iopamidol (Figure 4.40, **31**). Actually, iopamidol (Laboratory code B-15000 or MB 224D) had been synthesized in June of that year (Colombo 1974, 165). It showed elevated water solubility and low acute toxicity. Sufficient quantities for a more extensive testing were obtained in the following November (Brocchetta 1974, 12–15). Soon thereafter Piero Tirone, the head of the pharmacology department, reported his conclusions regarding the pharmaco-toxicological properties of iopamidol: "Compound of extremely low neurotoxicity

FIGURE 4.39   *Davide Pitrè.* (Courtesy of Simona Pitrè.)

CURRICULAM VITAE

Pitrè, Davide (b. March 4, 1923, Moltrasio, Prov. of Como, Kingd. of Italy //
d. September 27, 1991, Milan, Prov. of Milano, Italian Republic)

Liceo scietifico, Venice, Italy: High-school dipl. [maturità] (1941)/Univ.
Milano: Stud. of chem. (1941–1943)/Italian Wartime military serv. and
German prisoner of war (1943–1945)/Univ. Pavia: Stud. of chem. (1945–
June 1947); Laurea [Italian doctoral degree] in chem. (June 1947); Inst. of
Pharmaceut. Chem. and Toxicol.: Post-doctoral Fellow (March 1947–March
1950); Voluntary Asst. (March 1950–1951)/Schiapparelli Farmaceutici SpA,
Turin, Italy: Research Chem. (1951–1952)/Cilag Italiana SpA, Milan, Italy, in
1955 becoming Industria Chimica Dr. Fulvio Bracco SpA, in 1958 becoming
Bracco Industria Chimica SpA: Research Chem. (November 1952–1962); Head
of Chem. Dept. (1962–1969); Vice director of Research Laboratories (1969–
1984)/Univ. Pavia: Libera docenza [lecturer] in pharmaceut. chem. and toxicol
(June 1960); Second Laurea, in pharmacy (July 1974)/Univ. Milan, Faculty of
Pharmacy: part-time teaching assignment (November 1972–September 1991);
Assoc. Prof. (1985–September 1991).

and optimal systemic tolerability; shows pronounced urotropism"[25] (Tirone 1974). In
temporal coincidence, an initial patent application covering a family of compounds,
encompassing iopamidol, was filed on December 13, 1974 (Felder and Pitrè 1974a,b).
Switzerland was chosen for the initial filing since Italy did not allow patents for
pharmaceuticals until 1978. The patent was assigned to a wholly owned Swiss subsidiary of Bracco Industria Chimica SpA, called Savac AG, Chur, Switzerland. To
the inventors it appeared unlikely that the compound with which the search had
begun, iopamidol, should already embody the optimum. Hence, the quest for further
improved ones continued for a while unabated.

In the aforementioned talk of Felder, he specified that it was only some time
after July 1975 that he identified among numerous contrast agent candidates, iopamidol (Figure 4.40, **31**) as the best one for entering full-scale commercial development. It had favorable biological and pharmaceutical features, fulfilling almost all
of the requisites Almén had dreamt of. In part linked to its I/P ratio of 3.0, its acute
intravenous median lethal dose ($LD_{50, i.v.}$) in mice was the most favorable so far ever
achieved. Chemical degradation under autoclave sterilization was minimal, and it
formed solutions for injection with much lower osmolalities than the compounds of
the second-generation at the same concentration of iodine.

The contrast agent was eminently practical. Its manufacturing cost promised
to be lower than that of metrizamide. These characteristics were factors Felder
paid close attention to. In his career, he had accumulated pertinent experience both
while heading research and development and while in positions of responsibility
for chemical process development at Cilag AG, and for manufacturing at Bracco
Industria Chimica SpA. Patent strategy was reconsidered. Without losing the priority date of December 13, 1974 the pending Swiss application was substituted on

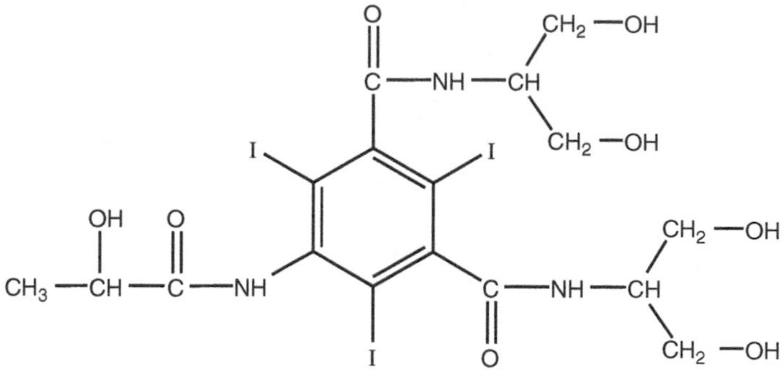

31

FIGURE 4.40 Iopamidol, one of the first two commercial nonionic uro-angiographic contrast agents.

31: Iopamidol (ISOVUE™, IOPAMIRO™, IOPAMIRON™, SOLUTRAST™, NIOPAM™, depending on the country, Bracco Industria Chimica. SpA, Italy). $I/P = 3.0$.

October 29, 1975 by two divisional applications, one covering the compound and the other a synthetic process for its manufacturing (Felder and Pitrè 1974a,b).[159] The contents of these patent applications became first accessible to the public through the corresponding single Deutsche Offenlegungsschrift of June 24, 1976 (Felder and Pitrè 1974a,b).

Iopamidol's chemistry and its performance in animal experiments were revealed at the 2nd Congress of the European Society of Cardiovascular Radiology on May 25–27, 1977 in Uppsala, Sweden (Felder, Pitrè, and Tirone 1977b; Belán et al. [1977] 1978). At the same congress ioxaglate meglumine/sodium made its début (Lautrou, Bergin, and Cardinal [1977] 1978) and the nonionic product ioglumide (P297) (Figure 4.36, **28**), synthesized at Laboratoires Guerbet SA was disclosed (Sovak et al. [1977] 1978, Tilly, Hardouin, and Lautrou 1973). The pleasant surprise with iopamidol was that toxicity, neurotoxicity above all, was far more reduced than one would have expected on the basis of osmolality reduction alone (Felder, Pitrè, and Tirone 1977a,b). In animal safety evaluations iopamidol outperformed dramatically all ionic contrast agents, it shared many of the advantages of metrizamide, and it even improved on others (Tirone and Boldrini 1981a,b, 1982). These observations found later clinical confirmation. In the case of neurotolerability preclinical tests, including intrathecal acute median lethal dose ($LD_{50, \text{i.th.}}$), turned out to be predictive for clinical outcome (Rosati, Morisetti, and Tirone 1992a; Rosati, Leto di Priolo, and Tirone 1992b).

It is interesting to understand how Felder and Pitrè arrived at the structure of iopamidol. The earlier experience at Eprova AG with the elevated solubility of the serinol derivative of methiodal prognosticated by Felder naturally favored experimentation with serinol. Serinol, which was not a widely available chemical, was provided by Eprova AG to Bracco Industria Chimica SpA for this purpose. The earlier mentioned difficulties with the synthesis of amides of *N*-acyl-5-amino-2,4,6-triiodoisophthalic acid derivatives were overcome by synthesizing intermediate acid chlorides bearing an unprotected anilinic function. As regards, the lactic acid moiety of iopamidol, Felder recalled having seen in the literature a significant improvement in water solubility afforded by lactoylation of p-phenetidine. By analogy the same was conjectured for an iodinated molecule carrying a lactoyl residue. Felder and Pitrè reasoned that the use of racemic lactic acid would increase solubility by contributing to the diversity of molecular configurations and conformations. In reality, the solubilizing effect of the lactoyl residue in p-phenetide was quite modest, and the use of enantiomerically pure lactic acid for the synthesis of a contrast agent yielded a compound with substantially higher solubility than was achievable with racemic lactic acid (Felder, Pitrè, and Tirone 1977b). Thus, the former premise was tenuous and the latter one was experimentally discredited. Worse, as already exposed, for good reasons solubility was then, and has remained since, notoriously unpredictable. Success under the aegis of a deficient theory is far from unusual, but it is rather rare that such an intimate glimpse into the processes behind important progress is possible. For shear necessity of keeping publication space to a minimum, scientific articles do not relate how progress was made but rather indicate how the progress *post hoc* can be rationalized using minimal printing space, i.e., communication with "deletion of modalities." A somewhat similar situation has been described above regarding the advent of acetrizoate sodium.

With the advice and coordination of Bonati, medical director at Bracco Industria Chimica SpA, the chairman of the Department of Radiology at the Institute of Hematology in Prague, Milan Svoboda, initiated experimental intravenous injection of iopamidol to humans in what we would today call phase I clinical trials. On the heel of these studies followed the first tests in man of iopamidol in angiography and urography. Giovanni Juliani, at the Mulinette Hospital in Turin, Italy, orally communicated his encouraging results first on January 21, 1977 (Juliani 1977). Vedran Nutrizio and collaborators at the Department of Radiology of the Clinical Hospital Center, Faculty of Medicine, University of Zagreb, Yugoslavia, communicated his by mid-June of the same year (Šimunić et al. 1977). The preclinical and preliminary clinical results were likewise disclosed at the XIV International Congress of Radiology, Rio de Janeiro, Brazil, in October 1977 (Felder, Pitrè, and Tirone 1977a; Bonati and Felder 1977). Full-scale clinical development as an angiographic and urographic contrast agent followed immediately. Initial experiments with intrathecal administration of iopamidol (B-15000) were first performed by Juliani, who added these studies to his aforementioned ones on intravenous injection (Bonati and Felder 1977), and by the Zagreb group (Nutrizio et al. 1981). The results led to inclusion of the clinical indication for myelography in the development of the product. Various other clinical indications followed.

In 1980, as part of the general chemical characterization of iopamidol substance and its solutions, solubility and crystal structures were analyzed (Pitrè and Felder 1980). It was then realized that iopamidol at the typical concentrations prepared for clinical use were supersaturated solutions. These solutions appeared to be stable if heat sterilized in a stoppered and sealed container. A concern that the compound might one day begin to spontaneously crystallize inside the vials turned into fear when in 1983 Bracco Industria Chimica SpA was informed of an impending academic publication describing such crystallization. By good fortune,

FIGURE 4.41   *Franco Bonati.* (Courtesy of Franco Bonati, Segrate, Italy.)

CURRICULAM VITAE

Bonati, Franco (b. April 15, 1921, Vigatto, Prov. of Parma, Kindom of Italy //
d. December 1, 2011, Segrate, Prov. of Lombardy, Italian Rep.)

Liceo Classico, Parma, Italy: High-school dipl. (Maturità) (1939)/Univ.
Parma: Stud. med. and surg. (1939–1945); MD (Aug. 1945); License to practice
(1946); Asst., Inst. of Med. Clinics, formally (1946–1963), factually (1946–
1956); Lecturer (Libera docenza) in Diagn. Med. (November 1958)/School
of Professional Nurses, Parma, Red Cross of Italy: Instructor in Pharmacol.
(1953–1956)/State licensed biologist (March 1969)/Industria Chimica Dr.
Fulvio Bracco, in 1958 becoming Bracco Industria Chimica SpA, Milan:
Head, Biological Laboratories (May 1956–December 1966); Medical Director
(January 1967–January 1982); consultant (February 1982–March 1991)/Assoc.
Ital. Pharm. Ind., Study Group for Technical-Scientific Problems: Vice-
president (1963–1972), President (1973–1981)/Eur. Fed. Assoc. Pharm. Ind.:
Techn. and Sci. Rep. of Ital. Assoc. Pharm. Ind. (1973–1986)/Univ. Cattolica,
Rome, Medical Faculty, School of Pharmacol. Specialization: Contract
Instructor in Deontol. and Sanitary Legislation (1984–1991)/Italian Soc. of
Pharmaceut. Sci.: Counsillor (1978-); Executive Secretary (1980–2001)/Italian
Soc. of Appl. Pharmacol. Sci.: Founder (1964); Councillor (1964–1967);
Honorary Member (1990–2011).

it became soon clear that crystallization could only initiate in opened vials, vials with a punctured stopper or those with glass microfissures owing to mechanical shock. Once the stopper was punctured, crystallization took hours to begin. The published report acknowledged these facts (Dawson, Pitfield, and Skinnemoen 1983). Some vials of iopamidol manufactured in the late 1970s are found in the archives of Bracco Imaging SpA. They contain still clear solutions. Since then it has been learned that, if industrially packaged, most nonionic contrast agents at their higher commercial concentrations are indeed permanently stable supersaturated solutions.[160] This can be understood in part as a consequence of the absence of surfaces on particles and container walls capable of catalyzing heterogeneous nucleation in a supersaturated solution obtained by cooling a particle-free, hot and highly concentrated regular solution (Servi and Turnbull 1966).[161] In the cases of many nonionic contrast agents, Almén's design theory had failed to lead to the wished-for elevated water solubility at 37 °C but had led to the equally usable supersaturation behavior.[162] A theory does not have to be correct to be stimulating. It suffices to be plausible and viable.

Whereas company internal resources could manage clinical development of the contrast agent for registration in Germany and in the home country of Italy, clinical development for the rest of the world markets required collaborations with other companies. Similarly, worldwide commercialization of the product could solely be envisioned through licensees. E. R. Squibb & Sons Inc. was predestined to be one of them, given the financial assistance it had provided for research and development and the exclusive license for the USA and semi exclusive rights to products for a number of other territories in the world it had thereby earned. *De facto*, it never exercised its rights outside the USA and Australia. When Schering AG inquired into the possibility of taking out licenses for iopamidol for the whole world, within legally available territories Bracco Industria Chimica SpA chose to be selective, granting only France, Australia, Japan, South Africa, and South America, to list the most important ones. For other territories, it preferred finding distributors that included in the deal rights for marketing, co-marketing and/or even pharmaceutical production of their therapeutic products in Italy.

After December 27, 1973, the publication date of the Austrian patent of Suter, Zutter, and Müller (1971), which covered methiodal amides, the competitor companies could have been alerted to the interest of Eprova AG and Bracco Industria Chimica SpA in nonionic contrast agents, but apparently, they were not alarmed. At Nyegaard & Co. A/S research on compounds that could follow metrizamide did proceed but was hampered by the heavy resource requirement of the development of metrizamide (Amdam and Sogner 1994). The follow-up development candidate failed toxicology tests in January 1974 and later that year it spontaneously crystallized out of the pharmaceutical solution for injection. Just as for the nonionic compound B 6590 mentioned earlier, at the time crystallization was thought to render it useless. A further development candidate, called C29 or compound 543[151] (Figure 4.42, **32**), was heat sterilizable in solution and had favorable overall biological attributes. On April 1, 1976, its solution too spontaneously crystallized. As a backup for C29 there had been identified the very similar compound 545. It later

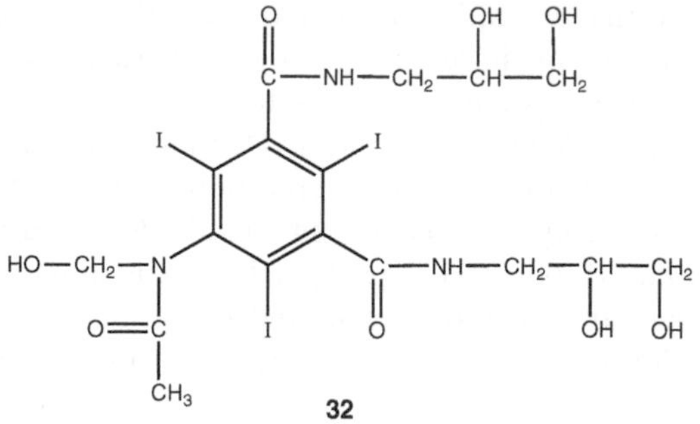

FIGURE 4.42 Iohexol, one of the first two commercial, nonionic, uro-angiographic contrast agents, and some congeners.

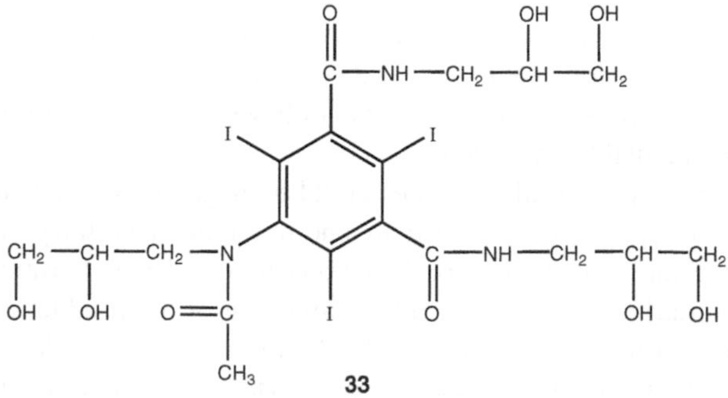

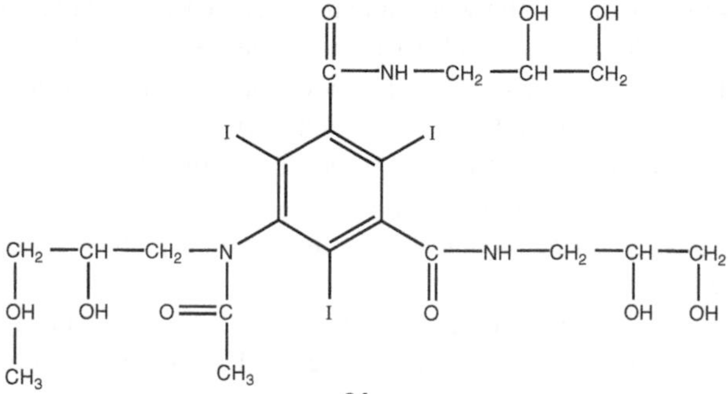

32: Experimental compound C29 (Nyegaard & Co. A/S, Norway). $I/P = 3.0$.
33: Iohexol (OMNIPAQUE™, Nyegaard & Co. A/S, Norway). $I/P = 3.0$.
34: Iopentol (IMAGOPAQUE™; IVÉPAQUE™, Nyegaard & Co. A/S, Norway). $I/P = 3.0$.

acquired the generic name iohexol (Figure 4.42, **33**). On June 11, 1976, i.e., before the investigators could have known anything about iopamidol, but just a few days before the German patent application was published (Felder and Pitrè 1974a,b) Nordal and Holtermann requested a patent covering among other compounds, iohexol (Nordal and Holtermann 1976).

When the German iopamidol patent application (Felder and Pitrè 1974a,b) was published on June 24, 1976, it became evident right away, even to the competition, that at last a nonionic contrast agent with excellent stability and solubility had been achieved (Amdam and Sogner 1994). C29 and iohexol were close relatives of iopamidol and formed solutions with similarly favorable osmolalities and viscosities (Haavaldsen 1980). Yet the team at Nyegaard & Co. A/S was worried by the fact that the compounds were complex mixtures of stereoisomers. It perceived this fact as a large regulatory hurdle for government drug approval. This probably explains the continued insistence on the slightly less complex C29 and the continued search of molecules with simpler stereochemistry, such as compound 593 (Haavaldsen 1980). The team was furthermore terrified by the crystallization problem, which Bracco Industria Chimica SpA with iopamidol was to encounter only later (Pitrè and Felder 1980; Dawson, Pitfield, and Skinnemoen 1983).

At the end of 1976, an exhausted Holtermann requested that he be relieved of his managerial duties and allowed to return to a more technical function. In consequence, the rudder of contrast agent research and development at Nyegaard & Co. A/S passed to the engineer Leif G. Haugen (Figure 4.29), whom Ulf Blix was to support by more direct involvement than before. The year 1977 was a year of terrible uncertainty. In May, investment in compound C29 (Almén 1980) was definitively dropped and by early 1978 there were discussions of abandoning research in the radiological field altogether (Amdam and Sogner 1994).

In 1978, the willingness of Sterling Drug, Inc.[79] to take iohexol under license for joint development and for distribution in the North and South Americas, Australia and many smaller Pacific Rim countries restored moral in a big way. People with lots of experience with the United States Food and Drug Administration assessed the problem of stereoisomers surmountable and had confidence in the long-term stability of supersaturated solutions, when heat sterilized in closed containers. With this refreshed optimism iohexol was developed in record time. The pharmacist Kamilla Dahlstrøm coordinated clinical trials as the head of the company's clinical department (Lindgren 1980).

In parallel, a number of additional candidate contrast agents were found in rapid succession (Amdam and Sogner 1994). By the end of 1978, it was iopentol (Figure 4.42, **34**) (Wille 1982) and within 1 year thereafter, the dimeric nonionic compound, iodixanol (Figure 4.44, **39**) (Amdam and Sogner 1994). Patenting of iodixanol followed only in 1982 (Hansen, Holtermann, and Wille 1982). Development of iodixanol and iopentol had to wait for a long while, since development of iohexol in times compatible with the necessity to compete with iopamidol, as well as ioxaglate meglumine/sodium, commanded all available resources for many years.

Like Bracco Industria Chimica SpA, also Nyegaard & Co. A/S lacked the means and experience needed for single-handed worldwide commercialization of a contrast agent. Thus, apart the licensee Sterling Drug, Inc., additional licensees were called for. Schering AG offered and was given licenses to sell iohexol in most territories for which it had failed to get a license for iopamidol from Bracco Industria Chimica SpA. Scandinavian countries, in which Nyegaard & Co. A/S itself had a solid commercial presence, were an exception. For Japan, which then and now constitutes a unique case, a license for iohexol was given to Daiichi Pharmaceutical Co. Ltd., Tokyo.

Starting in Germany, where most of the pivotal clinical trials with iopamidol had been performed, from 1982 onwards the product appeared in various countries under the brand names IOPAMIRO™, IOPAMIRON™, ISOVUE™, NIOPAM™, and SOLUTRAST™, depending on the licensee. On its heel, i.e., in 1983, followed iohexol with the worldwide trademark, OMNIPAQUE™. The initial response of the consumer substantially exceeded projections and for a while manufacturing capacity limited the expansion of supply to clinicians. Meanwhile, the two contrast agents account for about 51% of the world volume for nonionic iodinated products. By 2016, they have been injected about 1,350 million times. Many thousands of scientific publications document their diagnostic performance.

Both iopamidol and iohexol showed excellent performance in the clinic and turned out eminently practical in manufacturing and clinical application. They realized the vision of ionic uro-angiographic contrast agents that had led to the emergence of second-generation products (Table 4.1). Simultaneously they demonstrated the feasibility of products whose toxicity was improved to an extent not possible within the ionic category. With this, they completed the new technological vision, i.e., nonionic low-osmolal uro-angiographic contrast agents (Table 4.5). The vigorous efforts with which further compounds were explored, clinical applications were expanded, and commercial supply chains and distribution networks were built up, altogether testify to the effectiveness of the new vision.

The vision of low-osmolal uro-angiographic contrast agents in general and nonionic ones in particular, rapidly recruited followers. Companies (e.g., Schering AG, Sterling Drug, Inc., Mallinckrodt Inc., Squibb E. R. & Sons Inc.) not ready to compete head-on and straightaway through their own products with Bracco Industria Chimica SpA, Nyegaard & Co. A/S and Laboratoires Guerbet SA, yet sharing their vision, took licenses in various countries. The latter thereby overcame the difficulty of not being on their own prepared to be operative worldwide on a desirable time scale. Since various national subsidiaries of large corporations could autonomously find solutions to their needs, there emerged a complex international web of license and distribution agreements. This is not the place to describe them all, but the following situation may serve as an illustration. The subsidiary of Schering AG in Japan, Nihon Schering K.K., which initially supplied metrizamide under license from Nyegaard & Co. A/S, subsequently changed to iopamidol (NIOPAM) from Bracco Industria Chimica SpA, and continued to do so even after Schering AG had developed its own nonionic contrast agent, iopromide. The latter contrast agent in turn was given to Tanabe Seiyaku C., Ltd. for distribution in Japan under the name

---

TABLE 4.5
The Stepwise Emergence of the Vision for Nonionic Contrast Agents

1) Macromolecular contrast agents with blood pool behavior, efficient excretion, and low osmolality
2) Uro-angiographic contrast agent with low osmolality
3) Mediation of water solubility by hydroxyl-group-rich substituents instead of by carboxylate groups

The complete vision:

    Nonionic low-osmolal uro-angiographic contrast agents.

of PROSCOPE™. This all happened while E. R. Squibb & Sons Inc. had the nonexclusive license to bring iopamidol to Japan but did not exercise its right.

Schering AG pursued an astute worldwide strategy. Correcting for the fact that it had interrupted contrast agent research and development, it acquired licenses for iopamidol and iohexol for complementary territories, to cover most of the world, except the USA. With its pre-existing marketing and sales presence, it contributed in a major way to the affirmation of the new technology, without favoring either product, overall.

In the 1960s, the market for contrast agents, then all of the ionic variety, was perceived by many to become of the commodity type. Price began do dominate all other aspects. Hope of breakthrough chemicals entering the pipeline was dwindling. As a side effect, feedback to research and development functions from those most closely in contact with the customer, i.e., the marketing and sales functions, became scarce and poorly supported. As a result, the feedback contributed at best little to the advent of the new technology. With the arrival of iopamidol and iohexol this changed drastically. The vision behind these products finally captivated the business arms of various companies. The aforementioned elevated prices lost their deterrent effect. A full-scale involvement in the contrast agent business seemed again desirable. Research was renewed in various companies. At Schering AG, this brought forth iopromide in 1979 (Speck et al. 1979) and at Mallinckrodt Chemical Works, ioversol in 1982 (Lin 1982). In fact, since the arrival of iopamidol and iohexol the following similar agents have become available: i.e., iopromide [ULTRAVIST™, Schering, patented 1979, marketed 1985], ioversol [OPTIRAY™, Mallinckrodt, patented 1982, marketed 1989], iopentol [IMAGOPAQUE™, IVÉPAQUE™, Nyegaard, patented 1982, marketed 1993], iobitridol [XENETIX™, Guerbet, patented 1990, marketed 1994], iomeprol [IOMERON™, Bracco, patented 1979, marketed 1994], ioxilan [OXILAN™, Cook, patented 1985, marketed 1998].[163]

Nonionic contrast agents, iopamidol and iohexol, with their I/P ratio of 3.0, at equal iodine concentrations formed injection solutions of much lower osmolality than the ionic precursors with their I/P ratio of 1.5. For instance, the nonionic ones at the concentration of 300 mg(Iodine)/mL have values ranging from 520 to 700 mosmol/kg, as compared to the 1,600 mosmol/kg typical for monomeric ionic ones (ACR 1991; Krause et al. 1994a). The contrast medium with the highest iodine concentration sold today is iomeprol that, at 400 mg(Iodine)/mL, has an osmolality of still only 726 mosmol/kg. The solubility of iomeprol even allowed a formulation at 470 mg(Iodine)/mL but it was studied only in animals.

By the time of the search for nonionic uro-angiographic contrast agents the acute intravenous median lethal dose in rodents, $LD_{50, i.v.}$, had established itself empirically as a valuable selection criterion. Underlying the use of $LD_{50, i.v.}$ in contrast agent selection was the assumption that it constituted a predictor of severe adverse events in clinical practice. Indeed, the nonionic agents iopamidol and iohexol, whose $LD_{50, i.v.}$ were 2.5, respectively, 2.9 times larger than that of diatrizoate sodium produced in the clinic five to six times fewer severe adverse events than the older ionic forerunners. Scarce data do not allow yet conclusions regarding mortality of the third-generation of contrast agents, but they have certainly contributed to the overall fall of cases. In the United States, the mortality produced by all contrast agent combined in the years 1999–2001 is estimated at one in 1.1–1.2 million doses (Wysowski and Nourjah 2006), i.e., by a factor of about 17 better than the earlier given estimates for second-generation products (Hartman et al. 1982; Cashman, McCredie, and Henry 1991). The most immediately noticed benefit of third-generation contrast agents for the patient is the elimination of pain during injection (Palmer 1988; Katayama et al. 1990; Thomas, Williams, and Adam 1997). It can convincingly be explained by osmolality. Some vascular effects may have their origin in osmotic load delivered by a diagnostic dose. But in general, the mechanisms of adverse reactions are very complex and still poorly understood (Almén 1971; Morcos et al. 1998; Spinazzi 2005; Dawson 2006; Lasser 2011; Clement and Webb 2014). Their elaboration would go beyond the scope of the present treatise. Similarly, a discussion of small differences in adverse event frequencies observed with different nonionic monomeric contrast agents would only detract from the chosen main theme.

The monomeric nonionic contrast agents set an unprecedented safety record. Some of them additionally allowed myelographic examination of the entire subarachnoid space. Owing to their distinctively favorable properties they were hailed as the third-generation contrast agents. Ioxaglate meglumine/sodium that at 320 mg(Iodine)/mL has an osmolality of 580 mosmol/L, is sometimes subsumed under the group of third-generation products, without sharing all of the advantages of nonionic ones.

The monomeric nonionic iodinated contrast agents, whose ascent I have extensively described, constituted among the third-generation products those gaining the widest range of application and acceptance. At the needed concentrations, these products were still somewhat hyperosmolal, although not to a degree posing a significant clinical problem. Yet isoosmolality kept its lure. Increasing the I/P ratio offered a well-trodden path to further lowering of osmolality. The old diacidic dimer, iocarmate dimeglumine, and the monoacidic dimer, ioxaglate meglumine/sodium, by structural similarity suggested nonionic dimers, which can have an I/P ratio of 6.0. Such molecules were actually already conceived and covered in the patent of Almén, Nordal, and Haavaldsen that inaugurated the era of nonionic products (Almén, Nordal, and Haavaldsen 1969). But no evidence was adduced for elevated water solubility and acceptable toxicity of the only two compounds for which a synthesis was disclosed. In any event, it could be anticipated that solutions of nonionic dimers at desirable concentrations would have osmolalities low enough to allow addition of potentially beneficial salts up to isoosmolality with blood.

A decade after the research of nonionic contrast agents had really taken off, two independent industrial laboratories identified realistic nonionic dimers. Pfeiffer and Speck (1976) from Schering AG reported a family of compounds that included iodecol (Figure 4.43, **36**), and Mützel et al. (1979) another family that included iotasul (Figure 4.43, **35**). Since iotasul in solution showed phase separation upon temperature elevation from room temperature to 37 °C, just as some of Habicht's monomeric nonionic products had done earlier (Habicht and Zubiani 1963), it underwent clinical development as a lymphographic contrast agent for interstitial injection. While experiments, mostly in small animals but some also in man (e.g., Müller, Wenzel-Hora, and Addicks 1985) showed promise, usually the results were not gratifying.

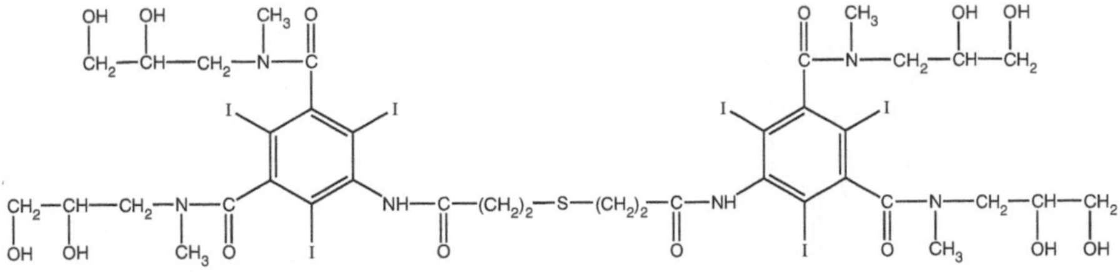

35

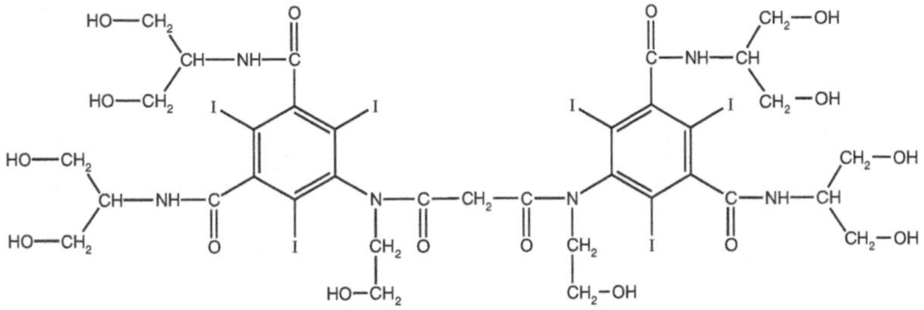

36

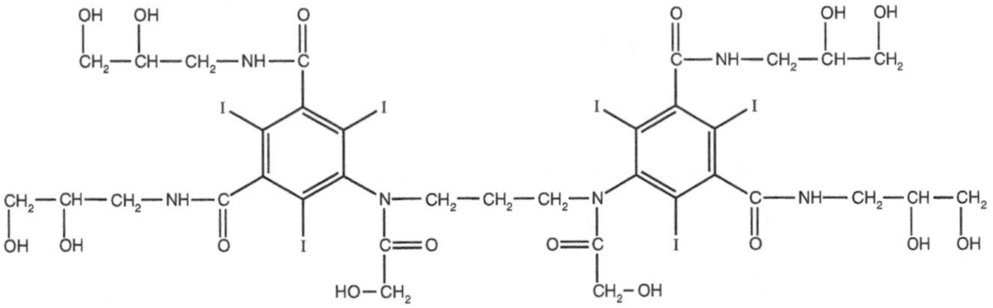

37

FIGURE 4.43 Experimental nonionic, dimeric contrast agents.

35: Iotasul (Mützel et al. 1979). $I/P = 6.0$.
36: Iodecol (Pfeiffer and Speck 1976). $I/P = 6.0$.
37: Bracco compound B-17500 (Felder and Pitrè 1979). $I/P = 6.0$.

The long travel distances from injection site to lymph nodes in man was to blame. Neither iodecol nor iotasul was ever developed to maturity.

In the laboratories of Bracco Industria Chimica SpA too the search for nonionic dimers was pursued. Felder and Pitrè (1979) had synthesized a family of dimers, among which the most interesting one was given the code number B-17500 (Figure 4.43, **37**). Initially it showed elevated solubility at all temperatures. It was developed almost to the level of clinical testing, with optimal results in animals when, to great surprise and disillusionment, spontaneous crystallization of the pharmaceutical preparation became a regular event. This fact caused the compound to be abandoned.

The nonionic dimer that finally overcame the solubility difficulties and was brought to market was iotrolan (Figure 4.44, **38**). This is how it came about. As covered in Section "Nonionic Contrast Agents", in the mid-1970s Sovak in his laboratory at the Radiology Department of the University of California in San Diego

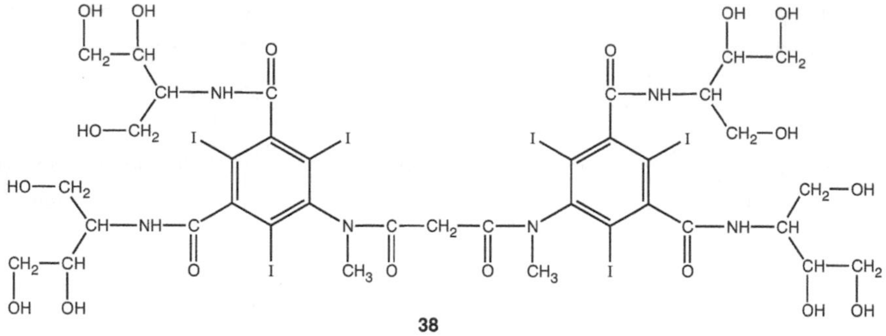

FIGURE 4.44 Commercialized nonionic, dimeric contrast agents.

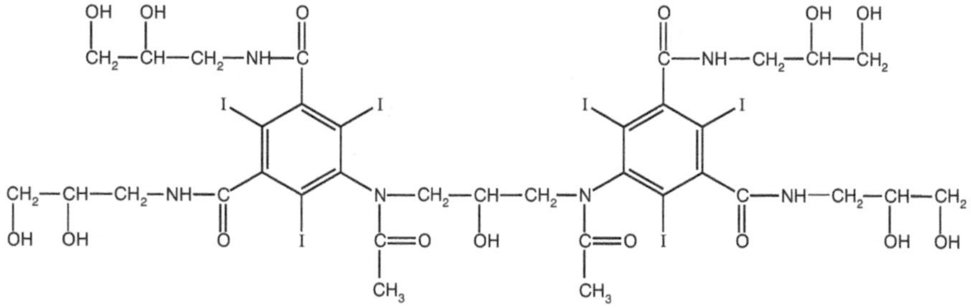

38: Iotrolan (ISOVIST™, Schering AG, Germany; OSMOVIST™, Berlex Laboratories, Inc. USA). $I/P = 6.0$.

39: Iodixanol (VISIPAQUE™, Nycomed Imsaging S/A, Norway). $I/P = 6.0$.

employed synthetic chemists in his pursuit of novel nonionic contrast agents (Sovak et al. 1975; Weitl, Sovak, and Ohno 1976; Weitl et al. 1976). In 1977 the chemist, Ramachandran S. Ranganathan, joined his group and the joint effort led first to the invention of the dimer, iotrolan, alternatively called iotrol (Sovak and Ranganathan 1980) and then to that of the monomeric nonionic molecule ioxilan (Sovak and Ranganathan 1985). Iotrolan at a concentration of 300 mg(Iodine)/mL showed an osmolality of 320 or 290 mosmol/kg, depending on the supplier's declaration, i.e., almost isoosmolal. Formulations at lower concentrations were rendered isoosmolal through the addition of sodium chloride and sodium bicarbonate.

Since his initial involvement with the search for improved products, Sovak, a professor of radiology, had sharpened his sense for business through extra-academic activities.[164] He realized that university departments are an unsuitable institutional environment for integrated research and development of drugs. For example, in such an environment the necessary experimentation to support and defend a patent beyond its application stage is difficult to justify. The need for secrecy before receipt of such protection is fundamentally incompatible with the encouragement of information exchange, which the good university training of young scientists and professionals demands. Yet, in the late 20th century without patent protection, there would be no commercially successful contrast agent. Nobody could responsibly take the financial risk of developing a product without the guaranteed duration of protection accorded by a patent. In order to facilitate work that without giving up his university position would allow him to pursue profitable products, Sovak founded Biophysica, Inc., La Jolla, California in 1981.

Iotrolan beat its dimeric competitors with entering the development phase. The compound was later taken under license by Schering AG, and was developed and, beginning in 1987, distributed under the trade name of ISOVIST™. It was formulated as isoosmolal solutions for injection at concentrations up to 300 mg(Iodine)/mL. Although carrying initially the clinical indications for myelography and for the visualization of various body cavities, in due time its clinical indication was extended to angiography. Minor adverse reactions, mostly delayed urticaria, forced the product off the market for angiography in 1995 (Moore 1995), but for myelography and for the visualization of various body cavities in various countries it remained available at least until 2012. It did not become a major product.

Roughly a couple of years after the invention of iotrolan, Hansen, Holtermann, and Wille (1982) at Nyegaard & Co. A/S patented the nonionic dimer iodixanol (Figure 4.44, **39**). The maximum iodine concentration in a solution for injection of nonionic dimers was limited by the dynamic viscosity, which had to remain in the range allowing facile rapid intravascular injection. At the time, it was feared that excessive viscosity had as well to be avoided because through shear stress it would produce endothelial damage during arteriography. Later that fear was shown to be unfounded (Hutcheson et al. 1999). In any case, for viscosity reasons, and perhaps because of limited solubility, the highest iodine concentration of iodixanol in use is 320 mg(Iodine)/mL, i.e., a concentration below the maximum concentrations of 370 or 400 mg/(Iodine)/mL offered for some monomers.

At the concentration of 320 mg(Iodine)/mL, a solution of iodixanol alone would have been hypoosmolal with regard to blood. Thus, sodium chloride and calcium chloride were added to establish isoosmolality. Iodixanol was introduced with angiographic indications in 1993 under the trade name of VISIPAQUE™ (Nycomed Imaging A/S[117]) and is currently the only available nonionic dimer for intravascular application. Except for the less than optimal maximum iodine concentration, with the dimeric products Almén's dream of the perfectly isoosmolal contrast agent appeared to have been realized. But did this isoosmolality translate into corresponding safety?

In the 1990s, the attention of researchers turned to one of the rare but potentially devastating adverse effects of intravascular injection of iodinated contrast agents, termed Contrast-Induced Nephropathy (CIN). This adverse effect manifests itself clinically as elevated serum creatinine values 2–3 days after contrast agent injection, indicative of a reduced glomerular filtration rate. The kidney damage can be irreversible. Recently, molecular markers have been identified, whose serum levels become abnormal before those of creatinine. Based on this finding, CIN has been renamed Contrast-Induced Acute Kidney Injury (CI-AKI). CI-AKI in patients with normal kidney function is rare and can be minimized by preparatory hydration of the patient, if necessary complemented by pharmacological interventions. At serious risk of CI-AKI are patients with impaired kidney function. Since the reduction of contrast agent osmolality on switching from ionic monomeric to nonionic monomeric products had reduced the risk of CI-AKI, it was reasonably hoped, isoosmolalality would bring further amelioration of the situation. But it was not to be so. A vast literature on studies both in animals and man has accumulated. Studies in animals, where the control over the experimental parameters is excellent, allowed deep insights into mechanisms behind the adverse effects to be gained (for review see Seeliger, Lenhard, and Persson 2014). A first clear result was that differences in osmolality of contrast agents alone could not adequately explain the observations. A second result was that a product's elevated viscosity boded ill for its kidney toxicity. As a matter of fact, low osmolality enhanced the detrimental effects of elevated viscosity. A principle mechanism explaining this fact involves the concentration of contrast agent in the renal tubules. Kidney tubule can concentrate solutes to some upper limit in osmolality. If they concentrate dimeric and monomeric contrast agents to the same osmolality, the dimeric agent shows dynamic viscosity much higher than the monomeric one. The former agent produces a viscosity so elevated as to impede tubular efflux and results in prolonged exposure of the organ to chemotoxic actions of the agent. Elevated viscosity too partakes in damage brought about by medullary hypoperfusion and hypoxia.

Clinical studies of CI-AKI turned out to be very difficult, in part due to serum creatinine levels being an unsatisfactory outcome measure, and in part due to the inherent difficulty of estimating differences in occurrence rates of rare adverse events. All studies involved comparison of products (for review see Jones and Babb 2008; Lenhard et al., 2012; Becker, Babb, and Serrano 2013; Bucher et al. 2014; Seeliger, Lenhard, and Persson 2014). Given the elevated safety of all currently used contrast agents, safety comparisons in a general patient population require extremely large numbers of patients, numbers only practical in retrospective analyses. The necessary number of patients in prospective safety trials may be reduced by focusing on patients with a high risk of complications associated with a given procedure. Accordingly, an avalanche of clinical trials has compared types and frequency of adverse events caused by isoosmolal iodixanol with those of various low-osmolal monomeric agent in high-risk patients. In addition, a comparison with the ionic dimer ioxaglate meglumine/sodium was made. Select phenomena observed in animals, e.g., prolonged presence of dimer in the kidney, were equally observed in man. A number of clinical trials failed to find a difference between monomeric and dimeric products. Others, depending on patient selection criteria, trial design, and safety parameter, found one or the other product class advantageous. This situation led to some cases of inadmissible use of clinical trials and findings for the commercial promotion of products. Interestingly, what has been said about the difficulties of comparing contrast agents for their causation of CI-AKI could be repeated as is for major cardiac adverse events (Jones and Babb 2008). It cannot be the present task to pursue this state of affairs further. Suffice it to say that even if any of the reported differences were taken as real, on average the product choices of clinicians and hospital buyers appear only marginally affected by them. One reason is that in many cases, risks associated with the surgical procedures for which the contrast agent is used are far greater than those ascribable to the injected product *per se*.

Limited maximal iodine concentrations and adverse effects have so far impeded nonionic dimers as a class of angiographic contrast agents to live up to the expectation set upon them. Although the one and only surviving angiographic agent of this class, iodixanol, has conquered a respectable place alongside the many monomeric top performers, it did not replace them in clinical practice. Nonionic dimers failed one of Schumpeter's criteria for an innovation, the creative destruction of old ways (Schumpeter [1912] 1997). For this reason, iodixanol is here classified together with the low-osmolal products as a member of the third-generation contrast agents.

An interesting concept, demonstrated, patented, and extensively tested in animals at Bracco SpA was that of achieving isoosmolal iodinated contrast agents of any useful iodine concentration, up to the highest one traditionally used, by mixing dimeric and monomeric agents in suitable proportions (Felder and de Haën 1992). Dynamic viscosities appeared to be favorable. The collected data for a particular formulation in animals supported the expectation of a performance superior to any of its predecessors. Recent studies gave reasons to believe that such a product would have had favorable kidney toxicity (Lenhard et al. 2012). It was quite unfortunate that the economic climate had changed. Thus, solely for reasons of a doubtful return on investment the further development of a product was discontinued.

A description of the history of the third-generation (nonionic, low-osmolal) contrast agents would be seriously incomplete without economic considerations. As already mentioned, the complexity of the issue allows here no more than a cursory treatment. In the 1960s, competition among producers of generic contrast agents of the second-generation (ionic, high-osmolal) had caused a price decay that had brought a dose of these sophisticated pharmaceuticals dangerously close to the level of a cup of ice cream. Profits were dwindling and many companies left the field. Products of the new, third-generation, were much more expensive to manufacture and increased regulatory burden demanded steadily growing investments. Favorably for the industry and its continued research and development activities, patent protection made it possible to set prices consonant with these conditions. At Nyegaard & Co. A/S, it was reputedly Holtermann's merit to have defended the metrizamide project against internal critics of the high cost of manufacturing by predicting the acceptability to the consumer of a correspondingly high price (Amdam and Sogner 1994). Metrizamide (AMIPAQUE) was indeed profitable, although the product volume sold was quite modest due to its limited clinical indications. However, conquering the primacy as myelographic contrast agent was certainly a prestige victory. Most importantly, the high price laid the foundation for the upcoming revolution in pricing and corresponding explosion of the global market value for contrast agents in general.

Faced with the problem of pricing of iopamidol, and exposed to contrasting opinions of advisors from inside her company and from licensees, the chief executive officer of Bracco Industria Chimica SpA, Diana Bracco, decided on, and succeeded, with a courageous high-price strategy that helped restore good profitability to the whole contrast agent industry. Thereby it permitted guaranteeing the future through continued major research investments into contrast agents, e.g., for magnetic resonance imaging and echography, diagnostic agents for nuclear imaging. Other companies followed suit pricing their own products similarly. In 1986, in the United States low-osmolal contrast agents cost 13–24 times more than the old, high-osmolal ones, whereas in Europe the factor ranged from 3.5 to 6.5 (Wilmot, Mehta, and Jha 2012). The spread in these factors reflects the mentioned complexity of price formation in different environments. Anyway, it is interesting to note that the initial price multiplicator from old to new technology in this situation was very similar to the one that had occurred when excretion urography with uroselectan sodium and methiodal sodium substituted the retrograde procedure with lithium iodide (UMBRENAL) (factor between 4 and 14) or with the same methiodal sodium (factor between 7 and 24) (see Section "From Sodium Iodide to Uroselectan Sodium and Methiodal Sodium").

In countries where the price differences between nonionic and ionic contrast agents were pronounced, in truth drastically distorted by under-pricing of the latter, e.g., in the United States, resistance to the new technology arose. For economic reasons, proposals were advanced and sometimes implemented to reserve the more expensive pharmaceuticals to patients at elevated risk of adverse reactions. A clinical trial designed to assess the validity of the concept showed that under such guidelines it was riskier to be classified "not at risk" and receive the old ionic products than to be classified "at risk" and receive the new nonionic products (Palmer 1988). This finding became known as the Australian paradox. It underscored a recognized, economy-related ethical and legal problem (Grainger 1986). The difficulty is illustrated by the following episode. At a symposium during the annual meeting of the RSNA at Chicago in 1990 (Almén et al. 1990), after several eloquent talks on why and how to regulate the new contrast agent technology in order to save costs to the health care system, speakers, and discussants were asked whether they would allow the use of old-generation ionic contrast agents on their family members and themselves. The individual speakers, including those that just had favored rationing of the new technology, in agreement with the audience as a whole, overwhelmingly responded negatively. Questionable double standards had become evident.

Notwithstanding a dramatically higher price of this generation of products as compared to the previous one, the clinical benefits led, within about a decade after introduction, professional organizations to recommend nonionic as replacement of ionic agents at least for patients at risk (e.g., Bettman 1989; ACR 1991). In some developed countries, to be precise Germany and Denmark, in the 1990s even some government agencies took the position that for certain major applications the use of the old ionic articles was no longer defendable (Sundhedsstyrelsen 1993; BfArM 2000).

Meanwhile, the relatively young scientific discipline of pharmacoeconomics gained momentum. This discipline examines the cost-effectiveness relationships for all stakeholders in the application of a pharmaceutical product, up to the health care system as a whole. Alas, as with the pricing issue, the problem turned out to be exceedingly complex and considering the great geographic variation in types and costs of contrast agents, medical practices, and services, valid pharmaco-economic analysis could at best be applied to nations. Another major difficulty lies in the fact that by the time an analysis is published and recommendations can be implemented, reality has often already changed enough to render action problematic. Lastly, in any regulatory intervention aimed at increased economy of resources, there are gainers and losers, and corresponding supporters and opponents. In any case, contrast agents promised to offer a well-circumscribed field in which experience with pharmaco-economic analysis could be gained. As an early representative of such an analysis stands Eddy (1992). Numerous further analyses have since followed without arriving at a consensus (e.g., Thomsen et al. 1997; Thomas, Williams, and Adam 1997; Wilmot, Mehta, and Jha 2012).

While the controversy over costs was not completely abated, nonionic contrast agents have steadily increased their share of total use of such radiological agents in the developed world, to reach about 94% in 2016. Thus, undeniably in the technology-shaping process a large step toward the vision of an effective but totally innocuous uro-angiographic contrast agent had been achieved.

THE DYNAMICS OF TECHNOLOGY SHAPING

The Failure of One-Dimensional Reductionism

Metrizamide formally opened the era of water-soluble, nonionic contrast agents. Almén's objective had been to achieve major improvements in the safety of generally usable uro-angiographic contrast agents. Measured against this objective, metrizamide failed to deliver. While it did reflect the chemical and physical design principles propagated by Almén (1969a,b), its costs of production were exorbitant and its phamaceutical presentation too cumbersome for general use. Instead, it found niche application in myelography, where to a large degree it displaced older products. With this "creative destruction" metrizamide for myelography did meet the Schumpeterian criteria for a technological innovation, an improvement innovation as it is called here. This does not apply to metrizamide for angiography. Hence, considering metrizamide as a prototype for a new generation of nonionic uro-angiographic products, on a par with what uroselectan sodium or methiodal sodium has been for first-generation products, is problematic. But metrizamide certainly was an extremely important stepping stone for finding compounds that would match expectations more fully. Metrizamide strengthened the trust in the feasibility of the targeted technology improvement in a way comparable to what PYELOGNOST had done for technology genesis.

About 8 years after the introduction of metrizamide, and in close succession, arrived on the market iopamidol and iohexol. These were the first contrast agents that fully met Almén's expectations. Iopamidol and iohexol were immediately perceived as the definitive new technology and became classified as the third generation of uro-angiographic products. The two products were so closely similar in pharmaceutical, pharmacological, toxicological, and radiological properties that their performance was not even considered worth separate analyses in a major Japanese survey of contrast agents-associated adverse events involving 337647 patients, organized by Hitoshi Katayama (Katayama et al. 1990).[165] Despite differences that can be documented mostly in animal experiments and in some cases in man, the many nonionic congeners that followed these two pioneers were frequently subsumed with the former two under a single group, e.g., in recommendations covering the use of the new products by government agencies and professional societies. Within a few years, the new generation of contrast agents displaced old-generation ones to a large extent. It evidently constituted not only reshaped technology, but with this creative destruction of old ways, it qualified as technological innovation in the Schumpeterian sense (Schumpeter [1912] 1997). The height of the innovative step seems smaller than in the first analyzed period on which sociological attention has been focused.

This reflects improvement innovation resulting from a certain technology-shaping phase as compared to fundamental innovation resulting from the genesis phase.

The histories of iopamidol and iohexol show a number of similarities. This is not unexpected given that both products were results of technology-shaping processes that started with comparable initial conditions. Their originators were medium-sized, family-owned companies, active in research, development, and production of earlier generation ionic contrast agents, with direct sales in a few countries and presence on major world markets through large distributors. They operated on shoestring research and development budgets and counted on international distributors for financial help. They strategically decided in the late 1960s to insist finding innovative contrast agents, just when most major players were getting out of that business. They privileged the search for high value niche products, and arrived instead, essentially at the same time, at main application ones. The histories of iopamidol and iohexol differed mainly in the typology of the niche products that preceded them and delayed their arrival through competition for development resources. In terms of industrial projects, iopamidol was preceded by ionic contrast agents for biliary imaging (iodoxamic and iopronic acid), whereas iohexol had a nonionic myelographic contrast agent (metrizamide) as precursor.

Analysis of the technology genesis phase has shown that the brilliant inventor is not a suitable key for understanding the genesis of technology. The same holds true for technology-shaping phases. Merit for major technological progress is not unambiguously assignable to a single group of collaborators, not to speak of a single individual. The fact is that Habicht has the primacy of announcing and demonstrating experimentally the concept of a water-soluble nonionic contrast agent that formed solutions of low osmolality. But the corresponding patent did not elicit entrepreneurial interests. Did the inventor lack the inner fire or the power of persuasion to convince others of the merits of the project or did he find himself simply in the wrong environment or at the wrong time?

Almén indubitably played a more crucial role in the formulation of the vision of an improved technology than any single person in the first historical period of actual special interest. His design theory suffered from inadmissible physicochemical extrapolation of viscosity theory of colloids to solutions of small molecular weight compounds, inclusion of ionic compounds in the list of purportedly nonionic ones, inclusion of neutral molecules devoid of hydrophilic groups, and the fact that even after his theoretical proposal concerning the water-solubilizing effect of aliphatic hydroxyl groups, the solubility of compounds remained quite generally unpredictable. Notwithstanding these deficiencies it had just the invigorating appeal that was called for at a time hopes to find a better contrast agent in most quarters had dwindled. But some luck and significant inventive steps were still needed before viable contrast agent candidates were identified. The pivotal actors to whom special credit can be given for materializing Almén's vision were the industrial chemists Holtermann, Nordal, and Haavaldsen at Nyegaard & Co. A/S and Felder and Pitrè at Bracco Industria Chimica SpA.

At a symposium in 1970, referring to low-osmolal contrast agents, and having in mind mostly ionic polymers, Sovak declared: Three spiritual fathers of this development, Drs. Hilal, Björk and Almén, are here. [Milos Sovak, quoted by Hilal (1970)].

Instead, with nonionic low-osmolal contrast agents in mind, Almén is meanwhile widely considered the only father of the new technology (Nyman, Ekberg, and Aspelin 2016a,b; Tweedle 2016). He himself did not a little to foster this image. To accept Almén as one of the fathers of nonionic contrast agents along with others from among the aforementioned individuals, i.e., Habicht, Holtermann, Nordal, Felder, and Pitrè, seems more fitting and in no way, diminishes his distinction. As a seminal father, he tirelessly tried to get his ideas accepted. As a caring father, he accompanied a first nonionic contrast agent through the development process, despite the fact that its principle application field did not match what he had originally in mind. As a champion of his ideas for products, Almén demonstrated a powerful inner drive that more than compensated for a superficial handicap in the form of an insecure demeanor. It also helped him to overcome his problematic interaction with Fondberg. In his drive, he resembled strongly Swick in the technology genesis phase.

Technological innovation can benefit greatly from academic luminaries as champions of a technology and/or products. Companies have their internal champions of ideas and products. In the course of the history of metrizamide and iohexol, Almén mutated from academic champion of a technological idea to product champion inside a company, and ultimately to technology and product champion in the outside world, qualified as academic luminary in the field. This quasi linear rendering of the career of a champion should not hide the fact that the company also needed champions of another type. Among this type of individuals occur some with strong, not always easy personalities. Some of them may disobey orders from above, as Habicht did[166] or be stubborn, as Holtermann was in his insistence on the development of novel contrast agents in an economically difficult environment. But stereotyping does not capture well the diversity found in real life. Typically, industrial champions of this type are given this title solely if, against all the odds, their deeds led to notably successful goods. They are typical material for "great man histories." I would like to emphasize again that there is much more to a new technology than the championship of a gifted individual.

Apart from industrial champions of ideas for products, industry needs champions of efficient project pursuit. The latter needs a leadership gift and acquired skills, supported by diversified knowledge. The champion of this kind is capable of performing well in the function of project leader or product manager. There exists a wealth of managerial know-how relevant for the selection of suitable candidates for these functions and their effective training. The names of this type of champion are rarely recognized outside the circle of industrial and academic circles participating in the development.[167]

All types of champions are in the end effective in technology genesis and shaping only as part of the network of actors and actions at the basis of the analytical model in evaluation. Thus, the concept of champion retains its usefulness in the description of the course of actions and events even in the model of Dierkes, Hoffmann, and Marz (1996), especially if further differentiated into champions of ideas for novel products and champions of efficient project pursuit regarding already identified products.

The technology-shaping phase was characterized by a great variety of pivotal actors forming a network of actors and actions. Most pivotal actors became members of clusters formed around companies. Within the clusters, interactions were particularly intense. Besides many interactions that turned out beneficial to technology improvement, there were some that hampered progress. Events transcending the technology-shaping issue, namely divestiture of companies (Cilag-Chemie AG, Erco Läkemedel AB, Eprova AG), were primarily to blame.

In the technology genesis phase consensus concerning the desirability of contrast agents for excretion urography and angiography was reached much ahead of consensus on feasibility brought the vision of the technology to maturity. The situation was not the same in the technology-shaping phase. For a long time, the desire for low-osmolality uro-angiographic contrast agents was not deeply felt by users and producers. Clinicians, in the first place those that had experienced the progress brought by the second-generation of products, could live with them and companies did not perceive a major market for novel uro-angiographic products in the face of the performance of existing ones and dramatically falling prices. A few individuals both within and outside companies instead felt otherwise (Hoey, Hilal, Habicht, Erikson, Ingelman, Fondberg, Almén), but possible research avenues generated little enthusiasm in the industry. Oligomeric compounds were pursued only with lukewarm support. Not even when Habicht and Zubiani at Cilag-Chemie AG demonstrated the feasibility of low-osmolal contrast agents based on nonionic molecules did interest grow significantly. Inside his company it did not suffice to convince the management under the new ownership, Johnson & Johnson [Corp.], of the worthiness of the effort. Attempts to sell or out-license the patents were similarly suspended. In the technology-shaping phase of present interest feasibility appeared on the vision horizon before consensus on desirability (medical need and market potential) was achieved. Evidently, medical need or market forces did not exert their influences similarly in technology genesis and shaping phases.

If one-dimensional reductionism in the history of contrast agent technology needed exemplification, no better one could be found than Almén's printed lecture entitled "Some memories from the 1960s on the origins of nonionic iodinated X-ray contrast media" (Almén [1994] 2001). Preoccupation with some causes of pain serves as a golden thread for linking successive stages of his personal development of ideas, efforts, and understanding. It provides the single overbearing motivation for the sequence of his actions. All begins with the sensibilization to the pain issue in childhood. It continues with the experience of pain and adverse events as an angiographer using ionic contrast agents and his recognition of their osmotic origin. Studies on contrast agent effects on batwings confirm the deleterious role of elevated osmolality. Ending it all are nonionic, low-osmolal contrast agents. Actual deviations from the path, pursuit of other types of projects, influences of competitors, unpleasant confrontations, and bothersome precedences, i.e., anything that could disturb the linearity of the narrative, were barely touched and more often actually suppressed. Yet only taking the suppressed information into account allowed detection of some erroneous chronologies. The present nonlinear history offers a corrective that is better suited for drawing generalizing conclusions on technology genesis and shaping.

Technology genesis and shaping phases resembled each other with regard to complexity of the actor's networks, absence of a uniquely enabling technological step, impossibility to arrange the events in a single causal sequence and heterogeneous driving forces. Like in the genesis case, the new technology could not be attributed to an individual. The widely practiced heroizing historiography risks teaching misleading generalizations about technological innovation. Any one-dimensional reductionism has to be rejected. In the technology-shaping phase, the analytical model of Dierkes, Hoffmann, and Marz (1996) passes its author's number 1 test. It is interesting that these authors based their conclusions on comparison of cases of completely independent technologies, namely the Diesel engine, the mechanical typewriter, and the mobile telephone, whereas the present conclusion is based upon a case of two materializations of the same fundamental technological innovation.

Technical Knowledge from Interference of Different Knowledge Cultures

Much of what has been said about the various knowledge cultures involved in the technology genesis phase, including their distinctive modes of knowledge production and reproduction, as well as documentation and conveyance, apply directly also to the technology-shaping phase selected for close examination, given that exponents from the same professional fields of activity took part. New in the phase scrutinized now was that all actors, individual, and industrial alike, had access to extensive shared knowledge of the technological field within which the search for technological progress made sense. Equally new was that pivotal actors from industry became numerically still more prominent than they had been in the earlier period.

As in the genesis phase, all pivotal actors pursuing novel technology produced and reproduced knowledge with primacy given to the object-reference dimension, but industrial actors differed strongly from their academic counterparts in their secondary reference dimension. For industrial actors (Habicht, Fondberg, Ingelman, Holtermann, Nordal, Suter, Zutter, Müller, Guy Tilly, Hardouin, Lautrou, Felder, Pitrè), the latter dimension, and in some cases, even the primary one, was the actor-reference dimension. Any of their efforts required negotiations between company employees and on occasions also with outsiders. Personal interest had to take second place to company ones. Academic scientists and physicians (Hilal, Björk, Erikson, Almén) instead were allowed to produce and reproduce knowledge in a more self-referential manner, cultivating curiosity, reputation, and career with higher priority. The differences in secondary reference caused interference of knowledge cultures, for technology first of the antiproductive, and subsequently of the productive variety. In the case involving Almén, Fondberg, Erco Läkemedel AB and Nyegaard & Co. A/B antiproductive interference may be blamed for a delayed start of a fruitful collaboration. Productive interference instead may be seen in the triumph over paralyzing economic pessimism prevailing at Nyegaard & Co. A/B in the case involving Almén, Holtermann, and Nordal.

The period under examination witnessed an interesting change in the knowledge production and reproduction of industrial scientists, chemists in particular. Felder characterized the change grossly in the following way: For decades drug discovery research had typically focused on identifying novel natural or synthetic chemicals within chosen realms and holes in existing patent coverage, with the hope of encountering clinically useful, and profitable classes of substances. Clinical indications were adjusted to the properties of available compounds based on broad biological screening. From the 1970s onward gradually, it became customary to start with clinical indications and desired features of a product and to design its structure rationally based on the clinical target and feedback from tailored biological tests (personal communication).[168] The new paradigm for finding molecules of interest rested on the possibility of constructing the outcome by rational manipulation of existing knowledge. The old one instead resembled more a card game with its randomly assorted hands analyzed rationally before taking action. Iodipamide dimeglumine and iocarmate dimeglumine are typical products born in the older tradition, metrizamide, iopamidol, and iohexol in the newer one. Those investigators who only in the 1960s became involved in compound design (Erikson, Almén, Fondberg) acted from the start within the newer tradition. They targeted angiography and low osmolality starting with theories on molecular structures, such as macromolecularity, molecular shape, nonionicity, and mediation of elevated water solubility by hydroxyl groups. In the knowledge production and reproduction, actors operating in the new way must be positioned closer to the manipulative/constructive pole, the others closer to the observational/analytical pole of the corresponding axis of differentiation. Researchers in most industries (e.g., Mallinckrodt Chemical Works, Laboratoires André Guerbet & Cie, Cilag-Chemie AG, Nyegaard & Co. A/B, Bracco Industria Chimica SpA) had grown up with the old way of proceeding but gradually changed to the new one.

Knowledge documentation consisted of two types. The first type consisted in descriptions, systematic chemical names and pictorial representations of organic chemical compounds, recipes for syntheses and pharmaceutical formulations, claims of invention, measurements of physical properties, product specifications, data on toxicology, safety pharmacology and pharmacokinetics, clinical indications, recommended clinical doses, administration modalities, interpreted radiographs, efficacy estimates, case histories, frequencies of adverse events, drug registration requests, market size estimates, production, distribution costs, etc. Within the analytical model, one would speak of a strong reliance on symbols for knowledge documentation. This kind of knowledge was preserved and readied for conveyance in the form of typed or printed vehicles, reports, publications, patents, production manuals, standard operation procedures, package inserts, radiographs, etc. The second type of knowledge documentation consisted of packaging design, labeling and logos, trademarks, expiration dates, patient consent forms, recommendations of precautionary measures and warnings in package inserts, advertisements, architecture of stands at exhibits, target prices, etc. Within the analytical model, they were symptoms and signals that documented and conveyed knowledge. Preservation of this knowledge relied mostly on archived samples and photos.

In the technology-shaping phase, an extreme conformity in the documentation behavior across academic and industrial actors and regarding various products could be observed. The conformity was much more pronounced than in the genesis phase. This may in part be explained by a closer similarity among actors, by the prevailing canon of science, and by the impact of dramatically increased government regulation. The exception was the trademarks. Whereas iohexol could be sold under a single trademark worldwide, unfortunately license agreements forced iopamidol to be distributed under a variety of different ones.

All participants in innovation used as vehicles of knowledge documentation and conveyance, among others, publications in journals, books, posters and oral presentations at congresses, company-sponsored symposia, company-sponsored brochures, directions for product use, exhibitions at stands, and personal contacts. Company employees and certain affiliated persons with secrecy obligations relied additionally on confidential vehicles, including internal reports, market analyses, study protocols, closed meetings, and patent applications. The way in which intellectual property was dealt with is again worthy of interpretation.

In the user community at large, patents were not part of the discourse. Even company employees in the course of interaction with technology users avoided reference to them. They did so for two reasons. First, patents had never entered the typical technology user's knowledge culture. Second, some users had an aversion to patenting of illness-related products.

In contradistinction, in the community of investigators immediately engaged in technological innovation, patents were obligatory elements of the discourse. They staked out the competitor's intellectual property that had to be circumvented in one's own pursuit of technological improvement innovation. Whereas in the technology genesis phase patents had assumed a role relatively late in the game, in the shaping phase they played a dominant role throughout. Patent disputes played no relevant role in progress. This was possible because in this case participants made effective use of appropriate documentation and conveyance modalities.

At the beginning of his career, Almén published papers on clinical issues (Almén 1966a, 1967), soon followed by some describing novel radiological devices (Almén 1965a, 1966a). While still being a trainee at a university hospital, and with its assistance, he applied for and was granted a patent covering a catheter steering device (Almén 1965b). With this action, he transcended the usual boundaries of knowledge documentation and conveyance of his profession and made a first foray into the world of technological innovation.[152] But he did not yet fully internalize the new modality, as evidence by his continued unwillingness to consider the patent literature. The affidavit signaled a change in attitude toward protection of intellectual property. In collaboration with Nyegaard & Co. on contrast agents, he produced numerous journal articles in addition to patents. He followed the latter for the rest of his working life, although rarely citing them. Almén's change in dealing with patents may be seen as the result of interference of industrial and academic knowledge cultures. Interference of knowledge cultures not only played an essential role in the technology genesis phase but also characterized the shaping phase. This is best illustrated by two cases, numbered 3 and 4.

CASE 3: Almén propagated with enthusiasm a theory for how to arrive at a class of angiographic contrast agents, showing elevated water solubility and improved toxicity because of low viscosity, diffusibility, and osmolality. Their myelographic application was outside of his focus. With the proposal of concrete chemical structures, the academic radiologist had ventured into the realm of synthetic chemistry and drug discovery under the new paradigm. The experienced industrial chemist, Holtermann, with a knowledge culture encompassing industrial strategy, risk taking, and leadership, was wise and humble enough to allow his decisions on chemistry research directions to be influenced by the proposals of a chemical layman with pronounced actor- and self-reference. Holtermann might have recognized Almén's great potential as a formidable champion for a new class of contrast agents, and as a flagship for his company should the proposals pave the way to commercial success.

The industrial chemists Nordal and Haavaldsen, acting predominantly within the old paradigm, synthesized a series of compounds varying in structure, within the limits of Almén's design principles. Without further theory, substituents were chosen and combined in an almost arbitrary way, bounded only by practicality and convenience (availability of chemical starting products, ease of synthesis). Because of the already running large-scale production of metrizoic acid, its iodinated core and precursors received preferential attention as starting materials. Broad biological screening revealed greatly varying neurotoxicity unrelated to osmolality. Old and new paradigms of drug discovery and company strategy had contributed through interference of corresponding knowledge cultures. The favorable neurotoxicity of metrizamide led to its selection for full-scale development as a contrast agent. But the primary clinical indication was changed to myelography instead of angiography. Osmolality of the contrast agent solution, emphasized by Almén and first accurately measured by Holtermann, on the way to metrizamide played no discernible role in the evaluation of candidate compounds. Despite the fact that the new clinical indication represented a deviation from his original objective, Almén espoused the new clinical indication with such enthusiasm that he even temporarily gave up his academic position as angiographer, becoming as company employee thoroughly involved in the pharmacological characterization necessary for drug registration purposes. He clearly internalized some aspects of industrial knowledge production.

Interference of various knowledge cultures did not only occur at the outer level, involving representative actors. Interference phenomena may be documented at the inner level too, i.e., within individual actors. When Almén connected his experience of ocular irritation due to the salinity of seawater to pain caused by hyperosmolal angiographic contrast agents, one may talk about a case of interference at the inner level between the knowledge culture of the sportsman and that of the radiologist. His design theory for low-osmolal contrast agents, together with his affidavit presenting chemical compounds, both composed single-handedly, are other splendid examples of the interference of different knowledge cultures within the individual. These cultures were as distinct as radiology, colloid physical chemistry and the chemistry of iodinated organic molecules, each having their own traditional modes of knowledge production and reproduction together with their modes of conveyance and documentation.

Physiology and pharmacology were the fields within which the hypotheses resulting from the interference were most easily tested. By diligence and some good luck, Almén found himself strategically at the optimal site of the interference through collaboration with Wiedeman at Temple University on microcirculation in bat wings (physiology) and participation at Nyegaard & Co. A/S in the preclinical development of metrizamide (pharmacology). Indeed, it was in Philadelphia that he progressed from rather immature ideas to the mature design theory for novel contrast agents, and in Oslo he saw his dream, initially in part only, later in its full splendor, become reality.

The patent covering metrizamide jointly authored by academic (Almén) and industrial (Nordal, Haavaldsen) scientists is a concrete trace of interference of knowledge cultures that differed in documentation and conveyance habits. Metrizamide, as a precursor of the new-generation contrast agents, was a technological artifact that showed the traces of interference of the knowledge cultures that gave rise to it. These conceptual traces took the form of compromises between original clinical goal (angiography) and practical limitations (stability, cost of production, complicated pharmaceutical presentation) of the best product candidate available at a decisive moment. As a consequence, Almén found himself participating in person in the development of metrizamide initially for myelography only, a contrast agent application in which excessive osmolality of contrast agents, his pet issue, was a problem of secondary importance.

An additional conceptual trace of knowledge culture interference in the advent of metrizamide is found in the variegated interactions accompanying the developments from dissatisfaction with available angiographic contrast agents to a myelographic contrast agents that for a while almost monopolized its market. Clinically explored typology of technical solutions and physical parameters to be mostly concerned with showed intertwining shifts. Industrial and academic rationales frequently interfered with each other, more often than not positively. Detailed examples include the passage from Almén's inspiration by the elevated water solubility of sugar to Nordal's awareness that his employer's own ionic contrast agent, metrizoic acid could easily be conjugated with a sugar, with advantageous chemical production cost. Another detailed example is the realization that Almén's proposed substituents rich in hydroxyl groups yielded some products with unexpectedly good neurotolerability, while failing to provide the desired general recipe for how to achieve elevated water solubility.

The interference of knowledge cultures had left concrete traces on AMIPAQUE at the time of commercial launch. A first such trace was its initial clinical indication, i.e., myelography. This clinical indication was the result of interference of Almén's academic knowledge culture, which was focused on improved uro-angiographic contrast agents, with the industrial one at Nyegaard & Co. A/S, including the subcultures concerned with chemistry, toxicity profiles, compound stability, chemical and pharmaceutical production, clinical doses, marketing and strategy. A related concrete trace was the burdensome pharmaceutical formulation of AMIPAQUE. It reflected in part the interference of the just mentioned industrial knowledge subcultures within the company. Actually, the generic and trade names, metrizamide and AMIPAQUE, may be included in the concrete traces of the intracompany interference. The first two syllables in the generic name (metriz-) and the last one in the trade name (-PAQUE) help the user to link the new product to the older one of the same supplier, metrizoate sodium (ISOPAQUE), thereby fostering customer retention, a marketing concern. The last two syllables in the generic name, which replace the −oate or −oic acid termination of all preceding contrast agents, and the lack of mention of a counter ion, remind the user of having to do with a nonionic product. This fosters product recognition and through this safety also, a concern of the regulatory affairs section of the producer, as well as the World Health Organization, which dictates the guidelines for creating generic drug names and manages approvals.

CASE 4: Metrizamide had generated the confidence that among compounds falling under the design principles that called for hydroxyl groups to substitute for carboxy groups as mediators of elevated water solubility, practical angiographic contrast agents could be found. In the search for products that exploited fully the potential of this technological solution, substituents selection oriented itself on theories (solubilizing effect observed in other drugs, effect of stereoisomerism on solubility and complexity of documentation for drug approval). For biological testing, metrizamide set standards of comparison and priorities, which streamlined the selection process. Osmolality assumed center stage in product profiling. Iopamidol and iohexol were the first products found by the new way of drug discovery, although elements of the old way still played some role.

In coming up with the chemistry of the two products interference of knowledge cultures was not striking. Straight confluence of different kinds of expertise sufficed. The crucial interference had already occurred on the way to metrizamide. In contrast to metrizamide, the game-changing exemplar of the new technology, iopamidol, and its close follower iohexol, lack any concrete traces that reveal how they came about. Fortunately, a frank description by Felder of the circumstances of iopamidol's conception has allowed a rare glimpse behind the curtains.

Examining instead the wider field of product development, interference of the knowledge cultures of the medicinal chemists (Pitrè, Felder and Suter), the manufacturing chemist (Felder), the entrepreneurs (F. Bracco, D. Bracco and Suter), and the intellectual property professional (Zutter) became evident. Actors with different preferred modes of data analysis participated. There were those that emphasized economic threats and opportunities (Felder, F. Bracco, D. Bracco) and those that focused more on understanding (Pitrè). In the case of the development of iohexol, participants expressing similarly divergent preferred modes of data analysis can easily be identified. The radiologist (Almén) and the medicinal chemists (Nordal, Haavaldsen), so to speak bench investigators, focused strongly on finding and understanding desired products, managers of company research (Holtermann, later Haugen) put additional emphasis on intellectual property protection (Haugen), and the entrepreneurs (U. Blix and others), just like the companies as organizations (Schering AG, Sterling Drug, Inc.), elaborated the events more in terms of economic threats and opportunities.

The issue of pricing of the new-generation contrast agents deserves consideration. Pricing integrates information about the product's clinical profile, its scientific and technical merits and limitations, its potential application dimension, the investment in its development and costs of manufacturing, distribution and promotion, with an estimate of what the market may bear or, in many countries, what governments may approve. But that is not all. It additionally reflects what meaning producer and various consumer types attribute to the performance of the product. By asking an elevated price for a nonionic contrast agent the entrepreneur expressed a corresponding confidence and pride in the product. For her/him its principle meaning was economic in nature and related to business development. For both Bracco Industria Chimica SpA and Nyegaard & Co. A/S nonionic contrast agents were correctly perceived as the great chance to evolve toward a fully integrated international firm, and to aspire at world leadership in a sector. To the patient with health insurance, the meaning of the new-generation product resided exclusively in reduced pain or discomfort during the diagnostic procedure.[169] He noticed technological progress especially if he had undergone the procedure with both old and new-generation products. In cases in which, for lack of insurance coverage and at their own expense, the patients were given a choice between old- and new-generation products, the positive meaning became mitigated by a higher cost. For the radiologist, the new-generation products meant among other benefits, less general, cardiac and renal side-effects, and less pain-induced motion of the patient. To the radiology technician it meant fewer instances of vomiting patients, causing unpleasant clean-up work, discontinuation of procedure, and patient schedule disruption. Lastly, for insurance companies of patients, physicians and hospitals, the meaning was associated with a reduction in adverse events requiring intensive care, cases of fatal outcome and related liability suits but as well an increased cost. This illustrates the interpretative flexibility of the meaning of a new technology. It is likely that all interpretations have contributed to the determination of how meaning justified the price. However, we are far from understanding how these interpretations complemented each other or, what is more, synergized in determining acceptance of the result, and thereby contributed to successful affirmation of the technology. We are even further away from being able to use our understanding to reliably predict acceptable prices for potential future contrast agents.

In light of these observations, the successful pricing of nonionic contrast agents, which allowed the new technology to conquer the market and the producers to blossom, has to be recognized as a positive contribution to technological innovation of its own. This contribution can be seen as the result of the interference of the knowledge cultures of the entrepreneur (D. Bracco, F. Bracco, U. Blix and others), the planning, marketing and sales people and that of the research and development personnel (Holtermann, Almén, Felder and Bonati), particularly with regard to conveying the clinical merits of the product).

In summary, the metaphor of new technology as the product of interference of different knowledge cultures found support in several episodes also in the second historical period examined. But the illustrations are somewhat less striking in this period (shaping phase) than in the first one (genesis phase). Most productive interactions of knowledge cultures were satisfactorily described as additive. Above I have proposed a modification of the analytical model that reduced the stringency of the requirement for interference. In this adjusted form, the model may be admitted having passed plausibility test number 2 of Dierkes, Hoffmann, and Marz (1996).

Interference as Communication and Individuation of Knowledge

Habicht and Zubiani failed to obtain the cooperation of Johnson & Johnson Corp., which had at the time of their discovery of nonionic contrast agents, just acquired Cilag-Chemie AG, their working place. Without entrepreneurial support, even the best idea for a product is usually condemned to oblivion. Habicht and Zubiani published their findings only in the form of a patent. Thus, they had negligible impact on the academic research community, which regretfully tends to ignore patents. The case illustrates that in the absence of adequate communication, inventions remain ineffective in terms of technology genesis.

The epitome of bad communication was that involving Fondberg, Bjäringer, Holtermann, Blix, Bjørnson and Almén up to the time the latter became a team member at Nyegaard & Co. A/S. Fondberg's solitary research initiative was so purely communicated and documented that almost every part of it today can reasonably be doubted. Almén's initial credulity is just another manifestation of neglected communication. Cooperation failed.

Secrecy dictated by ongoing negotiations about the divestment of the Erco company certainly played a further role detrimental to cooperation. Notwithstanding the passage of oral and written words, nothing ensured that the receiver understood what was meant; cooperation failed to materialize. Whatever valid ideas and experience may have been made available, they at least temporarily evaporated, leaving behind nothing but bad feelings.

This story splendidly illustrates the fundamental role of communication in its fullest sense in technology genesis and shaping. No undisputed key to an understanding of the causes of the communication failure could be identified, but the continued professional successes of all the participants of the affair suggest that something had been learned. In the case of Almén's manner of dealing with industry the affidavit constitutes tangible evidence.

Like Swick before, Almén carried an inner mission. Not intimidated by numerous initial difficulties he pursued the search for collaboration with industry until he had found the right partner in Holtermann from Nyegaard & Co. A/S. Despite the fact that communication with Nyegaard & Co. A/S had set off on the wrong foot, sustained cooperation was soon achieved, when Holtermann was enabled to rebuild a contrast agent research group and took the lead in communicating with Almén. Holtermann used face-to-face meetings in a neutral place (New York) to instill his interlocutor with trust in the company. Visits to Oslo by Almén were used to build confidence in the collaboration. Convincing proposals by Nordal on how to test Almén's ideas efficiently did the rest.

Although it was to its disadvantage Nyegaard & Co. A/S did not interfere with Almén's academic vocation to rapidly publish his insights. This tolerance must be seen as the price paid by industry for cooperation between an academic and industry with a difficult past. These actions constitute textbook-ripe illustrations of communication strategies judicious for the case. They were highly successful and for a while Almén became a true member of the research and development team at Nyegaard & Co. A/S. He maintained collaborations with the company for the rest of his working live; reproduction of cooperation was achieved.

As observed in the first historical period examined, coordination of sustained cooperation was very much the natural realm of company research directors. Holtermann's crucial role has been emphasized. Felder as research director at Bracco Industria Chimica SpA was responsible for implementing his company's changed innovation strategy and excelled in his role as coordinator of sustained cooperation. He coordinated the activities at Eprova AG and Bracco Industria Chimica SpA by initially assigning separate research areas. When shocked by becoming aware that the competition (Nyegaard & Co. A/S) was ahead in a technology held to be desirable but a long way off, it was he who recognized, earlier and still more astutely than Holtermann, the enormous business potential of substantially improved uroangiographic contrast agents. Given the high priority he now assigned to the corresponding research, as soon as resources could be liberated, he let the team in Milan join the effort of Eprova AG. Advice, experience, and compounds (e.g., serinol that became a solubilizing substituent of iopamidol) were exchanged. With iopamidol, he, together with Pitrè, had realized Almén's dream completely. With the help of Bonati for the organization of the clinical aspects, he managed the extremely rapid development of the product for registration with the health authorities. Through assistance to people responsible for engineering, manufacturing, and licensing he contributed to the build-up of a supply chain.

The role of coordinator of sustained cooperation is equally the role of the champion of efficient project pursuit, or project leader, outlined above. It requires radiating confidence, made believable by a successful track record and the reaching of milestones, such as patent filings and approvals, permission to initiate clinical experimentation, timely clinical successes in man, registration of new drugs, etc. It involves obtaining cooperation from members of many different knowledge cultures, from radiology, chemistry, and pharmacology to engineering, manufacturing, marketing, and general management. Holtermann and Felder were optimally prepared for that role by long and varied experiences. Communication understood in the sense of the analytical model was one of the forms in which interference of different knowledge cultures at the outer level became manifest. Where communication failed, even the best of ideas evaporated.

At the inner or individuation level, extraordinary accomplishments in internalization of distinct knowledge cultures are not difficult to find. As a radiologist, Almén mastered the transformation of considerations about clinical adverse reactions and physicochemical arguments about contrast agents and their pharmacological effects, into propositions involving a class of synthetic molecules, members of which eventually arrived in the hands of the practicing clinician. To this end, he did not shy away from appropriating the language of the patent chemist when he illustrated his affidavit with Markush formulas. In this respect, he went a big step further than Swick in the technology genesis phase, who only gave useless advice on compound structures to the chemists.

Almén never reached the sophistication of Binz and Swick in adapting his language to the various audiences he addressed, i.e., chemists, radiologists, industrial personnel, and entrepreneurs. In the beginning of Almén's involvement in contrast agent research the magic expressions around which his thoughts and proposals revolved were "low diffusion rate," "water-soluble macromolecules," "low osmotic pressure" (Almén 1966a), "low osmolality" (Almén and Wiedeman 1968b), "low osmolarity," "oligomers and polymers," "low viscosity," "axial ratio," "hydrophilic radical," and "nonelectrolytic" (Almén 1969a), the last of the listed expression meaning what today is usually called "nonionic." Remember, the expression "hydroxyl groups" appeared not before his affidavit (Figure 4.35), and for public reception, not before the published patent application (Almén, Nordal, and Haavaldsen 1969). To the radiologist the concepts of tonicity or osmolarity[170] were more familiar than non-electrolyticity or nonionicity. For the synthetic chemist, the opposite was the case, because the latter terms refer to properties of molecules, whereas the former ones have a meaning exclusively for their solutions in relation to physiology. Almén used some terms interchangeably and in various combinations irrespective of the audience. It might be argued that any good scientist could deal with all these concepts and immediately see their connections. In principle, this is true, and hindsight lets such mental flexibility appear trivial. Yet, it is evident that Almén's diffuse message was not easily received.

The fact is that a number of company representatives Almén initially had spoken to never followed up on the contact. Some major radiological journals did not believe his message to be of sufficient interest for publication. Even Holtermann and Felder initially qualified Almén's proposals to be of merely marginal relevance. Almén was a weak communicator. The weakness is likewise encountered in his choice of communication means. A telephone call is about as inopportune a means to propose a novel idea to a company as there exists, a typical mismatch between message and chosen means of communication (Stork and Sapienza 1992).

While in writing Almén appeared as self-assured, in direct contact with people he projected a rather shy and insecure personality. He lacked self-assuredness, the most preferred personality traits in negotiators. For this reason, one must admire all the more the tenacity with which he was repeatedly willing to face negotiating situations in the course of his numerous attempts to build collaboration with industry. To be an effective negotiator he would have been required to slip into a guise foreign to him but a better match to the knowledge culture of his vis-à-vis (internalization). In this respect, it appears his performance was poor. Instead he overcame the problem mostly by tenacity. Tenacity is one of his character traits, which became strikingly manifest also in projects he had pursued up to old age. Weaknesses in the area of communication may have been in part responsible for the very long time it took for his theory to evolve into a widely shared vision of a new technology. It is probably one of the contributing factors to his unpleasant confrontation with Fondberg.

Individuation of knowledge cultures is a demanding activity. No amount of experience let it become mere routine, as illustrated by the case of Holtermann. He had to obtain the cooperation of the company's top management in the form of the decision to industrially develop metrizamide. He had to get permission of concomitant research on follow-up products. He had to convince it of continued commitment to contrast agents in general, even in the face of serious research setbacks (reproduction of cooperation). Holtermann became so exhausted, that after metrizamide had reached the market and a patent covering iohexol had been applied for, but before the development of iohexol was decided, he renounced his management position in favor of a more technical one. He reduced the field in which he had to cultivate interference of knowledge cultures. His exhaustion reflects mostly individuation taking the form of strong, perhaps excessive, identification (internalization) of the individual with his task.

As the technology genesis phase did before, the technology-shaping phase provides evidence for the centrality for technological innovation of communication and individuation processes among actors representing diverse knowledge cultures. Actually, they may be identified with interference of knowledge cultures, in agreement with the analytical model (plausibility test number 3).

The Emergence of a Technological Vision

Against all skepticism of the radiological community in the 1960s regarding the desirability of better uro-angiographic contrast agents, if this would mean increased costs, several research groups kept up their search. A first group of investigators (Priewe, Hoey, Wiegert, Hilal) desired water-soluble uro-angiographic contrast agents with osmolalities lower than those of available products. A second group (Erikson, Ingelman, Fondberg, Almén) desired contrast agent showing vascular confinement. Both groups identified oligo- or polymeric ionic contrast agents as possible answers to their respective objectives. At the same time, each group recognized in the property pursued by the other a beneficial side. In this way, the first two steps toward a new vision were made (Table 4.5). We may let the recognition that excessive osmolality was a property of contrast agents to overcome, especially for angiographic ones, mark the beginning of a new technology-shaping phase.

With the exception of the group of researchers collaborating with or for Pharmacia AB (Erikson, Ingelman, Granath, Lindberg), or Erco Läkemedel AB (Almén, Fondberg), most experts in the field did not consider macromolecular contrast agents a realistic solution to outstanding problems, particularly since polyacidic dimeric and trimeric molecules had already been scrutinized at length and found to hold little promise. Experimental low-osmolal contrast agents composed of water-soluble nonionic compounds that would reduce cellular damage were explored (Habicht, Zubiani) years before any involvement of Almén in contrast agent design. Because the results were only communicated in the form of patents, this step initially went unrecognized by most of those who should have taken notice. For those who did take note (Suter, Zutter, Felder), the knowledge was to remain latent only temporarily. Yet in the absence of a broad impact, the earliest ideas and experiences with nonionic contrast agents contributed little to the formation of a new vision.

After many trials and tribulations with the ideas about oligo- or polymeric contrast agents and first experimental experience with measured osmolalities, Almén caught that glimpse of a novel vision for uro-angiographic contrast agents that eventually captured enough attention to begin revitalizing the sector. He drew on an analogy between the stinging ocular pain caused by seawater and the vascular pain encountered during injection of hyperosmolal contrast agents. He abandoned his previously assumed causation of toxic effects of contrast agents by hyperosmolality-induced increases in blood viscosity in favor of a mechanism akin to that causing ocular pain. He thought of reducing the *I/P* ratio of monomeric compounds, and thereby their osmolality, through abolition of charged groups. He imagined nonionic contrast agents with the necessary elevated water solubility achieved by replacing the carboxylate groups of ionic agents first with neutral hydrophilic substituents, substituents bearing many hydroxyl groups in particular. To justify this component of the vision he called upon the analogy to ordinary sugar and its elevated solubility. The third step in the formation of the new vision was made (Table 4.5). Some components of earlier visions continued to be implicitly endorsed. So were some of the associated design principles. Nobody expected a fundamental revolution, as, for example, a change in X-ray-absorbing atom would have brought. Clearly an improvement innovation was pursued.

However, before a new vision could be completed, some additional hurdles had to be taken. Primarily it was essential to gain confidence in the feasibility of nonionic compounds with sufficient water solubility. This confidence was initially lacking even among those who knew of Habicht's promising results and those that had synthesized nonionic compounds obeying Almén's design principles before they ever had become accessible to them (Felder, Suter). Even Holtermann counted among the doubters. The early success of Nordal with finding nonionic compounds possessing elevated water solubility provided exactly the evidence needed to convert one of the crucial skeptics (Holtermann) into a believer. Later, metrizamide, despite its limited chemical stability, provided similar evidence for investigators outside of Nyegaard & Co. A/S (e.g., Felder, Pitrè, Guy Tilly, Hardouin, Lautrou, Sovak). The newly gained confidence in the feasibility of the technology went a long way toward completing a new vision.

Still lacking was a widespread conviction that nonionic contrast agents would benefit angiography (desirability). Curiously, what constituted the point of departure for Almén turned out to be the last major stumbling block that prevented his ideas from becoming a widely shared vision. This obstacle was partially overcome when the low toxicity of iopamidol and iohexol were discovered. It was completely over-come only when clinical studies revealed their superior safety compared to ionic contrast agents.

One line of extrapolation based on the experience with solubility, chemical sta-bility, and industrial producibility of numerous contrast agent candidates during the quest that led through metrizamide, iopamidol and iohexol to the future, suggesting that novel improved compounds can still be found (feasibility) (Figure 1.1). A second line of extrapolation based on the toxicity of the same two to three compounds prom-ised, and in time offered, clinical benefits and thus also enlarged markets (desirabil-ity). The intersection of the two lines of extrapolation provided a vanishing point on the horizon, a point, which succeeded in orientating all actors, independently of their professional extraction, position, and academic or industrial affiliation. At last, the new vision of nonionic, low-osmolal uro-angiographic contrast agents, often termed third-generation contrast agents, was complete (Table 4.5). As previously demonstrated for the vision of contrast agents in general (Table 3.1), and ionic uro-angiographic ones in particular (Table 4.1), the vision of low-osmolal nonionic ver-sions for uro-angiographic applications too came about in multiple steps (Table 4.5).

In parallel to the described activities, Lautrou and collaborators at Laboratoires André Guerbet & Cie explored monoacidic dimeric contrast agents for angiography. The product they created, HEXABRIX, appeared on the market in 1980, i.e., at least 2 years before iopamidol and iohexol did. It responded to the medical wish for low osmolality identified by Almén, but the chemical structure constituted a hybrid between Hoey's early and Almén's later proposals. Despite the product's respectable success in the cardiological niche market, it remained the unique embodiment of the type of molecule. The dominant vision, rightly or wrongly, called for nonionic com-pounds. Belatedly even the producer of HEXABRIX, Guerbet SA, through the com-mercialization in the late 1990s of the nonionic compound iobitridol (XENETIX) in Europe and ioxilan (OXILAN in the USA, IMAGENIL™ in Japan) has become a convert to the common vision.

By the mid-1970s, Almén's theory had definitely captured a critical mass of actors. The vision had been completed. Additional companies with experience in the field of contrast agents fell under its spell and began research. After a long hiatus in significant progress regarding contrast agents, clinical researchers had again new products to study. By exploring their application limits and elaboration of optimal protocols for their utilization, they contributed crucially with their knowledge culture to the shaping of the new technology. Company's marketing and sales forces too adopted the vision. The multitude of producers of generic drugs that attempted their luck at entering in competition as newcomers testifies that the vision had captured not only the establishment.

Again, technological innovation, this time regarding technological improvement during the technology-shaping phase, occurred under the influence of a vision. This vision had emerged in a stepwise fashion. The roles that analogies have played in the emergence of the vision have been pointed out, lending plausibility to the proposed role of analogical arguments in the genesis and shaping of technology in general.

Synchronization and Interference

As was already the case in the first historical period of special interest, also in the second period scientific meetings constituted the principle vehicle for synchronized communication and individuation. Of particular relevance were the regular reunions of the ISOPAQUE reformulation group, the meetings of the Swedish Association of Medical Radiology in Stockholm, and the 2nd Congress of the European Society of Cardiovascular Radiology. Outstanding examples of communication and individuation concerned properties of contrast agents as chemical products (degree of polymerization, electrical charges) and in solution (water solubility, osmolality, viscosity). These aspects were previously outside the focus of discussions. Interference of industrial and academic knowledge cultures lead to novel classes of candidate products (Fondberg, resp. Almén) and altered Almén's manner of dealing with intellectual property and job negotiations. Synchronous presentations at the aforementioned European congress of three novel and promising uro-angiographic contrast agents (iopamidol, ioxaglate meglumine/sodium, ioglumide) recruited new adherents to the vision, eventually unleashing an impressive competitive race for follow-up products and contrast agent market shares.

Comparable company histories and essentially synchronous but independent management decisions were made such that in the second half of the 1960s Nyegaard & Co. A/S and Bracco Industria Chimica SpA found themselves as contrast agent industries in strikingly similar circumstances. The contrast agent business was considered strategic, but niche products were pursued. The observed synchronicity up to that point in time finds no illumination by the vision of present interest.

Adoption of the vision and its synchronization function can instead illuminate the flurry of activity in the late 1970s and 1980s that, beginning with metrizamide, iopamidol, and iohexol, resulted in several additional uro-angiographic low-osmolal contrast agents. Before products were in use for a sufficient duration, no one could anticipate how much clinically superior the new product generation would really be. Similarly, no one could know how large a market the products would command. Up to that time the clinical benefits and economic potential for improved uro-angiographic products had been estimated as modest at best. The situation changed when the vision of uro-angiographic contrast agents leapt to investigative radiologists on the one hand and to marketing and sales, as well as business development functions, on the other. The former developed intense clinical research and publication activities. The latter switched from just serving an existing market, to creating an enlarged one, with measures that included courageous pricing. A complex worldwide web of licenses, cross-licenses, and distribution agreements formed. This required a great deal of synchronized communication by numerous individual and institutional actors. It offered many occasions for interference of knowledge cultures, especially of national ones (e.g., product importation, regulation, health providers, opinion leaders, pricing, insurance, etc.). Early on there was not yet any economic feedback from the market. Therefore, the early synchronization is better attributed to the vision than to the "invisible hand of the market."

Synchronicities both with and without a connection to a vision of contrast agents were revealed. As in the first historical period, in the second period the analytical model in its slightly simplified form passed its author's plausibility test number 4.

The Function of the Technological Vision

The new vision provided industrial and academic efforts with a clear orientation, namely, nonionic, low-osmolal uro-angiographic contrast agents as replacement for the old-generation ionic products. It continued to do so for decades. This orientation was understandable to actors from widely differing professional specialties and roles.

The vision furthermore exerted a motivating function. It motivated industry to invest in contrast agent innovation by fostering hope for a substantially enlarged market, to be opened up by products that had both improved clinical performance and profitability. Radiological investigators became motivated by the opportunity to delineate the utility limits of innovative products in established clinical applications and to explore new ones. Through the vision common ground was found even across competing companies. The complex web of license and distribution agreements among competitors in various countries for products satisfying the vision is a monument to the orienting and motivating functions of the vision.

Dissection of the function of a vision only into orientation and motivation subfunctions, as proposed in the analytical model modified by Kahlenborn et al. (1995), was easy to justify by the observations on the second historical period of interest. The analytical model in the modified form thus passed plausibility test number 5 of Dierkes, Hoffmann, and Marz (1996).

Technology Espousal and Resistance

Low-osmolal contrast agents responded to a medical desire, the importance of which was not widely agreed upon. As luck would have it, the new technology offered more benefits than had been expected, enabling it to blossom even in the face of only a limited utility delivered by a prototypical product (metrizamide). When iopamidol and iohexol finally arrived, their safety and tolerability was so strikingly superior to old-generation products that there was little discussion of their intrinsic merits. The development of a number of contrast agents so similar to the first two, that their differentiation on the market has been difficult, testifies to the espousal of the reshaped technology by producers and users alike.

Nonionic dimers were envisioned already in the first intimation of the design principles behind the third-generation contrast agents. Neither on theoretical nor on experimental grounds do they merit treatment as results of a separate technology-shaping phase. They have and will not substitute the monomers on a dominant scale, i.e., they fail Schumpeter's criterion for technological innovation relative to monomers.

Were it not for their substantially increased price, nonionic contrast agents would have been accepted immediately without hesitation by the clinician. As in the first period of special interest, elevated price differences, and to some extent also limited manufacturing capacities, slowed the conversion from old to new-generation products. Unlike in the first case, in the second one there arose some efforts to restrict the costlier products, with the accompanying difficulties of weighing in benefits to patients and service providers. The patient selection generated ethical and legal dilemmas for the physician and hospitals. In contempt of continued rhetoric, the new technology has meanwhile asserted itself in all developed countries. It constitutes technological innovation in the Schumpeterian sense. With continued price decays that are typical of all aging products, the same can soon be expected for the remainder of the world. The new technology will then have become the norm.

TECHNOLOGY CONSOLIDATION AND ECONOMIZATION PHASE

HISTORY

In the 1970s, in parallel with the development of nonionic contrast agents, dramatic developments in the X-ray imaging equipment took place. Computer tomography (CT) appeared on the scene (Webb 1990). Widespread the expectation took hold, that CT would render contrast agents superfluous. In the course of a lecture at a Berlex Symposium in 1984, Burton Drayer showed a slide entitled Famous First Word 1974/1984, with which he reminded his audience that in all preceding instrument developments analogous evaluations had proven wrong. And right he was. Classical CT created unprecedented demand for contrast agents.

Not far behind the arrival of CT followed the introduction of magnetic resonance imaging. This technique, progress in nuclear imaging and especially in ultrasonography, together with the improvements in CT without contrast agents, conspired in the 1980s to erode the role of X-ray-based excretion urography (Pollack and Banner 1985; Amis 1999). Even in the field of angiography these imaging modalities threatened to make significant inroads into the domain hitherto dominated by contrast-enhanced X-ray imaging. Some skeptics prognosticated the decline of the importance of X-ray technology in general.

Then, suddenly the tables turned. Accompanying the assertion of third-generation contrast agents in the 1990s another revolution in X-ray imaging occurred, the introduction of spiral and multislice spiral CT.[171] These technologies allowed acquisition of three-dimensional information at unprecedented speeds. Third-generation contrast agents on their own but also in combination with new imaging equipment, offered rich new areas of clinical exploration for academics. To date more than 2,000 articles have been published solely involving iopamidol. While some examinations became possible without previously required contrast agents, or with less of them, new ones absolutely necessitating them were added. The total number of imaging procedures performed continued to grow. To date the momentum has not yet diminished

(Lusic and Grinstaff 2013). It looks like X-ray-based radiography will continue as the major diagnostic imaging modality for the foreseeable future. Propitiously for the contrast agent producers, the technological advances in the equipment sector and the associated diversified clinical applications have increased the demand for their products. The market for X-ray contrast agents in 2015 was upward from three billion dollars and growing.

More than 30 years have passed since iopamidol and iohexol, followed by similar third-generation products, shaped radiological practice, and contrast agent business with a saltatory improvement innovation, nonionic uro-angiographic contrast agents. In comparison with the clear improvement for the patient these two products had brought, additional improvements delivered by members of the category that followed, both monomeric and dimeric, remained very modest, if demonstrable at all. The chemical structures of all members of the category reflected the same design principles. Numerous laboratory products designed on the basis of alternative principles have failed the practicality test. In 1998, with ioxilan the last member of that generation of products was commercially launched. Since then, i.e., for at least 17 to 20 years, no new product has been introduced and none are expected in short order. Contrast agent technology had entered a frozen state, reminiscent of the mechanical typewriter example referred to in Section "The Analytical Model of Technology Genesis and Shaping." This hiatus strongly indicates that the technology-shaping phase had come to a closure and the consolidation and economization phase of technology had initiated. Actually, the change in phase may be dated to the mid-1990s and the expiration of patent coverage of the most successful products.

As a further sign of the consolidation and economization phase the competitive arena became driven by reductions in all types of costs, introduction of manufacturing processes with reduced risks and environmental impact (de Haën 2009), shifts in marketing strategy from an exclusive emphasis on products performance to one of bundling contrast agents with logistic services, disease management, training of clinical personnel, information technology-based patient management solutions, research contracts, as well as imaging and injection equipment (Wild and Puig 2004). In the economization phase, several third-generation contrast agents increasingly became perceived as interchangeable, and economic considerations dominated buyer's choices. In the health care industry, strong consolidation of service providers occurred and these could negotiate lower prices (Gopal 1996; Ascher 1996). Although the economization phase has brought drastic reductions in profitability for contrast agent producers, their business is still going strong.

Toward the end of the 20th century, the economic climate for further innovation in X-ray contrast agents had deteriorated. Various changes in ownership of producing industries[30,37,118,164] led to the loss of experienced R&D personnel, reductions in the sizes of discovery research groups, or their redirection to other imaging modalities, in conjunction with program interruptions and terminations. Most importantly, the failure of isoosmolal products to deliver the expected benefits was discouraging. In spite of this, not all research came to a halt.

Repeated efforts in the 1980s to collect customer's wishes regarding contrast agents turned out to be disappointing. Most respondents felt no need for new products and wished merely more of the same at lower prices. The sparse suggestions for new products that could be collected showed little consensus. Within the industries likewise, there was little consensus as to the most desirable drug discovery objectives. A wide variety of distinct research avenues were and continue to be pursued in industry and especially in academia. Contrast agents with novel qualities that can be produced in the laboratory and show some potentially useful imaging effects in animals, abound (Lusic and Grinstaff 2013). Alas, almost all are completely unrealistic in one or more of the following respects: large-scale manufacturing, stability, elimination from the body, existence of old patents, size of potential market, and possible return on investment.

Among the many avenues of exploration, I have chosen only two for short considerations. Just enough details are given to illustrate the difficulties for technological progress in the consolidation and economization phase. A scientific, fully referenced analysis of the situation would go far beyond the scope of the present treatise.

A first avenue of exploration aimed at products with reduced toxicities among relatives of the iodinated contrast agents of the third-generation. Some researchers held it still possible to find, guided by the prevailing technological vision and following trodden paths, compounds and pharmaceutical formulations for which it would be possible to demonstrate relevant clinical advantages over existing products. The fact that the toxicity of the dimers, iotrolan and iodixanol, differed in a clinically relevant manner nurtured that hope. Apparently already in the period of existence of Nycomed Amersham plc (1997–2001)[117] the nonionic dimer ioforminol with the code number AN-113111 was invented. After the company's acquisition by GE Healthcare Inc., patent protection was sought (e.g., Thaning 2007) and the code number was changed to GE-145. The compound stood out by allowing solutions of lower osmolality than iodixanol, thus enabling isoosmolal formulations containing higher concentrations of electrolytes, in particular sodium chloride and calcium chloride. Reminiscent of what the company's predecessor, Nyegaard & Co. A/S, had pursued much earlier in its metrizoate salt reformulation program (see Section "Perfecting Ionic Uro-Angiographic Contrast Agents"), optimization of electrolyte composition was examined as a way to improve cardiotoxicity, foremost the reduction of ventricular fibrillations and hemodynamic effects during coronary angiography.

Studies in rats failed to reveal a toxicological advantage of GE-145 over the company's own product, iodixanol (Wistrand et al. 2010). In pigs, favorable cardiotoxicity of GE-145 injection could be shown (Chai et al. 2010). In 2014 ioforminol injection 320 mg(Iodine)/mL was the object of a phase 2 clinical safety study in elderly subjects undergoing coronary procedures. The incidence of renal biomarker-based contrast-induced acute kidney injury (CI-AKI) was a designated safety issue of the trials. Besides this clinical effort, a patent was requested covering X-ray diagnostic procedures with contrast media having low contrast agent and elevated electrolyte concentrations (Newton et al. 2012). It looks like GE Healthcare is gambling on discovering a clinically relevant performance of ioforminol injection that would make it sufficiently stand out from currently available contrast agents, including its own iodixanol injection, to make a return of investment plausible. All developments appear to be discontinued.

In 1989 Sovak (see above), during a period in which he was collaborating with Schering AG, Berlin, on clinical applications of iotrolan, invented a class of nonionic dimers among which is found iosimenol. Iosimenol at typical iodine concentrations allowed isoosmolal formulations that in addition showed a dynamic viscosity around 30% lower than possible with other dimers. At early times in the history of iosimenol low viscosity was considered desirable, principally because it would allow the use of small injection needles. In the past, elevated viscosity was also occasionally speculated, as it appears now wrongly so (Hutcheson et al. 1999), to cause shear stress mediated effects on vascular endothelium when contrast agent temporarily substituted blood completely in vessels during coronary angiography. Sovak obtained for his invention a patent (Sovak 1989). Schering AG became the assignee of the patent, but it and its successor company, Bayer Pharma AG, showed no inclination to develop the new compound.

Undeterred by an initial lack of interest by industry, Sovak on his own set out to develop iosimenol as the first low viscosity and isoosmolal contrast agent. He performed preclinical studies at the Department of Radiology, University of California, San Diego Medical School, La Jolla, CA (Sovak et al. 2004), and convinced himself to have at hand the first example of a new generation of products (Sovak et al. 2005). His company Interpharma Praha AS, Prague, Czechia, with experience in the manufacturing of generic products,[172] by 2003 organized clinical trials up the phase 2/3 level.

As explained earlier, dimers suffer from prolonged retention in the kidney. This is owed mostly to excessive viscosity at the elevated concentration in renal tubules made possible by low osmolality. Gradually it was realized that the pursuit of isoosmolality alone held little promise. Eventually it became seen even as misleading (Seeliger, Lenhard, and Persson 2014). Based on these insights, contrast agents that form isoosmolal solutions of low viscosity attracted interest. Meeting these requirements, in the midst of development, iosimenol acquired a new justification.

In the period from 2004 to 2008 Mallinckrodt, Inc., as applicant deposited patents both for a chemical manufacturing process (Bailey et al. 2008) and a pharmaceutical formulation for injection (Periasamy and Doty 2004) of iosimenol. Most likely Mallinckrodt had acquired a product license from Schering AG. After this nothing more on iosimenol was heard from Mallinckrodt. Then, as a surprise, a newcomer to the contrast agent business appeared on the scene. On August 26, 2008, the large corporate group, Otsuka Pharmaceutical Co. Ltd, Tokyo, Japan,[173] announced the acquisition of Interpharma Praha AS, and with it the rights to iosimenol. Sadly, half a year later Sovak died (Anonymous 2009) and with him iosimenol lost its champion. International Otsuka subsidiaries took over the organization of clinical trials. In 2014 reports on phase 2 results were published (e.g., Meurer et al. 2015), but evidence at this time for ongoing clinical trials is missing. It appears development of iosimenol has been discontinued.

The case of iosimenol splendidly illustrates what happens when a reigning technological vision has lost its lure and a new one is not yet available. Collective pursuit of a goal becomes spotty and half-hearted. The case serves as an example of the opposite of the effective technology genesis and shaping under the spell of a vision in earlier periods. Thereby it bolsters the sociological theory under scrutiny.

A second avenue of exploration resurrected the dream of the blood pool contrast agent, which, except for long-term safety problems, had been splendidly fulfilled by colloidal thorium dioxide particles, THOROTRAST. Foremost among numerous researchers, Harry W. Fischer had dedicated an important portion of his lifetime of research to the identification of an equally well-performing but safe follow-up product (Fischer 1990). Alternative, nonradioactive metal oxide colloids, e.g., tantalum oxide (Thomas, Henry, and Kaplan 1951), stannic oxide (Fischer 1957), and various types of apatites were examined (Deutsch et al. 1994). Extremely prolonged persistence in the body of all these solid particulate agents constituted risks for patients.

One way of avoiding prolonged persistence in the body while maintaining blood pool agent behavior was accorded by the preparation of various solid colloidal iodinated organic products that in the body could be biodegraded to established water-soluble contrast agents (e.g., Fischer 1977a,b; Lauteala, Kormano, and Violante 1984). Unluckily, from an excretion and safety standpoint, their intravascular use never looked promising, not to speak of their excessive potential cost of manufacturing. The same can be said about metal cluster compounds, envisioned more recently (Yu and Watson 1999; Berger et al. 2011). Since the disappearance of THOROTRAST no company has taken the risks associated with the development of X-ray contrast agents based on colloidal solid particles.

The last really propitious attempt at blood pool contrast agents initiated at E. R. Squibb & Sons, Inc., Princeton, NJ, and involved liposomes filled with ionic or nonionic contrast agents (Mackaness and Hou 1978). Such products additionally promised to be accumulated in organs rich in macrophages, through this facilitating the imaging of liver, spleen, and bone marrow. The two decades following the first publication of the concept saw a flurry of both university- and industry-based studies in this field and the imaging potential of such products in animals was rapidly demonstrated [for review see Sachse (2008)]. For certain liposome formulations, industrial laboratories purportedly have overcome the major hurdle of manufacturing on a large scale (e.g., Minchey et al. 1988; Schneider, Tournier, and Lamy 1987; Adzamli et al. 1990; Schneider et al. 1990; Tournier and Lamy 1995; Sachse, Rossling, and Leike 1993). Some clinical trials could be performed (e.g., Spinazzi et al. 2000; Leander et al. 2001). Taking into consideration all aspects, the results did not bode well for a profitable product. Experienced contrast agent developers and producers appear to have abandoned further efforts.

A few years ago, the start-up company Marval Biosciences, Inc., San Diego, CA and Houston, TX, was granted puzzling interrelated patents for liposomes filled with soluble iodinated contrast agents and bearing on their surface polymer chains, for use in imaging of leaky vasculature (e.g., Annapragada et al. 2004). A particular implementation of the invention consists of iodixanol encapsulating liposomes bearing on their outside polyethylene glycol appendices that protect them from being efficiently captured by macrophages and maintains them well within healthy vasculature. They leak from the defective vasculature in tumors that can for this reason be detected by X-rays. The product called PEGylated liposomal iodixanol injection, with code NCTX, in 2014 has entered phase I clinical trials, i.e., its safety is being tested in healthy volunteers. The technical obstacles, such as shelf life and leakage of contrast agent from the liposomes, as well as the economic hurdles are enormous. The development has apparently been discontinued.

Given the large number of companies and public institutions, too large to be listed here, at which research on particulate contrast agents had been performed at least into the late 1990s, the question arises spontaneously: Was this dogged persistence meritorious behavior of the kind often needed to break an impasse, or was it simply a reflection of a lack of innovative ideas? The fact is, neither these efforts nor renewed searches for macromolecular soluble iodinated contrast agents (Meyer and Legreneur 1991; Doucet et al. 1991) have identified products whose clinical utility would promise a reasonable return on investment. Furthermore, the developments in X-ray instrumentation, foremost the increase in three-dimensional image acquisition speed, have rendered blood pool contrast agents obsolete. A recent review of iodinated blood pool contrast agents restricted its attention to preclinical aspects (Hallouard et al. 2010), thereby tacitly admitting that clinical use is at best a remote possibility. If contrast agent technology needs innovation, it has to come from an altogether different corner.

In parallel with the described efforts third-generation contrast agents suffered the passage from proprietary to generic products. Experience in the pharmaceutical industry had led to the credo that immediately after the expiration of patent protection of a proprietary product generic drug companies have a splendid opportunity to enter the field with drastically reduced prices made possible by low development costs and lean company structures. By the end of 1994, the basic patents covering iopamidol, and mid-1996 that covering iohexol, expired. The market for nonionic contrast agents whetted the appetite of genericists. A raft of companies, newcomers to the field, announced their intention to bring generic iopamidol to market, e.g., in the USA alone Abbott Laboratories in Abbott Park, Faulding, Inc. in Elisabeth, ESI-Lederle in Philadelphia, Division of Wyeth-Ayerst Laboratories and Brightstone Pharma, Inc. in Cary. Some did get their product approved by the US Food and Drug Administration. But reality rapidly caught up with them. Few products were ever commercially launched on the US market. One reason was that by that time prices already had dropped strongly (Wilmot, Mehta, and Jha 2012), in part because of competition among at least eight similar products. Another reason was that the market had become very cost-competitive. On September 11 and December 18, 1996, the electronic magazine *Diagnostic Imaging* (*UBM Medica*) came to this conclusion considering the arrival on the market of OXILAN and the approval by the FDA of Faulding's generic iopamidol. Decades of experience with large-scale chemical manufacturing (plant scale for hundreds of tons per year requiring 10–30 times those amounts in raw materials) and continued process research had honed established manufacturer's ability to produce economically. For a newcomer to be competitive meant the building of costly product-specific plants. This constituted a major barrier to both entry into and exit from the activity. Last but not least, the customer was no more satisfied with bottled products alone. He expected diversified pharmaceutical presentations, various contrast agent associated services and research opportunities. This could only be delivered by a marketing and sales force and R&D personnel with a tradition in tune with the imaging specialist, not by one trained on other medicinal areas. The new-coming supplier of generic products would have to build up his proper worldwide sales force, a massive and risky investment.

Despite the difficulties in many parts of the world iopamidol and/or iohexol, in one case even ioversol, are today offered on the market by companies without proper experience in contrast agent innovation, including Agfa HealthCare Imaging Agents GmbH, Cologne, Germany, iMax Diagnostic Imaging, subsidiary of Hovione FarmaCiencia SA, Loures, Portugal, Tae Joon Pharmaceutical Co. Ltd., Seoul, South Korea, Zhejiang Hisin Pharmaceutical Co. Ltd., Taizhou, China, SRS Pharmaceutical Pvd. Ltd., Mumbai, India and Teva Seiyaku K.K., Nagoya, Japan, Sanochemia Pharmazeutika AG, Vienna, Austria. They have to compete with the traditional innovators, which have invested heavily in manufacturing facilities and distribution channels located worldwide. The multitude of newly entering companies should not obscure the fact that so far, their market share has remained small.[174] The principle competitors in the field, except for changes in ownership and some acquisitions of companies active in contrast agent related fields, have remained essentially the same few that had developed proprietary third-generation contrast agents. Their revenues allow continued investments in research and development of contrast agents, although mostly outside the field of X-rays.

Aside from making substantial investments in manufacturing and distribution, major industrial players adapted to the loss of proprietary status of their products through extension of clinical indications of existing products, pharmaceutical formulations for new applications, e.g., formulation for oral administration in cases of suspected gastrointestinal perforation (Bell, McKinstry and Mills 1987), the adaptation of vial sizes to specific applications and the development of user-friendly primary packaging, such as bulk packages, plastic bottles and sacks, and importantly, prefilled syringes for power injectors. This development is a sign of a mature technology and market.

Over most of the period of contrast media technology genesis and shaping the modalities of intravascular injection to patients had remained essentially fixed, namely the injection via handheld, freshly filled syringes. This changed with the arrival of fast computer tomographs and the improved diagnostic approaches they brought. The newly desired elevated injection rates and fractionated volume control could only be met by power injectors. Initially, dedicated firms promoted the power injector technology, but with time most of them were absorbed by contrast media producers. In part, this reflects the fact that the latter, with their pharmaceutical experience, were better prepared for the development and production of prefilled syringes. For certain applications prefilled syringes for such injectors were developed which, with their enormous size and resistance to pressure, marked a milestone in the genesis and shaping of injection technology. Innovation in this technology could compensate for lack of innovation in the chemical compound area.

The Search for a New Vision

The last chapter has illustrated the widespread perception that with the assertion of the third-generation contrast agents the vision that had guided technology genesis and shaping up to that time had exhausted its appeal. The few holdouts that still search along trodden paths are the exception that proves the rule. The reigning lack of shared research orientation and the curtailing of resources in the industry shows what it means to be without a vision. Unarticulated was the hope by some players of coming up with a new, or at least substantially altered technological vision. It was clear that some elements of the old vision would have to be included into the new one, not necessarily in explicit fashion. Most wanted were new kernels of a vision.

New elements of visions suitable to complement the previous ones have sprung up in various forms since the beginning of the technology consolidation and econo-mization phase. As already discussed, "Blood pool contrast agents" that would allow dramatic reduction of the invasiveness of numerous procedures seemed at one point a noble enough objective to become the kernel of a revised form of the existing vision, and this despite the fact that few believed such products could ever substitute the current ones to a major extent. The return on investments simply seemed unat-tainable. "Contrast agents as cheap as water" may instead be the potential kernel for a vision readable between the lines of articles concerned with health care cost containment. Unfortunately, it runs contrary to continued innovation that faces ever-increasing development costs. These increased costs are in good part explainable by two causes, namely increasingly stringent regulatory requirements by health author-ities and changes in cost reimbursement policies by health insurers. Taken together, they require extended clinical trials aimed at separate proof of efficacy for each additional clinical indication, even minor ones, and trials that demonstrate a positive cost–benefit balance. Increased environmental protection constitutes an issue that is more likely than not to increase costs further. "User-friendly contrast agents" may be the proposed vision element that has led to prefilled syringes reaching the market. "Contrast agents from the tap" appear to be the vision element behind multidose packages and automated administration devices based on variable and multidose peristaltic pumps. Hospital labor costs versus environmental impact of packaging materials, cost to the patient and health care system versus safety for the patient, and legal protection for the drug producers are examples of the opposing demands in need of reconciliation under the last two proposed elements of a vision.

Although activities addressing the various aspects captured in the aforementioned elements of new visions are ongoing, none of the latter has so far evolved into a full-fledged vision. They failed to achieve the broad consensus on the combination of desirability and feasibility, economic feasibility included, which singles out success-ful visions. In the face of these realities, it can only be hoped that the economization phase of the discussed technology will not deprive the contrast agents sector in its entirety of the financial incentives to innovate, whether in the very area of X-rays or in adjacent areas of interest, such as those of magnetic resonance imaging, echogra-phy, various imaging methods based on radioisotopes or light imaging with fluores-cent imaging agents.

Hope for a promising future for the contrast agent sector usually springs from developments that occurred while this book was in writing. In the wake of the dramatic developments in molecular biology and genetics and the instrumentation needed in their pursuit was born the technological vision of "personalized medi-cine," a form a medicine wherein therapy in all its forms is tailored to the genetic and epigenetic constitution of the individual patient. This constitution of a patient may be analyzed by *in vitro* experiments on nucleic acid extracts from easily obtain-able cells. In contrast, detection, localization and molecular specification of tumors and sites of infections, tissue degeneration, or drug action can benefit greatly from imaging in vivo. Thus, essentially in synchrony with the vision of "personalized medicine" was born that of "molecular imaging" (Nunn 2007). No radiological congress these days passes without a major lecture on this subject, and this despite the fact that for radiologists, prospects for being major protagonists and beneficiaries are much slimmer than for their competitors, the specialists in nuclear medicine or ultrasonography. In particular, the limited sensitivity of X-ray technology, with or without contrast agents, renders it generally ill-suited for molecular imaging. The exception is the earlier mentioned and time-honored contrast-enhanced imaging of the liver (Chapter 3). At best X-rays may contribute within a combination technology, e.g., a single machine in which X-ray CT provides a high-resolution anatomical image and scintigraphy provides the molecular identification and characterization of lesions. Magnetic resonance imaging unfortunately does not fare much better. Although echography has the requisite sensitivity, the nature of contrast agents, i.e., micronsized stabilized gas bubbles, limits its applicability mostly to intravascular targets. Instead, for nuclear medicine imaging the sky seems to be the limit. The embrace of the vision "molecular imaging" even by the radiologist illustrates the strength of visions. Visions emerge, have a career, are substituted, and eventually become superfluous. It can fairly be concluded that with the arrival of the vision "molecular imaging" the roughly 100-year-old one on X-ray contrast agents, the focal point of this book, has ceded its role.

5 Conclusions

The present investigation produced lessons in history of contrast agents, historiography of technology, and sociology of technological innovation. Extensive amplification of the contrast agent database with details, including retrieval of "deleted modalities," has generated a wealth of new insights into historical processes and connections. As the first and principle lesson in history, the study leaves a narrative about the advent of uro-angiographic contrast agent technology that does much more than merely refine currently prevailing ones. It contradicts them often drastically and identifies documents that in any further elaboration of the subject would require consideration.

A second lesson regards historical cycles. Division of the contrast agent history into phases suggested itself naturally. On a large scale, the case of uro-angiographic contrast agents, from its beginning to present days, easily fitted the version of the technology life cycle with genesis, shaping, and consolidation/economization phases. The shaping phase invited subdivision into consecutive subphases with their own technology life cycle. For example, the shaping phase was found already in the literature to be divided into subphases associated with so-called first-, second-, and third-generation contrast agents. Interestingly, the subphases repeated to some extent on a downsized scale the phases of the large scale. The observation leads to the hypothesis that nested technology life cycles are a characteristic of technology genesis and shaping in general.

Besides historical insights, the present analysis also offers lessons for historiography of technology. A first one derives from the recognition of the failure of "great man historiography" to capture convincingly the observed processes of technological innovation. Purely product-oriented narratives that describe technological progress through a succession of commercial products, winners *ad interim*, that come and go just like magic, cannot aspire to do better. More promising turned out an approach that gave social processes in the history of contrast agent technology an upgraded appreciation.

The frequent parallelism in genesis of ideas and realization of projects, observed in the case of contrast agents, offered empirical support for the opinion that even without the identified historical protagonists, technology genesis and shaping would have taken a similar course. What would have changed is the number and character of the individual actors and institutions, as well as the distribution of their roles but not the game as a whole. This situation is certainly not unique to the present case. It may actually be a frequent occurrence.

In a historical approach oriented on chaos theory (Shermer 1995), the case of contrast agent technology may be seen as a complex dynamic system characterized by a bifocal strange attractor. The trajectories of the technology alternated between proximity to the innovation focus and proximity to the consolidation focus, thereby delineating the shaping phase and its subphases. In this model, historical conditions may differ within certain limits from one period to the next, while the system continues on very similar trajectories.

A second lesson in historiography of technology regards patents. With few exceptions, historians of contrast agent technology, amateurs, and professionals alike have ignored in their chronicles the role of patents and neglected to cite them. Industries that owned them and developed products based on their teaching suffered similar neglect. Unfortunately, this situation is by no means rare. Remarkably, even industry-based historical writers rarely give insights into the circumstances that brought about progress of the kind embedded primarily in the form of patents. They are comforted in their behavior by the fact that the scientific journal literature cites patents rarely or ignores them altogether. Given the rigorous priorities patents convey and the essential role they play in technological innovation, as emphasized by Braudel (1982), one of the most eminent 20th-century historians, their neglect is utterly astounding and regrettable.

The following are some reasons for the situation: Ignorance of the functioning of the patent system is widespread. Means for following the corresponding literature (abstracting/alerting services) are costly. Until in the late 20th century patents became electronically accessible, searching for them was difficult. Assessing their pertinence is nontrivial. The academic community tendentially holds a deprecating attitude toward patenting and associated profitability, particularly in the context of illness-related inventions, although this is changing (Li, Azoulay, and Sampat 2017). Moreover, the deprecating attitude toward patenting is institutionalized in two ways. First, at least up to the times of special focus here, most academic scientists involved in contrast agent research never had received education in the field of patenting. They have never read, let alone written, a patent in their entire career. Obviously, most professors were unable to teach any aspect of pertaining to them. To some degree this continues to be the case. Second, instructions to authors and reviewers of manuscripts for scientific journals are deficient in instructions for how to deal with patents. Reviewers of manuscripts are asked to check that the earlier literature is adequately cited. Given the low standing of patents, their consideration falls traditionally outside the obligations of a referee. Plagiarism of their contents and false novelty claims in the scientific journal literature is one of the results. Neglect of patents reflects also a weak spot in the quality control of academic research. The manner in which they are considered in the literature in general, and in historical writing in particular, needs to be modernized. A small first practical step proposed here regards a purposeful citation style for patents in historical papers.

Citing patents in a meaningful manner is inherently less straightforward than citing other types of publications. Scientists, engineers, lawyers, and historians have quite different needs. The citation style versions for patents recommended by the manual of the US National Library of Medicine (Patrias [2007] updated 2015) come currently closest to meeting the requirements of historians of technology. I propose and practice in this publication, like in an earlier one (de Haën 2001), a slight adaptation thereof. It meets still more closely the historian's needs. Details are given in the introductory comment to the reference section and are illustrated in the list of references. Worth special mention is the choice to use the priority date, rather than the publication date, for referencing in the text and for ordering the references in lists. It informs the reader of the latest possible date of the invention.[175] It helps avoid totally confusing chronologies of events resulting from the reliance on publication or issuing dates.[176] The date, country, and document number of the earliest publication of the patent content worldwide, be that an application only or a patent issued in another jurisdiction, are included. The earliest time the public could have learned about the invention is communicated.

Now to the lessons learned with regard to the sociology of technological innovation. Elucidation of the genesis and shaping of technology in terms of social dynamics is still a young area of endeavor. Satisfactory consensus on terminologies, methods, and generalizations has yet to be achieved. In order to overcome this difficulty, this study has sought orientation mainly in a single school of sociological thought, namely that of Dierkes, Hoffmann, and Marz (1996). As hoped for, the restriction to these researchers' analytical model has rendered the effort manageable for this author and hopefully will do so likewise for the reader untrained in sociology.

The choice to structure a historical account in relation to a particular sociological model, and the attempt to come to normative conclusions, implicitly expresses this author's belief that history is allowed to aim for more than what idealist and post-modernist philosophers are willing to concede, respectively, the mere construction of narratives about series of contingencies and the attribution to such narratives of meaning just as dependent on literary conventions as on empirical foundations. It is widely accepted that historiography oriented by the search for patterns akin to mechanical laws has remained disappointing. The power of the present approach lies in the avoidance of the shortcomings suffered by some that concentrate on the identification of causes and causative agents. Whatever insights on technological innovation processes are gained here lead instead to a nonquantitative, probabilistic description for conditions under which repeatedly distinct but similar and empirically verifiable phenomena will occur.

The analytical model operates principally with two metaphors. One of them introduces interference of knowledge cultures as a source of genesis and shaping of technology. The other one endows a vision, emerged in coincidence with the inception of a new technology effort, with the capacity to orient its thrust, while concomitantly motivating those endorsing it. The two are connected by the role the interference of knowledge cultures plays in the formation of a vision. The distinctiveness of knowledge cultures partaking in the genesis and shaping of contrast agent technology was self-evident, but the systematization proposed by the analytical model deepened appreciation of the cultural variety beyond professional specialization. The history of contrast agents suggested the pertinence of reference dimensions for knowledge production, reproduction, and conveyance additional to those of the original model. A first additional dimension regards the preferential use of the physiological senses. A second additional dimension covers preferred modalities of data acquisition and analysis. The newly added ones do not conceptually complicate the model, and they do not touch the core of it.

The examination of the contrast agent history for cases of productive interference of knowledge cultures, as opposed to cases of plain additivity thereof, was successful but required diligence. Both conceptual and concrete traces for interference reflecting technological progress were pinpointed. This brought support to the plausibility of the analytical model. But it failed to demonstrate interference as compulsory for technological innovation. In the contrast agent history protagonists differing with respect to preferred use of senses abounded, but interference involving knowledge cultures of this kind was predominantly of the antiproductive kind.

The contrast agent history allowed convincing identification of an evolving vision with a fixed core presiding over two historical periods that were completely separate in terms of times and actors. Furthermore, it led to the proposition that visions are the result of intuition at the inner, and analogical reasoning at the outer level, in concurrence with interference of knowledge cultures.

In the first place, the analytical model opened a window for a highly edifying perspective on a case of technology history. It offered powerful instruments for rendering intelligible the complexity of the genesis and shaping of contrast agent technology in the two historical periods to which special attention had been paid. In particular, it helped reveal striking similarities of social processes in them. Remember that the model had previously described well the genesis and shaping of the technologies of the Diesel engine, the mechanical typewriter and the cellular telephone, as exposed by Dierkes, Hoffmann, and Marz (1996). Now it captured with similar success two periods in the case of contrast agents. At least in the types of social processes involved, the model has demonstrated predictive power. Integration of the original analytical model by the additional reference dimensions for knowledge cultures and by the proposed role of intuition and analogical reasoning in the emergence of a vision showed the model to be upgradable. This was already expected by its creators. The science behind it revealed its fertility.

Given that the original analytical model in its slightly relaxed form fitted smoothly two distinct periods of major progress in contrast agent technology, it has passed the five plausibility tests proposed by its creators. Its application to the contrast agent history as a whole was less illuminating than in the case of the two historical periods contained therein. Instead, the mentioned version of the technology life cycle could easily be married with the analytical model, to the benefit of both.

Although many of the analytical model's facets found their illustration in the contrast agent history, not a few failed to do so, at least in my search. This may reflect that the model's concept map is very strongly, and may be even excessively, ramified. Production and reproduction of knowledge are clearly distinct activities. So are cooperation and reproduction of cooperation as well as internalization and reproduction of internalization. Little convincing are distinctions between coordination of reproduction of cooperation and coordination of cooperation or that between coordination of reproduction of internalization and coordination of internalization. Unclear is what the concepts coordination of reproduction of cooperation and coordination of reproduction of internalization are supposed to bring over what coordination of cooperation and coordination of internalization cannot capture. The infelicitous longer terms serve more the esthetics of a hierarchically structured concept map than explanatory power and may profitably be replaced by the shorter ones. Also combining production and reproduction of cooperation, or of internalization, in single production categories would affect the model only in a minor way. These examples illustrate how the analytical model might be slightly streamlined.

While concentrating on the network of actors and actions intimately and immediately involved in the genesis and shaping of a new technology, the analytical model neglects the roles of other components of reality. Little attention is paid to the consequences for the final success of a technology caused by competing technologies that arise and develop in parallel but with little or no interconnections. The role played by competition for resources between completely separate technologies receives scarce attention. Large-scale political structure, change and turmoil, rapid changes in belief and value systems, position of the time frame within macroeconomic cycles, as well as the nature, size, organization and quality of the institutional assets involved in research and development are just a few of the aspects neglected. While it is difficult enough to try to influence technology genesis and shaping on the basis of the insights crystallized in the analytical model, it would be just about impossible to influence most of the neglected components of reality. Thus, at least from an interventional perspective, the analytical model retains its justification even when its blind spots are admitted.

Now, does the present sociological analysis offer lessons for how to improve processes of technological innovation? Some concrete recommendations of modest profundity and limited novelty are indeed possible. For optimal communication, precondition for productive interference of knowledge cultures, types and styles of presented documents should in the first place be tailored to the typical mode of knowledge production, reproduction, and conveyance of the recipients, obviously without compromising the validity of the message. The role played by preferred senses in knowledge production, reproduction, and conveyance should be explicitly taken into account in such activities as communication with collaborators, team building, career planning, and job reassignment during reorganizations. This can be expected to reduce misunderstandings and friction caused by them, motivate personnel, and foster interference of knowledge cultures. Excellent books give guidance for how the success of language-based technical communication can be enhanced by optimal visual display of quantitative information (Tufte 1983). Additional benefits may come from involvement of other senses. For example, to a mixed audience called to make a decision regarding a technology it may be of advantage to present the same message in the form of a data table that includes error estimates and in the pie chart format, possibly accompanied by a story illustrating an emblematic data point, and finally an embodiment of the technology as a model to be handed around. A convincing example regarding medical risk communication has recently been published (Operskalski and Barbey 2016).

The analyzed history illustrated both the inevitable and legitimate practice of "deletion of modalities" in scientific and technical communication and its illegitimate counterpart. It is recommended that researchers be made conscious of this practice and its delimitation from misbehavior, including fraud. The differences in the practice in scientific publications and patents should be pointed out. Technology genesis and shaping could profit from enhanced familiarity of the academic community with scopes and importance of patenting and with its distinctive communication necessities, "deletion of modalities" included.

Questions deeper than those about types and styles of presentations regard the possibility of intervention on the core concepts of the analytical model, "interference of knowledge cultures" and "technological vision." Do these concepts merely help rendering historically documented technological innovation intelligible or does their success in achieving that, justify normative conclusions, a hope expressed earlier? Let us begin with pondering "interference of knowledge culture" and the possibility to elicit it. Uniting people in a common organization and/or location can be counted on to result in additivity of knowledge cultures. In this manner, the probability of interference occurring might be increased but not guaranteed. In any case, formation of interdisciplinary structures is a measure widely adopted by industrial and academic institutions with the intent to foster innovation. The history of contrast agents included a few examples of interference wherein actors profited from belonging to the same organizational unit, but as a whole, it produced scarce support for an importance of this condition.

Venturing into new areas has been recommended as a means for enhancing creativity (Pariser 1998), interference of knowledge cultures being one mediator thereof. In the contrast agent history, there are cases, where such interference caused actors to enter a new field and just as frequent cases of inverted order. Hence, no generalization is called for.

Lessons with regard to "technological visions" seem much more problematic than those regarding interference of knowledge cultures. One may distinguish between management of a vision and creation of a new one. Regarding the former, it is certainly possible for management to position an organization within a reigning vision and make corresponding efforts at promotion. The history of contrast agent technology documents the success of companies joining a common vision. But it also illustrates the difficulty of identifying the right moment for escaping its spell. Awareness of the phenomenon of technological vision in general and of the one presiding over ongoing efforts may help circumnavigate that difficulty.

As opposed to managing a vision, creation of one de novo seems to be a quite different challenge. Dierkes, Hoffmann, and Marz (1996) have struggled with this issue. Yet it is creation that is called for at the inception of an innovation period. According to the original analytical model, a vision originates by interference of knowledge cultures, and interference is fostered by a vision. In order to avoid circular reasoning, I have proposed here that the elementary step of interference involves proceeding by analogy, at the individual's internal level appearing as intuition, and at the external level as analogical reasoning, often cast in the form of metaphors. Metaphors are the material from which visions may emerge. Viewed this way, initial interference of knowledge cultures precedes formation of a vision, but in subsequent moments the two processes may shape the further evolution as intertwined and concurrent processes.

The contrast agent history clearly supports the positive role of visions in technological innovation. The question then arises whether it teaches as well how their formation can be encouraged. It does not do so explicitly. But if the proposed roles of intuition, analogical thinking and metaphors are admitted, creativity stimulating tools, such as the Six thinking hats-method of de Bono (1999) or other such techniques (Luther 2013), may be useful. In practice, however, such methods generally fail to gain sustained application in favor of rational discussions.

Given that visions can be as powerful in directing all the actors involved in the pursuit of a technology, as assumed in the analytical model, it is not astonishing that interest groups, such as industry, academia, government, regulatory agencies, political parties, and consumer associations, have tried to stimulate genesis and shaping of technology through the creation of visions by committee. Formulation of Technological Vision 2020 by the US chemical industry in 1996 (Schulz 1996) exemplifies such an attempt. Although the effort is likely to have some usefulness, I doubt that visions, as broadly based as this one, will show similar effectiveness as the kind of visions that have guided genesis and shaping of contrast agent technology.

It is conceivable that a thorough study of the emergence and development of visions may also lead to their more efficient utilization as instruments of societal guidance. Inevitably such instrumentalization would occur both for the good and for the bad, as it did in the case of Gustave Le Bon's insights into mass psychology. Promotions, which attempt to create visions, are of only limited efficacy, unless they magically hit the predisposed sensibilities of a sufficiently large population of addressees. As a consequence, manipulation of human endeavor and technological progress alone by a few forgers of a vision, be they exponents of businesses, religious groups, or political powers, seems intrinsically limited. What a blessing!

References

Given the central role congress presentations have played in getting correct chronologies for the present history, here the references are ordered according to the earliest date of presentation, given in square brackets. Outside the brackets follows the date of appearance in print. Square brackets around a first name initial of an author indicate that in this specific publication, his initial, although known, was not provided. Whenever possible, citations of patents make reference to the primary one. If the patent application (App.) that established the priority date has been withdrawn, for example, because a granted supranational patent superseded it, an exactly equivalent version granted in another state is cited. Most patents are part of an international family, but information about different members of this family is given only if they offer the sole way of retrieving useful information, for example, on inventors and applicant companies, or evolving issues entering the documents during the international patenting process. In order to best serve historical inquiry, references to patents give the following information in order: (a) list of inventors, set in upright type and closed with a semicolon. (b) List of applicants/assignees, as persons or institutions, set in oblique type and closed with a period. If the patent does not identify the inventors, in their place only the applicants are provided, and these may actually include the inventors. If the inventor is not identified but known from other sources, for example, an equivalent patent in a different national jurisdiction, the information is added in roman type within a parenthesis following the list of applicants. (c) Earliest priority date(s) (Prior.), with indication of the relevant country or international authority, if different from that of the cited patent. (d) Publication date (Publ.) If an earlier publication of the patent application in any jurisdiction is relevant to the cited issue, the type of document, its number, and the publication date are given within parenthesis at the end of the reference. (e) Patent title. (f) Patent number, including the country code indicating the country or international authority. From among the members of an international patent family, the one to be cited preferably reflects the patent issuing authority closest in relevance to the argument the citation pertains to. Short references to patents in the text and in the footnotes list inventor(s) or applicant(s), followed by the priority year (not publication year!). The order of entries into the list of references also considers priority year and adheres otherwise to the same rules regarding ordering as for other publications.

Abel, J. J., and L. G. Rowntree. 1909. On the pharmacological action of some phthaleins and their derivatives, with especial reference to their behavior as purgatives. I. *J. Pharmacol. Exp. Ther.* 1(2): 231–264.

Abrams, H. L. 1996. Cardiac radiology. In *A History of the Radiological Sciences. Diagnosis*, eds. R. A. Gagliardi, and B. L. McClennan, 255–269. Reston, VA: Radiological Centennial.

ACR (American College of Radiology. Committee on Drugs and Contrast Media). 1991. *Manual on Iodinated Contrast Media*. Chicago, IL: American College of Radiology.

Adzamli, I. K., S. E. Seltzer, M. Slifkin, M. Blau, and D. F. Adams. 1990. Production and characterization of improved liposomes containing radiographic contrast media. *Invest. Radiol.* 25(11): 1217–1223.

Ahlgren, P. 1973. Long-term side effects after myelography with watersoluble contrast media: Conturex, Conray Meglumin 282 and Dimer-X. *Neuroradiology* 6(4): 206–211.

Akhundov, T. S., A. D. Tairov, M. V. Imanova, and A. I. Iskenderov. 1989. Thermal equation of state for aqueous sodium iodide solutions [in Russian]. *Izv. Vyssh. Uchebn. Zaved./ Energ.* (6): 97.

Allardt, H.-G.; *Schering-Kahlbaum AG, Berlin.* Austrian Prior. 1930, June 12. Publ. 1932, July 19. Methods for Producing Alkali Salts of Iodomethyl Sulphonic Acid. Patent US1867793.

Allardt, H.-G.; *Schering-Kahlbaum AG, Berlin.* Prior. 1931, Dec. 2. Publ. 1933, May 4. Methods of Preparation of Brominated and Iodinated Alkyl Sulfonic Acids and Their Salts [in German]. Patent DE575678.

Allardt, H.-G., and O. von Schickh; *Schering-Kahlbaum AG, Berlin.* Prior. 1930, Sept. 17. Publ. 1932, Nov. 2. Methods of Preparation of 3,5-Dihalogen-2-pyridone-N-alkylcarboxylic- and -sulfonic Acids, Respectively Their Salts [in German]. Patent DE563131.

Almén, T. 1965a. Simple device for bowel examinations of patients with a colostomy. *Br. J. Radiol.* 38(445): 75–76. doi:10.1259/0007-1285-38-445-75.

Almén, T. H. O.; *Torsten Hakan Oskar Almén, Lockarp, Sweden.* Swedish Prior. 1965b, July 5. Publ. 1970, Mar. 17. Medical Probe for Injecting X-ray Contrast Medium into the Body. Patent US3500820.

Almén, T. 1966a, Sept. 1. A steering device for selective angiography and some vascular and enzymatic reactions observed in its clinical application. Chap. XI "Development of a water-soluble contrast medium with low diffusion rate." Translated from Swedish by L. James Brown. *Acta Radiol. Diagn.* 5(Suppl. 260): 141–151. doi:10.1177/0284185166005S26014.

Almén, T. 1966b, Dec. 10. *A Steering Device for Selective Angiography and Some Vascular and Enzymatic Reactions Observed in its Clinical Application.* (Akademisk Avhandling) 151 p. Malmö: Lundgrens Söner Boktryckeri.

Almén, T. [1966] 1967. Transpeptidase activities after celiakography [in Swedish]. Presentation at the Meeting of the Swed. Assoc. Med. Radiol., Jönköping, Sept. 17, 1966. *Nord. Med.* 77(6): 200.

Almén, T. 1967. Milk-of-calcium in a renal cyst diagnosed by angiography and puncture. *J. Urol.* 98(2): 175–176. doi:10.1016/S0022-5347(17)62848-9.

Almén, T. 1969a, Aug. Contrast agent design. Some aspects on the synthesis of water soluble contrast agents of low osmolality. *J. Theor. Biol.* 24(2): 216–226. doi:10.1016/S0022-5193(69)80047-0.

Almén, T. 1969b, Dec. 29. Contrast agent design. Some aspects on the synthesis of water soluble contrast agents of low osmolality. *Chem. Abstr.* 71(26): 128684j.

Almén, T. 1971. Toxicity of radiocontrast agents. Chap. 13 In *International Encyclopedia of Pharmacology and Therapeutics.* Section 76, Radiocontrast Agents, vol. 2, ed. P. K. Knoefel, 443–450. Oxford: Pergamon Press.

Almén, T. 1980. Experience from 10 years of development of water-soluble nonionic contrast media. *Invest. Radiol.* 15(Suppl. 6): S283–S288.

Almén, T. 1985. Development of nonionic contrast media. *Invest. Radiol.* 20(Suppl. 1): S2–S9.

Almén, T. [1994] 2001. Some memories from the 1960s on the origins of nonionic iodinated X-ray contrast media: Lecture to the Mälardalen's radiology club, Nov. 10, 1994. Little Chalfont UK: Amersham Health. In Swedish: www.seldinger.se/hedersmedlemar/pdf/Torsten%20Almen.pdf (Accessed Apr. 26, 2018).

Almén, T. 1995. Visipaque—A step forward. A historical review. *Acta Radiol.* 36(Suppl. 399): 2–18. doi:10.1177/0284185195036S39902.

Almén, T., and L. Fondberg. [1966] 1967. Macromolecular contrast agents for angiography [in Swedish]. Presentation at the Meet. Swed. Assoc. Med. Radiol., Stockholm, Dec. 2–3, 1966. *Nord. Med.* 77(18): 580.

Almén, T., and G. Nylander. 1962. Serial phlebography of the normal lower leg during muscular contraction and relaxation. *Acta Radiol.* 57(4): 264–272.

Almén, T., and M. P. Wiedeman. 1968a, May/June. Application of contrast media to the external surface of the vasculature. Effects on microcirculation in the bat wing. *Invest. Radiol.* 3(3): 151–158. doi:10.1097/00004424-196805000-00002.

Almén, T., and M. P. Wiedeman. 1968b, Nov./Dec. Application of monomers and polymers to the external surface of the vasculature. Effects on microcirculation in the bat wing. *Invest. Radiol.* 3(6): 408–413.

Almén, T., V. Nordal, and J. Haavaldsen; Nyegaard & Co. A/S, Oslo. Prior. 1969, June 27 and 1970, Feb. 9. Publ. 1973, June 27. Nonionic Iodinated X-ray Contrast Agents. Patent GB1321591. (German disclosure DE2031724. 1971, Jan. 7).

Almén, T., J. Haavaldsen, and V. Nordal; *Nyegaard & Co. A/S, Oslo.* 1971, May 10. Nonionic iodized X-ray contrast media. *Chem. Abstr.* 74(19): 99662e.

Almén, T., E. P. Steinberg, P. D. Jacobson, and J. E. Thomas. 1990. Low-osmolality contrast media—Issues and answers. Imaging symposium at the 76th Sci. Annu. Meet. Radiol. Soc. North Am., Chicago, Nov. 30, 1990. *Radiology* 177(P)(Suppl.): 34–35.

Altshuller, G. S. 1988. *Creativity as an Exact Science: The Theory of the Solution of Inventive Problems.* New York: Gordon and Breach.

Amdam, R. P., and K. Sogner. 1994. *Wealth of Contrasts. Nyegaard & Co. —A Norwegian Pharmaceutical Company 1874–1985.* Oslo: Ad Notam Gyldendal.

Amis, E. S. 1999. Epitaph for the urogram. *Radiology* 213(3): 639–640. doi:10.1148/radiology.213.3.r99dc47639.

Amundsen, P., O. P. Foss, H. C. Godal, and S. Nitter-Hauge. 1973. Intravenous injections of metrizamide into human volunteers. *Acta Radiol. Diagn.* 14(Suppl. 335): 339–345. doi:10.1177/0284185173014S33542.

van Andel, P. 1994. Anatomy of the unsought finding. Serendipity: Origin, history, domains, traditions, appearances, patterns and programmability. *Br. J. Philos. Sci.* 45: 631–648.

Andersson, M., K. Juel, and H. H. Storm. 1993. Pattern of mortality among Danish Thorotrast patients. *J. Clin. Epidemiol.* 46(7): 637–644. doi:10.1016/0895-4356(93)90037-2.

Andrew, E., K. Dahlstrøm, K. Sveen, and T. Renaa. 1981. Amipaque (metrizamide) in vascular use and use in body cavities: A survey of the initial clinical trials. *Invest. Radiol.* 16(6): 455–465.

Annapragada, A., R. V. Bellamkonda, E. Karathanasis, and R. M. Lebovitz; *Marval Pharma, Inc., New York.* Prior. 2004, Apr. 21. Publ. 2014, Dec. 16. Nano-Scale Contrast Agents and Methods of Use. Patent US8911708.

Anonymous. 1930. Discussion concerning uroselectan [in German]. *Med. Mitt.* 2(2): 44–46.

Anonymous. 1932. Award of gold medals. *Radiology* 18(1): 154. doi:10.1148/18.1.154a.

Anonymous. 2009. Deaths. Sovak, Milos. *New York Times* 158(54571): B10, col. 7.

Arnell, S., and F. Lindström. 1931. Myelography with sinodan (Abrodil). *Acta Radiol.* os-12(3): 287–288 & Table XXVIII. doi:10.1177/028418513101200307.

Arns, R. G. 1997. The high-vacuum X-ray-tube: Technological change in social context. *Technol. Cult.* 38(4): 852–890.

Ascher, B. 1996. X-ray contrast agents market matures into generic market. *Clinica* 736/737: 8–9.

Aspelin, P. 1979. Effect of ionic and nonionic contrast media on red cell deformability in vitro. *Acta Radiol.* 20(1A): 1–12. doi:10.1177/028418517902001A01.

Bachem, C., and H. Günther. 1910. Barium sulfate as shadow forming contrast agent in Roentgen examinations [in German]. *Z. Roentgenkunde Radiumforsch.* [Leipzig] 12(11): 369–376 & Table XIV.

Backer, H. J. 1926. The dihalogen-methanesulfonic acids [in French]. *Rec. Trav. Chim. Pay-Bas.* 45(11): 830–837. doi:10.1002/recl.19260451109.

Bailey, A. R., M. M. Jones, M. T. Kneller et al.; *Mallinckrodt LLC, Hazelwood.* Prior. 2008, Jan. 14. Publ. 2012, July 4. Process for the Preparation of Iosimenol. Patent EP2240432.

Ball, P. 2017. The power of the blackboard. *Phys. World* 30(6): 32. doi:10.1088/2058-7058/30/6/43.

Barke, R. 1970. *Röntgenkontrastmittel.* Leipzig: VEB Georg Thieme.

Barry, C. N., and D. K. Rose. 1953. Urokon Sodium 70 percent in excretion urography. *J. Urol.* 69(6): 849–855. doi:10.1016/S0022-5347(17)68159-X.

Battelli, A., and A. Garbasso. 1896. About Roentgen rays [in Italian]. Presentation at the Physics Institute, Univ. of Pisa, Italy, Jan. 25, 1896. *Nuovo Cimento,* 4th ser. 3(1): 40–61 & pl. V.

Bauer, M. 1995. Resistance to new technology and its effects on nuclear power, information technology and biotechnology. In *Resistance to New Technology. Nuclear Power, Information Technology and Biotechnology,* ed. M. Bauer, 1–42. Cambridge, UK: Cambridge University Press. doi:10.1017/CBO9780511563706.002.

Bayer AG, Leverkusen. 1930s Sales Crop Protection [in German]. Unpublished manuscript. WV—ZD—Archiv, Signature 1/6.6.26. Leverkusen: Bayer AG, Corporate History & Archives.

Bayer-Meister Lucius. Pharmaceutical Sales-Center, I. G. Farbenindustrie AG, Leverkusen. 1930. Suppl. No. 4 to the Price List No. 4 for Wholesalers [in German]. WV—ZD—Archiv. Leverkusen: Bayer AG, Corporate History & Archives.

Becher, W. 1896. Regarding the application of Roentgen's procedure in medicine [in German]. *Dtsch. Med. Wochenschr.* 22(13): 202–203. doi:10.1055/s-0028-1139548.

Becker, J., J. Babb, and M. Serrano. 2013. Glomerular filtration rate in evaluation of the effect of iodinated contrast media on renal function. *AJR Am. J. Roentgenol.* 200(4): 822–826. doi:10.2214/AJR.12.8871.

Belán, A., K. Benda, J. Fabián, and J. Bláha. [1977] 1978. Advantages of a new nonionic contrast medium. The results of animal experiments. Presentation at the 2nd Congr. Eur. Soc. Cardio-vascular Radiol., Uppsala, Sweden, May 25–27, 1977. *Ann. Radiol.* (Paris) 21(4–5): 279–282.

Bell, A. G. 1880. On the production and reproduction of sound by light. *Am. J. Sci.,* 3rd ser. 20(118): 305–324. doi:10.2475/ajs.s3-20.118.305.

Bell, K. E., C. S. McKinstry, and J. O. Mills. 1987. Iopamidol in the diagnosis of suspected upper gastro-intestinal perforation. *Clin. Radiol.* 38(2): 165–168.

Belt, E. 1974. Von Lichtenberg: A memoir. *Urology* 3(3): 385–387.

Berberich, J., and S. Hirsch. 1923. The roentgenographic representation of the arteries and veins in living humans [in German]. Presentation at the Frankfurt Med. Assoc. Oct. 15, 1923. *Klin. Wochenschr.* 2(49): 2226–2228. doi:10.1007/BF01712032.

Berger, M., H. Schmitt-Willich, D. Sülzle et al.; *Bayer Intellectual Property GmbH, Monheim; Bayer Pharma AG, Berlin.* Prior. 2011, June 27. Publ. 2013, Jan. 3. Bis Tridentate W_3O_2 Clusters for X-ray Imaging. Patent App. WO2013000970A1.

Bernstein, E. F., C. R. Reller, and T. B. Grage. 1962. Experimental studies of Angio-Conray—A new angiographic agent. *Radiology* 79(3): 389–394. doi:10.1148/79.3.389.

Bettendorf, G. 1995. Schoeller, Walter Julius Viktor [in German]. In *Zur Geschichte der Endokrinologie und Reproduktionsmedizin: 256 Biographien und Berichte,* ed. G. Bettendorf, 488–490, Berlin: Springer. doi:10.1007/978-3-642-79152-9_200.

Bettmann, M. A. 1989. Guidelines for use of low-osmolality contrast agents. *Radiology* 172(3): 901–903. doi:10.1148/172.3.901.

BfArM (Bundesinstitut für Arzneimittel und Medizinprodukte, Berlin, Germany). 2000. Revocation of Ionic Radiographic Contrast Agents (Amidotrizoic Acid, Iotalamic Acid) for Intravascular Use [in German]. www.bfarm.de/SharedDocs/Risikoinformationen/Pharmakovigilanz/DE/RI/2002minus/RI-asi-ion-roentgenkontrastm-widerruf.html (Accessed Apr. 28, 2018).

Bijker, W. E. 1987. The social construction of Bakelite: Towards a theory of invention. In *The Social Construction of Technology Systems: New Directions in the Sociology and History of Technology*, eds. W. E. Bijker, T. P. Hughes, and T. Pinch, 159–187. Cambridge, MA: MIT Press.

Bijker, W. E. 1995. *Of Bicicles, Bakelites and Bulbs: Towards a Theory of Sociotechnical Change*. Cambridge, MA: MIT Press.

Binz, A. Berlin. 1928, Dec. 3. Letter to the Rektor der Landwirtschaflichen Hochschule Berlin [in German]. Universitätsarchiv der Humboldt Universität, Document UK-Pers., R 9, Bd. 2, Bl. 14.

Binz, A. 1930. Abstract of discussion to Jaches (1930). *J. Am. Med. Assoc.* 95(19): 1411–1412.

Binz, A. [1930] 1931. The chemistry of uroselectan. Presentation at the Annu. Meet. Am. Urol. Assoc., New York City, June 1930. *J. Urol.* [Baltimore] 25(3): 297–301.

Binz, A.: Prior. 1935a, Jan. 28. Publ. 1936, July 28. Process for the Manufacturing of Halogen-Methane Sulphonic Acids and Their Salts. Patent GB451034.

Binz, A. 1935b. Biochemical and medical significance of newer pyridine derivatives [in German]. *Angew. Chem.* 48(28): 425–429.

Binz, A. 1937a. The history of uroselectan [in German]. *Z. Urol.* 31(2): 73–84.

Binz, A. 1937b. Chemical research and medical applications [in German]. *Ber. Dtsch. Chem. Ges.* 70/A(7): 127–140. doi: 10.1002/cber.19370700733.

Binz, A., and H. Maier-Bode. 1933. The action of iodopyridone derivatives on streptococci [in German]. *Biochem. Z.* 257: 351–360.

Binz, A., and C. Räth; *Schering-Kahlbaum AG, Berlin*. Prior. 1923, Nov. 20. Publ. 1931, June 11. Methods of Preparation of Pyridine arsenic Acids [in German]. Patent DE528113.

Binz, A., and C. Räth; *Schering-Kahlbaum AG, Berlin*. Prior. 1924a, July 1. Publ. 1930, Oct. 23. Methods of Preparation of 2-Oxopyridine-5-stilbinic Acid [in German]. Patent DE530495.

Binz, A. and C. Räth. Prior. 1924b, Oct. 30. Publ. 1926, Apr. 30. Process for the Production of Iodized Pyridine Derivatives. Patent GB251578.

Binz, A., and C. Räth; *Schering-Kahlbaum AG, Berlin*. Prior. 1925, Apr. 12. Publ. 1930, Jan. 15. Methods of Preparation of Unsymmetric Arsenocompounds [in German]. Patent DE488574.

Binz, A., and C. Räth. 1927. Chemotherapeutics from the pyridine series [in German]. *Tierarztl. Rundsch.* 33(48): 891–892.

Binz, A., and C. Räth. 1930a. The chemistry of uroselectan [in German]. *Med. Mitt.* 2(2): 32–35.

Binz, A., and C. Räth. 1930b. The chemistry of uroselectan [in German]. *Klin. Wochenschr.* 9(49): 2297–2298. doi:10.1007/BF01847005.

Binz, A., and O. von Schickh; *Chemische Fabrik von Heyden AG, Radebeul*. Prior. 1933, Sept. 5. Publ. 1935, Nov. 26. Methods of Preparation of 3-Aminopyridine [in German]. Patent DE622345.

Binz, A., C. Räth, and A. von Lichtenberg. 1930. The rendering of kidneys and the urinary tract in the X-ray image with the help of iodopyridone derivatives [in German]. *Z. Angew. Chem.* 43(23): 452–455.

Binz, A., C. Räth, and K. Junkmann. 1930. Concerning the biochemistry of iodinated and arsenated pyridone derivatives [in German]. *Biochem. Z.* 227: 200–204.

Björk, L. [1964, 1966] 1967. Isopaque B in angiocardiography and coronary angiography. A comparative study. (Proc. of symposia held in Copenhagen, Denmark, Nov. 1964 and Sandefjord, Norway, 1966.) In *Cations in Intravascular Contrast Media and Development of Specific Metrizoate Formulas. Pharmacological and Clinical Studies*, eds. P. Lindgren, K. Dahlstrøm, and L. Fondberg. *Acta Radiol. Diagn.* 6 (Suppl. 270): 112–115. doi:10.1177/0284185167006S27012.

Björk, L. 1966. The osmotic effects of Urografin 76 percent and Isopaque 60 percent in angiocardiography. *Am. J. Roentgenol. Radium Ther. Nucl. Med.* 98(4): 922–926. doi:10.2214/ajr.98.4.922.

Björk, L., J. Ekholm, U. Eriks(s)on, and S. Skoglund. [1963] 1964. A new technique for cardiovascular Roentgen examinations [in Swedish]. Presentation at the Swed. Assoc. Med. Radiol., Stockholm, Nov. 30, 1963. *Nord. Med.* 71(25): 782.

Björk, L., U. E. Erikson, K. A. Granath, B. G.-A. Ingelman, and B. J. Lindberg; *Pharmacia AB, Uppsala.* Prior. 1966a, Dec. 2. Publ. 1972, Aug. 28. X-ray Contrast Agents Whose Active Ingredient is a Water-Soluble 2,4,6-Triiodobenzoic Acid-Containing Polymer [in Swedish]. Patent SE348110. (App. publically accessible 1968, July 1).

Björk, L., U. E. Erikson, B. G.-A. Ingelman, and B. J. Lindberg; *Pharmacia AB, Uppsala.* Prior. 1966b, Dec. 13. Publ. 1972, May 23. Triiodobenzoic Acid Derivatives for Use as X-ray Contrast Agents [in Swedish]. Patent SE345198. (App. publically accessible 1968, July 1).

Björk, L., U. E. Erikson, K. A. Granath, B. G.-A. Ingelman, and B. J. Lindberg; *Pharmacia AB, Uppsala.* Prior. 1967, Nov. 16. Publ. 1972, Aug. 28. X-ray Contrast Agent Whose Active Ingredient is a Water-Soluble 2,4,6-Triiodobenzoic Acid-Containing Polymer [in Swedish]. Patent SE348111. (App. publically accessible 1969, May 17).

Björk, L., U. Erikson, and B. Ingelman. 1969a. Clinical experience with a new type of contrast medium in peripheral arteriography. *Am. J. Roentgenol. Radium Ther. Nucl. Med.* 106(2): 418–424. doi:10.2214/ajr.106.2.418.

Björk, L., U. Erikson, and B. Ingelman. 1969b. Clinical experience with a new type of contrast medium in carotid arteriography. *Am. J. Roentgenol. Radium Ther. Nucl. Med.* 107(3): 637–640. doi:10.2214/ajr.107.3.637.

Björk, L., U. Erikson, and B. Ingelman. 1970. Polymeric contrast media for roentgenologic examination of gastrointestinal tract. Preliminary report. *Invest. Radiol.* 5(3): 142–148. doi:10.1097/00004424-197005000-00002.

Björk, L., U. Erikson, and B. Ingelman. 1976. Preliminary report on angiography with polymeric contrast agents in rabbits and dogs. *Ups. J. Med. Sci.* 81(3): 183–187. doi:10.3109/03009737609179046.

Black, M. 1962. *Models and Metaphors; Studies in Language and Philosophy.* Ithaca, NY: Cornell University Press.

Blascow, N. W., and A. Wade, eds. 1972. Diodone injection. In *Martindale: The Extra Pharmacopoeia*, 26th ed., 459–460. London: Pharmaceutical Press.

Bleichröder, F. 1912. Intraarterial therapy (with remarks by E. Unger and W. Löb) [in German]. Demonstration offered at the. Hufeland Gesellschaft., Berlin, May 9, 1912. *Berl. Klin. Wochenschr.* 49(32): 1503–1505.

Bleyer, J. M. 1896. Researches on rapid photography by means of Edison's kinetograph and the animated reproduction with the kinetoscope. Showing also by this method the present and future possibilities of taking and re-producing certain animated physiologic movements as the larynx, heart, intestines, etc., and certain diseases having visible symptoms, etc., with a historic review of fifty years of photography. *J. Am. Med. Assoc.* 26(15): 727–730. doi:10.1001/jama.1896.02430660023003h.

Blos, D. 1942. The Schering specialties [in German]. *Schering-Bl./Werkz. Betriebsgemeinschaft Schering A.G.*, n. ser. 5(1/2): 2–9.

Blühbaum, T., K. Frik, and H. Kalkbrenner. 1928. A new type of application of colloids in the X-ray diagnostics [in German]. *Fortschr. Geb. Rontgenstr.* 37(1): 18–29.

Blum, V., F. Eisler, and T. Hryntschak. 1920. Cystoradioscopy [in German]. *Wien. Klin. Wochenschr.* 33(31): 677–680.

Böhler, E. 1965. *Der Mythos in Wirtschaft und Wissenschaft.* Freiburg i. B.: Verlag Rombach.

Bohannon, J. 1910. Calling all dancing scientists! *Science* 328(5983): 1286. doi:10.1126/science.328.5983.1226-b. www.sciencemag.org/projects/dance-your-phd/official-rules (Accessed Apr. 28, 2018).

Boijsen, E. 1971. The future development of angiography [in German]. In *Röntgendiagnostik. Technologie, Untersuchungsmethodik, Organisation*, eds. W. A. Fuchs, and E. Voegeli, 40–47. Bern: Hans Huber.

Boijsen, E. 1996. Swedish radiology as reflected in Acta Radiologica 1921–1996. *Acta Radiol.* 37(3P2): 412–440. doi:10.1177/02841851960373P206.

Bonati, F. 1995. From the pharmacist artisans to the chemical industries: The contrast agents [in Italian]. In *Immagini e segni dell'uomo. Storia della radiologia italiana*, ed. A. E. Cardinale, 239–257. Naples: Idelson.

Bonati, F., and E. Felder. 1977. Iopamidol: A New Nonionic Contrast Medium. Preliminary Clinical Results. Presentation at the 14th. Int. Congr. Radiol., Rio de Janeiro, Brasil, Oct. 23–29, 1977. (Abstr. and Proc. were never published) Doc. IOP/1000. Archives of Bracco Imaging SpA, Milan, Italy.

Bonnemain, B. 2014. *Guerbet: une aventure de plus d'un siècle au service de la santé*. Paris: Pharmathèmes.

Bonnemain, B., and M. Guerbet. 1995a. The history of lipiodol (1901–1994) or how a medical drug can evolve with its time [in French]. *Rev. Hist. Pharm.* (Paris) 42(305): 159–170.

Bonnemain, B., and M. Guerbet. 1995b. The discovery of Ténébryl, the first French water soluble uro-angiographic product [in French]. *Rev. Hist. Pharm.* (Paris) 42(306): 279–284.

Børdalen, B. E., H. Wang, and H. Holtermann. 1970. Osmotic properties of some contrast media. *Invest. Radiol.* 5(6): 559–565. doi:10.1097/00004424-197011000-00023.

Boriani, A., and G. Boriani. 1940. Experimental research around the direct actions on renal function brought about by the preparations commonly used for röntgendiagnostic purposes [in Italian]. *Scritti Ital. Radiobiol.* 7(2): 109–126.

Boveri, M. 1996. *Verzweigungen: Eine Autobiographie: Herausgegeben von Uwe Johnson*. Frankfurt a. M.: Suhrkamp.

Bozzini, P. 1807. *The Light Conductor or Description of a Simple Instrument and its Use for Illuminating Inner Cavities and Interstices of the Living Animal Body* [in German]. Weimar: Landes-Industrie-Comptoir.

Braasch, W. F. 1925. Urography. *Surg. Gynecol. Obstet.* 41(1): 114–115.

Braasch, W. F. [1930] 1931. The value of uroselectan in the renal lithiasis. Presentation at Symp. Uroselectan, Am. Urol. Assoc., New York, June 10–12, 1930. *J. Urol.* [Baltimore] 25(3): 265–274. doi:10.1016/S0022-5347(17)72843-1.

Braudel, F. 1982. *The Wheels of Commerce*. New York City: Harper & Row.

Brocchetta, M. 1974. *Chemistry Laboratory Notebook* no. 535, Oct. 29–Dec. 11 [in Italian]. Archives of Bracco Imaging SpA, Milan, Italy.

Bröer, R. 2002. Legend or reality? Werner Forssmann and heart catheterization [in German]. *Dtsch. Med. Wochenschr.* 127(41): 2151–2154. doi:10.1055/s-2002-34642.

Broman, T., and O. Olsson. 1949. Experimental study of contrast media for cerebral angiography with reference to possible injurious effects on the cerebral blood vessels. *Acta Radiol.* os-31(4): 321–334. doi:10.1177/028418514903100404.

Broman, T., and O. Olsson. 1956. Experimental comparison of iodonum with sodium acetrizoate with reference to possible injurious effects on the blood-brain barrier. *Acta Radiol.* os-46(1/2): 346–350. doi:10.1177/028418515604600144.

Bronner, H. 1929. Short summary of the general urological X-ray technology for the medical practice [in German]. *Rontgenprax. Diagn. Rontgen. Radium. Lichtther.* 1(14): 626–643.

Bronner, H., and J. Schüller. 1930. Excretion pyelography with 'Abrodil' [in German]. *Dtsch. Med. Wochenschr.* 56(33): 1393–1394. doi:10.1055/s-0028-1125874.

Bronner, H., G. Hecht, and J. Schüller. 1930. Excretion-pyelographie with 'Abrodil' [in German]. *Fortschr. Geb. Rontgenstr.* 42(2): 206–218.

Brooks, B. 1924. Intraarterial injection of sodium iodide. Preliminary report. *J. Am. Med. Assoc.* 82(13): 1016–1019. doi:10.1001/jama.1924.02650390006002.

Brown, T. L. 2003. *Making Truth: Metaphor in Science*. Urbana: University of Illinois Press.

Brown, R., S. H. Rahimtoola, G. D. Davis, and H. J. C. Swan. 1965. The effect of angiocardiographic contrast medium on circulatory dynamics in man. Cardiac output during angiocardiography. *Circulation* 31(2): 234–240. doi:10.1161/01.CIR.31.2.234.

Bruwer, A. J., ed. 1964. *Classic Descriptions in Diagnostic Roentgenology*, vol. 2. Springfield, IL: Charles C. Thomas.

Bucher, A. M., C. N. De Cecco, U. J. Schoepf et al. 2014. Is contrast medium osmolality a causal factor for contrast-induced nephropathy? *Biomed. Res. Int.* 2014. 931413. doi: 10.1155/2014/931413.

Buka, F. 1896. About the direct observation of internal body parts by means of X-rays [in German]. *Dtsch. Med. Wochenschr.* 22(19): 304. doi:10.1055/s-0028-1139615.

Burhop, C. 2009. Pharmaceutical research in Wilhelmine Germany: The case of E. Merck. *Bus. Hist. Rev.* 83(3): 475–503. doi:10.1017/S000768050000297X.

Burns, J. E. 1915. Thorium: A new agent for pyelography. Preliminary report. *J. Am. Med. Assoc.* 64(26): 2126–2127. doi:10.1001/jama.1915.02570520020005.

Burns, J. E. 1916. Thorium—A new agent for pyelography. *Bull. Johns Hopkins Hosp.* 27(304): 157–164.

Burns, J. E. 1917. Further observations on the use of thorium in pyelography. *J. Am. Med. Assoc.* 68(7): 533–535. doi:10.1001/jama.1917.04270020197010.

Burrows, E. H. 1986. *Pioneers and Early Years: A History of British Radiology*. St. Anne, Channel Islands, UK: Colophon.

Busch, W. [1884] 1965. *Klecksel the Painter*, transl. Max Born. New York: Ungar.

Butler, C. 1995. FO1A/PA Unit, United States Immigration and Naturalization Service, Washington DC. Information Control Number 95004618.

Cabanis, E. A., and M.-T. Iba-Zizena Cabanis, eds. 1994. Laboratoires Guerbet/specialists in medical imaging [in French]. In *Contribution to the History of European Neuroradiology*, 2nd ed. Paris: Edition Pradel.

Callon, M. 1992. The dynamics of techno-economic networks. In *Technological Change and Company Strategies*, eds. R. Coombs, P. Saviotti, and V. Walsh, 72–102. London: Harcourt Brace Jovanovich.

Callsen, J.; *I. G. Farbenindustrie AG, Frankfurt a. M.* Prior. 1926, June 20. Publ. 1930, Jan. 15. Methods for the Preparation of Halogen Containing Alcohols from Halogen Containing Aldehydes [in German]. Patent DE489281.

Cameron, D. F. 1918. Aqueous solutions of potassium and sodium iodides as opaque mediums in roentgenography. Preliminary report. *J. Am. Med. Assoc.* 70(11): 754–755. doi:10.1001/jama.1918.02600110012002.

Cameron, D. F., and C. C. Grandy. 1918. Sodium and potassium iodide in roentgenography. *J. Am. Med. Assoc.* 70(21): 1516–1517. doi:10.1001/jama.1918.02600210006002.

Cannon, W.B. 1914. Early use of the Roentgen ray in the study of the alimentary canal. *J. Am. Med. Assoc.* 62(1): 1–3.

Canzler, W. 1997. The success of the automobile and the sorcerer's apprentice syndrome [in German]. In *Technikgenese. Befunde aus einem Forschungsprogramm*, ed. M. Dierkes, 99–129. Berlin: Edition Sigma, Rainer Bohn Verlag.

Cartwright, L. 1995. *Screening the Body. Tracing Medicine's Visual Culture*. Minneapolis: University of Minnesota Press.

Cashman, J. D., J. McCredie, and D. A. Henry. 1991. Intravenous contrast media: Use and associated mortality. *Med. J. Aust.* 155(9): 618–623.

Casper, J. 1967. The introduction in 1928–29 of thorium dioxide in diagnostic radiology. *Ann. N. Y. Acad. Sci.* 145(3): 527–529. doi:10.1111/j.1749-6632.1967.tb50255.x.

Castellanos, A., R. Pereiras, and A. García. 1937. The radio-opaque angio-cardiography [in Spanish]. *Arch. Soc. Estud. Clin. La Habana* [Havana, Cuba] 31(9/10): 523–573.

Ceccarelli, L. 2001. *Shaping Science with Rhetoric: The Cases of Dobzhansky, Schrödinger, and Wilson.* Chicago: University of Chicago Press. doi.org/10.1086/428980.

Chai, C.-M., H. Rasmussen, M. Eriksen et al. 2010. Predicting cardiotoxicity propensity of the novel iodinated contrast medium GE-145: Ventricular fibrillation during left coronary arteriography in pigs. *Acta Radiol.* 51(9): 1007–1013. doi:10.3109/02841 851.2010.504743.

Champagne, M. 2016. Diagrams of the past: How timelines can aid the growth of historical knowledge. *Cogn. Semiotics* [Berlin] 9(1): 11–44. doi:10.1515/cogsem-2016-0002.

Chemische Fabrik auf Actien (vormals E. Schering), Berlin. Prior. 1884, Mar. 7. Publ. 1884, Dec. 3. Methods for the Preparation of Iodoform, Bromoform and Chloroform by Means of Electrolysis [in German]. Patent DE29771.

Cho, A. 2016. Will Nobel Prize overlook LIGO's master builder? *Science* 353(6307): 1478–1479. doi:10.1126/science.353.6307.1478.

Chwalla, R. 1930. Our recent experiences with intravenous pyelography. *Br. J. Urol.* 2(3): 256–267. doi:10.1111/j.1464-410X.1930.tb00004.x.

Ciamician, G., and P. Silber, Rome. Prior. 1885, June 9. Publ. 1886, Apr. 7. Methods of Preparation of Tetraiodopyrrole, Named Jodol [in German]. DE35130.

Clement, O., and J. A. W. Webb. 2014. Acute adverse reactions to contrast media. Mechanisms and prevention. In *Contrast Media. Safety Issues and ESUR Guidelines*, eds. H. S. Thomsen, and J. A. W. Webb, 51–60. Berlin: Springer.

Cole collaborators. 1932. Part I. Roentgenologic exploration of the mucosa of the gastro-intestinal tract. *Radiology* 18(2): 221–263. doi:10.1148/18.2.221; Part II. Important anatomical data of the digestive tract. *Radiology* 18(3): 471–520. doi:10.1148/18.3.471; Part III. Findings Observed in the gastro-intestinal tract. *Radiology* 18(5): 886–941. doi:10.1148/18.5.886.

Colombo, A. 1974. *Chemistry Laboratory Notebook* no. 500, June 27 [in Italian]. Archives of Bracco Imaging SpA, Milan, Italy.

Congressional Record—Senate. 1975. Dr. Swick Gets His German Kudos—45 Years After. *Proc. and Debates 94th U.S. Congress, 1st Session*, 121 (Part 19): 24904.

Constant II, E. W. 1980. *The Origin of the Turbojet Revolution* (Johns Hopkins Studies in the History of Technology, new ser. no. 5). Baltimore, MD: Johns Hopkins University Press. doi.org/10.1086/352945.

Cotrim, E. S. 1954. Cardiac, blood pressure and respiratory effects of some contrast media. *Acta Radiol.* os-42(Suppl. 116): 58–74. doi:10.1177/0284185154042S11609.

Coulston, F., and J. O. Hoppe. 1959. The pathologist and toxicologist in the evaluation of the safety and methods of development of radiodiagnostic compounds. *Ann. N. Y. Acad. Sci.* 78(3): 740–755. doi:10.1111/j.1749–6632.1959.tb56060.x.

Council A. M. A. 1932. Council on pharmacy and chemistry. Thorotrast. *J. Am. Med. Assoc.* 99(26): 2183–2185. doi:10.1001/jama.1932.02740780035010.

Council A. M. A. 1935. New and nonofficial remedies: Containing descriptions of the articles which stand accepted by the Council on Pharmacy and Chemistry of the American Medical Association. *Am. J. Publ. Health Nations Health* 25(9): 1058–1059.

Daft, R., and R. Lengel. 1986. Organizational information requirements, media richness and structural design. *Manag. Sci.* 32(5): 554–571. doi:10.1287/mnsc.32.5.554.

Dahmen, H. 1927. Therapeutic experiences and experiments with preparations from the pyridine series [in German]. *Tierarztl. Rundsch.* 33(48): 894–897.

Danto, A. C. 2007. *Narration and Knowledge, Including the Integral Text of Analytical Philosophy of History.* New York: Columbia University Press.

Dawson, P. 1997. Röntgen's other experiment. *Br. J. Radiol.* 70(836): 809–816. doi:10.1259/bjr.70.836.9486045.

Dawson, P., P. Pitfield, and K. Skinnemoen. 1983. Isomeric purity and supersaturation of iopamidol. *Br. J. Radiol.* 56(670): 711–713. doi:10.1259/0007-1285-56-670-711.

Dawson, P. 2006. Adverse reactions to intravascular contrast agents. *Br. Med. J.* 333(7570): 663–664. doi:10.1136/bmj.38981.652118.DE.

de Bono, E. 1999. *Six Thinking Hats*. New York: Back Bay Books.

de Haën, C. 1995. The Earliest Days of X-ray Contrast Agents. Presentation at the Röntgen Centenary Congr. 1995, Birmingham, UK, June 12–16. Book of Abstracts, 272.

de Haën, C. 2001. Conception of the first magnetic resonance imaging contrast agents: A brief history. *Top. Magn. Reson. Imaging* 12(4): 221–230.

de Haën, C. 2009. Knowledge culture conditioned bounded rationality and human artificial neural network processes (HANNP Theory): How risks of accidents and environmental impact of a new chemical production process and plant site have entered decisions. *Saf. Sci.* 47(6): 843–852. doi:10.1016/j.ssci.2008.10.011.

de Haën, C., F. Uggeri, O. Gazzotti, and M. Brocchetta; *Dibra SpA, Milan*. Prior. 1993, Dec. 23. Publ. 1996, Dec. 24. Tetraiodopyrrole Derivatives as Contrast Medium and as Antiseptic. Patent US5587487.

Del Re, G. 1996. The 'Dreams of Kékulé' and the birth of modern chemistry [in Italian]. *Kos* [Milan], n. ser. 128: 18–21.

Del Rio, A. 1974. Von Lichtenberg in Mexico. *Urology* 3(3): 388–389.

Demel, R., V. Kollert, and M. Sgalitzer. 1930. Histological examination of the arterial wall after contrast filling of the arteries with uroselectan [in German]. *Zentralbl. Chir.* 57(48): 2964–2966.

Demel, R., M. Sgalitzer, and V. Kollert. 1931. The clinical results of arteriography in diseases of peripheral arteries [in German]. *Mitt. Grenzgeb. Med. Chir.* 42: 357–391.

Dennett, D. C. 1991. Real patterns. *J. Philos.* 88(1): 27–51. doi:10.2307/2027085.

de Regt, H. W. 1996. Philosophy and the kinetic theory of gases. *Br. J. Philos. Sci.* 47(1): 31–62.

Deutike, P. 1952. Prof. Dr. Theodor Hryntschak [in German]. *Wien. Med. Wochenschr.* 102(34): 669.

Deutsch, E. A., K. F. Deutsch, W. P. Cacheris et al.; *Mallinckrodt Medical, Inc., St. Louis*. Prior. 1994, July 6. Publ. 1995, Nov. 21. Treated Apatite Particles for X-ray Diagnostic Imaging. Patent US5468465.

Deutsche Gesellschaft für Urologie. 1926. Agenda of the VII. Meeting. Vienna, Sept. 30–Oct. 2, 1926 [in German]. *Wien. Med. Wochenschr.* 76(38): 1125–1126.

Deutsche Gesellschaft für Urologie. 1929. List of members of the German Society of Urology (New members accepted at 8th Congr., Berlin, Sept. 26–29, 1928) [in German]. *Verh. Dtsch. Ges. Urol./Z. Urol. Sonderband* 8: 7–19.

Deutsche Gold- und Silber-Scheideanstalt vormals Roessler, Frankfurt a. M. German Prior. 1924, June 30. Publ. 1926, Dec. 1. Process for the Preparation of a Novel Pyridine Substitution Product [in German]. Patent CH117895.

Deutsche Gold- und Silber-Scheideanstalt vormals Roessler, Frankfurt a. M. Austrian Prior. 1925, Oct. 17. Publ. 1927, May 20. Preparation Process for New Pyridine Derivatives [in French]. Patent FR621989.

Deutsche Gold- und Silber-Scheideanstalt vormals Roessler, Frankfurt a. M., and *Albert, A.* Prior. 1924, Nov. 28. Publ. 1926, Mar. 29. A Process for the Production of New Derivatives of Organic Arseno-Compounds. Patent GB249588.

Dieffenbach, J. F. 1834. *Physiologisch-chirurgische Beobachtungen bei Cholera-Kranken*, 2nd amplified ed. Güstrow: F. Opitz.

Dierkes, M., U. Hoffman, and L. Marz. 1996. *Visions of Technology. Social and Institutional Factors Shaping the Development of Technologies*. New York: St. Martin's Press.

Diernhofer, K. 1928. About mastitis treatment with selectan [in German]. *Tierarztl. Rundsch.* 34(28): 523–527.

Doby, T. 1976. *Development of Angiography and Cardiac Catheterization.* Littleton, MA: Publishing Sciences Group.

Dohrn, M., and H. Horsters; *Chemische Fabrik auf Actien (vormals E. Schering), Berlin.* Prior. 1922, Dec. 29. Publ. 1924, Aug. 4. Methods of Preparation of 2-Amino-5-chlorpyridine and 2-Amino-3,5-dichlorpyridine [in German]. Patent DE400191.

Dohrn, M., and P. Diedrich; *Schering-Kahlbaum AG, Berlin.* Prior. 1930, Dec. 19. Publ. 1932, Mar. 7. Methods of Preparation of N-Alkyldiiodochelidamic Acid [in German]. Patent DE545916.

Dolbear, A. E. 1894. Electricity and photography. *Cosmopolitan* [New York] 16(6): 765–766.

Dommann, M. 1999. X-ray vision must be learned in the sweat of observation: About the semiotics of shadow images [in German]. *Traverse* 6(3): 114–130. doi:10.5169/seals-17730.

Dommann, M. 2003. *Durchsicht, Einsicht, Vorsicht. Eine Geschichte der Röntgenstrahlen 1896–1963.* Zürich: Chronos Verlag. doi:10.3929/ethz-a-004459534.

Dorn, H. 1957. X-ray contrast agents [in German]. *Pharmazie* [Berlin-East] 12(6): 315–322. (7): 415–431 & (8): 499–514.

Dotter, C. T., and F. S. Jackson. 1950. Death following angiocardiography. *Radiology* 54(4): 527–533. doi:10.1148/54.4.527.

Doucet, D., D. Meyer, C. Chambon, and B. Bonnemain. 1991. Blood-pool X-ray contrast agents: Evaluation of a new iodinated polymer. *Invest. Radiol.* 26(Suppl. 1): S53–S54.

Draaisma, D. 2001. The tracks of thought. *Nature* 414(6860): 153. doi:10.1038/35102645.

Dünner, L. 1923. Representation of the lung pattern in the radiograph of the living [in German]. Presentation at the Berlin Med. Soc., Mar. 14, 1923. *Dtsch. Med. Wochenschr.* 49(14): 459.

Dünner, L., and A. Calm. 1923. The radiology of the vessels, especially the pulmonary vessels of living humans [in German]. *Fortschr. Geb. Rontgenstr.* 31(5/6): 635–636.

E. Merck in Darmstadt (Hugo Winternitz). Prior. 1897a, Apr. 8. Publ. 1898, Mar. 2. Methods of Preparation of Durable Iodo- and Bromo-Fats [in German]. Patent DE96495.

E. Merck in Darmstadt (Hugo Winternitz). Prior. 1897b, Apr. 8. Publ. 1902, Oct. 22. Methods of Preparation of Durable Iodo- and Bromo-Fats [in German]. Patent DE135835.

Eberhard-Metzger C., C. Thor-McCarthy, A. Tuffs, L. Wagner-Roos, C. Wassman, U. Soeder and R. Schley. 1993. From the Shadow Image to the Spinning Atom. *Research Report Schering: Diagnostics*: 6–11, Berlin: Schering Corporate Communications.

Eddy, D. M. 1992. Applying cost-effectiveness analysis. The inside story. *JAMA* 268(18): 2575–2582. doi:10.1001/jama.1992.03490180107037.

Eichholtz, F. 1927. About rectal anesthesia with Avertin (E107). Pharmacological section. *Dtsch. Med. Wochenschr.* 53(17): 710–712. doi:10.1055/s-0028-1165256.

Eisenberg, R. L. 1992. *Radiology, An Illustrated History.* St. Louis, MO: Mosby-Year Book, Inc.

Ekholm, J., U. Erikson, and S. Skoglund. 1964. An improved method for experimental angiography. *Br. J. Radiol.* 37(443): 839–843. doi:10.1259/0007-1285-37-443-839.

Ekstrand, T. K. I. B., A. Munksgaard, and B. V. Wickberg; *Erco-Läkemedel AB, Stockholm.* Prior. 1970, Mar. 31. Publ. 1975, Aug. 18. X-ray Contrast Medium Containing as a Contrast Producing Substance a 2,4,6-Triiodobenzoic Acid Derivative [in Swedish]. Patent SE378065. (German disclosure DE2111127. Publ. 1971, Oct. 2).

Elke, M., in collaboration with H.-E. Schmitt, E. Felder, J. M. Fröhlich, C. Gückel, E. W. Radü, D. Scheidegger, and U. Speck. 1992. *Kontrastmittel in der radiologischen Diagnostik: Eigenschaften, Nebenwirkungen, Behandlung,* 3rd ed. Stuttgart: Georg Thieme.

Ellegast, H., and B. Thurnher. 1962. The beginnings of roentgenology in Vienna [in German]. *RöFo* 96(1): 145–158.

Elzen, B., B. Enserink, and W. A. Smit. 1996. Socio-technical networks: How a technology studies approach may help to solve problems related to technical change. *Soc. Stud. Sci.* 26: 95–141. doi:10.1177/030631296026001006.

Eprova AG, Schaffhausen, Switzerland (1968, March 1): Invoice to Bracco Industria Chimica SpA, Milan, Italy, for "Execution of experiments regarding the synthesis of amides of diiodofumaric acid".

Erbach, K. 1923a. About the Trönsegaard Cleavage of Proteins [in German]. Med. diss., Ruprecht-Karls-Univ., Heidelberg.

Erbach, K. 1923b. Curriculum vitae [in German]. Document H-III-862/39 in Arch. Univ. Heidelberg.

Erbach, O. 1932. Familial Occurrence of Cutis Hyperelastica (and Curriculum Vitae) [in German]. Med. diss., Hamburgische Univ., Hamburg.

Erikson, U. 1961. Arteriography on amputated legs [in Swedish]. *Nord. Med.* 66(33): 1134–1135.

Erikson, U. 1965. Circulation in traumatic amputation stumps. An angiographical and physiological investigation. *Acta Radiol. Diagn.* 3(Suppl. 238): 1–122. doi:10.1177/0284185165003S23801 to doi:10.1177/0284185165003S23806.

Erikson, U., and A. Hulth. 1962. Circulation of amputation stumps. Arteriographic and skin temperature studies. *Acta Orthop. Scand.* 32(1–4): 159–170. doi:10.3109/17453676208989570.

Erikson, U., and S. Olerud. 1966. Healing of amputation stumps, with special reference to vascularity and bone. *Acta Orthop. Scand.* 37(1): 20–28. doi:10.3109/17453676608989400.

Fedeli, F. 1964. *Chemistry Laboratory Notebook* no. 253, Sept. 7 [in Italian]. Archives of Bracco Imaging SpA, Milan, Italy.

Felder, E. Bracco Industria Chimica SpA, Milan. 1969, Mar. 3. X-ray Contrast Agents: State of the Art, Development Trends, Situation of Own Work. (1st version) [in German. Company-internal document]. Company Archives of Bracco Imaging SpA, Milan, Italy.

Felder, E. 1970. Perspectives of research in the field of contrast agents [in Italian]. Presentation at the Italian-Sovietic Symp. New Developments in the Field of Contrast Media, Milan, June 15–17, 1970. *Radiol. Med.* 56(9): 844–851.

Felder, E. 1986. The evolution of contrast agents [in Italian]. In *Ionici o nonionici, i molteplici aspetti dei mezzi di contrasto*, ed. A. Chiesa, 11–19. Brescia: Clas Editoriale.

Felder, E., and C. de Haën; *Bracco SpA, Milan*. Italian Prior. 1992, Dec. 24. Publ. 1997, Dec. 9. Aqueous Injectable Formulations Useful for Radiodiagnosis Comprising Iodinated Aromatic Compounds Used as X-ray Contrast Media. Patent US5695742.

Felder, E., and D. Pitrè; *Bracco Industria Chimica SpA, Milano*. Prior. 1968, May 2 (corrected). Publ. Mar. 15, 1971. New X-ray Contrast Agents and Methods for their Preparation [in German]. Patent CH502105.

Felder, E., and D. Pitrè; *Savac AG, Chur*. Prior. 1974a, Dec. 13. Publ. 1978, Dec. 29. A Readily Water-Soluble, Nonionic X-ray Contrast Agent [in German]. Patent CH608189. (German disclosure DE2547789. Publ. 1976, June 24).

Felder, E., and D. Pitrè; *Bracco Industria Chimica SpA, Milan*. Prior. 1974b, Dec. 13. Publ. 1980, Mar. 31. Methods of Preparation of Readily Water-Soluble 2,4,6-Triiodo-isophthalic Acid Amides. Readily Water-Soluble, Nonionic X-ray Contrast Agents [in German]. Patent CH616403. (German disclosure DE2547789. Publ. 1976, June 24).

Felder, E., and D. Pitrè; *Bracco Industria Chimica SpA, Milan*. Italian Prior. 1979, Aug. 9. Publ. 1982, Oct. 5. Derivatives of 2,4,6-Triiodo-isophthalic Acid, Processes for Their Synthesis and X-ray Contrasting Materials Containing These. Patent US4352788.

Felder, E., D. Pitrè, and Tirone P. 1977a. Preclinical Studies with Iopamidol: A New Nonionic Water-soluble Contrast Agent. Presentation at the 14th Int. Congr. Radiol., Rio de Janeiro, Brazil. Oct. 23–29, 1977. (Abstr. & Proc. never published).

Felder, E., D. Pitrè, and P. Tirone. 1977b. Radiopaque contrast media XLIV—Preclinical studies with a new nonionic contrast agent. Presentation at the 2nd Congr. Eur. Soc. Cardio-vascular Radiol., Uppsala, Sweden, May 25–27, 1977. *Farmaco Sci.* 32(11): 835–844.

Fey, V. R., and E. I. Rivin. 1997. *The Science of Innovation: A Managerial Overview of the TRIZ Methodology*. Southfield, MI: The TRIZ Group.

Feynman, R. 1965. *The Character of Physical Law*. Cambridge, MA: MIT Press.

Feynman, R. P. 1982. Simulating physics with computers. *Int. J. Theor. Phys.* 21(6/7): 467–488.

Finby, N., N. Poker, and J. A. Evans. 1956. Ninety percent Hypaque for rapid intravenous roentgenography. *Radiology* 67(2): 244–246. doi:10.1148/67.2.244.

Fischer, H. W. 1957. Colloidal stannic oxide: Animal studies on a new hepatolienographic agent. *Radiology* 68(4): 488–498. doi:10.1148/68.4.488.

Fischer, H. W. 1977a. Iothalamate ethyl ester as hepatolienographic agent. *Invest. Radiol.* 12(1): 96–100.

Fischer, H. W. 1977b. Improvement in radiographic contrast media through the development of colloidal or particulate media: An analysis. *J. Theor. Biol.* 67(4): 653–670. doi:10.1016/0022-5193(77)90252-1.

Fischer, H. W. 1986. Catalogue of intravascular contrast media. *Radiology* 159(2): 561–563. doi:10.1148/159.2.561.

Fischer, H. W. 1987. Historical aspects of contrast media development. In *Contrast Media from Past to the Future*, ed. R. Felix, 3–18. Stuttgart: Georg Thieme.

Fischer, H. W. 1990. Particles. *Invest. Radiol.* 25(Suppl. 1): S2–S6.

Fischer, H. W., and J. W. Eckstein. 1961. Comparison of cerebral angiographic contrast media by their circulatory effects. An experimental study. *Am. J. Roentgenol. Radium Ther. Nucl. Med.* 86(1): 166–177.

Fischer, H. W., and S. H. Cornell. 1967. Toxicity study of sodium metrizoate containing calcium and magnesium. *Acta Radiol. Diagn.* 6(2): 126–132. doi:10.1177/028418516700600204.

Fischer, W. 1972. Pharmacokinetics of Iodamide as Methylglucamine and Sodium Salt after Intravenous and Intraarterial Administration in Humans [in German]. Med. diss., Freie Univ., Berlin.

Fleck, L. 1983. Look, See, Know [in German]. In *Erfahrung und Tatsache*, ed. L. Fleck, 147–174. Frankfurt a. M.: Suhrkamp.

Fleck, L., with a foreword by T. S. Kuhn. 1981. *Genesis and Development of a Scientific Fact*, eds. T. J. Trennn, and R. K. Merton, trans. F. Bradley, and T. J. Trenn. Chicago, IL: University of Chicago Press.

Fölsing, A. 1993. *Albert Einstein. Eine Biographie*. Frankfurt a. M.: Suhrkamp.

Fondberg, L. [1966] 1967. The effect of calcium and magnesium ions in attempts to balance contrast agents [in Swedish]. Presentation at the Meet. Swed. Assoc. Med. Radiologists, Jönköping, Sweden, Sept. 17, 1966. *Nord. Med.* 77(6): 199–200.

Forssmann, W. 1929. Probing the right heart [in German]. *Klin. Wochenschr.* 8(45): 2085–2086. doi:10.1007/BF01875120.

Forssmann, W. 1931a. On the contrast representation of the cavities of the living right heart and the pulmonary artery [in German]. *Munch. Med. Wochenschr.* 78/I(12): 489–492.

Forssmann, W. 1931b. Methodology of contrast representation of the central circulatory organs [in German]. Presentation at the 55th Meet. German Soc. Surg., Berlin, Apr. 8–11, 1931. *Arch. Klin. Chir.* 167: 787–790.

Forssmann, W. 1972. *Selbstversuch. Erinnerungen eines Chirurgen*. Düsseldorf: Droste.

Forster, A. 1896. *Radiographische Aufnahmen ausgeführt mit Röntgenschen Strahlen im Physikalischen Institut der Universität Bern*. Bern: Stämpfli.

Fouillet, X., H. Tournier, H. Khan et al. 1995. Enhancement of computer tomography liver contrast using iomeprol-containing liposomes and detection of small liver tumors in rats. *Acad. Radiol.* 2(7): 576–583. doi:10.1016/S1076-6332(05)80118-7.

Fränkel, W. K. 1930. A new non-irritating contrast agent for perivesical outpatient pyelography [in German]. *Munch. Med. Wochenschr.* 77/II(34): 1447–1448.

Frings, M. Bayer AG, Leverkusen, Germany. 1994, June 23. Letter to Christoph de Haën, Bracco SpA, Milan, Italy, with information from Bayer Archive Document WV-ZD.

Fritsch, K. 1911. Jodipin in X-ray images [in German]. *Beitr. Klin. Chir.* 75(1/2): 168–183.

Fuchs, F. 1928. Urological experiences with the halogen solutions according to Albrecht-Ulzer [in German]. *Wien. Klin. Wochenschr.* 41(5): 160–161.

Gaedicke, J. 1896. Studies on Roentgen rays [in German]. *Photogr. Arch.* [Berlin] 37(10/790): 147–154.

Gagliardi, R. A., and B. L. McClennan. 1996. *A History of the Radiological Sciences. Diagnosis.* Reston, VA: Radiological Centennial.

Galison, P. L. 1997. *Image and Logic: A Material Culture of Microphysics.* Chicago, IL: University of Chicago Press.

Garthe, E. 1929. On a New Preparation Method for 2-Chlorpyridines [in German]. PhD diss., Friedrich-Wilhelms-Univ., Berlin.

Gehe & Co. AG, Dresden, and *Roseno, A.* Prior. 1929, Feb. 3. Publ. 1930, Oct. 7. Methods of Preparation of Inclusion Compounds from Sodium Iodide and Urea [in German]. Patent DE509265.

Gehes Codex. 1914, 2nd ed. Dresden: Gehe & Co., AG.

Geissler, [P.] 1896. The diagnosis of bone foci by X-rays [in German]. Presentation at the 25th Congr. German Soc. Surg., Berlin, May 27–30, 1896. *Verh. Dtsch. Ges. Chir.* 25: 28–29.

Glasser, O. [1931] 1993. *Wilhelm Conrad Röntgen and the Early History of the Roentgen Rays.* San Francisco, CA: Norman.

Goffman, E. 1974. *Frame Analysis: An Essay on the Organization of Experience.* New York City: Harper & Row.

Goldstein, A. E., and B. S. Abeshouse. 1935. A historical and practical consideration of pyelographic media. *Am. J. Roentgenol. Radium Ther.* 33(2): 165–175.

Gonsette, R. E. 1973a. Biologic tolerance of the central nervous system to metrizamide. *Acta Radiol. Diagn.* 14(Suppl. 335): 25–44. doi:10.1177/0284185173014S33506.

Gonsette, R. E. 1973b. Metrizamide as contrast medium for myelography and ventriculography. Preliminary clinical experiences. *Acta Radiol. Diagn.* 14(Suppl. 335): 346–358. doi:10.1177/0284185173014S33543.

Gonsette, R., and G. André-Balisaux. 1969. A New Hydrosoluble and Resorbable Contrast Material for Myelography and Ventriculography. Presentation at the 12th Int. Congr. Radiol., Tokyo, Japan Oct. 6–11, 1969. Book of Abstracts. Tokio: Hitachi Printing Solutions.

Gopal, K. 1996. Market forces drive down X-ray contrast prices. *Diagn. Imaging Europe* 45–52.

Gortan, M., and C. Ravasini. [1929] 1930. The radiographic picture of the urinary tract with intravenous injection of uroselectan [in German]. Presentation at the 9th Italian Congr. Medical Radiol., Trieste, Nov. 17, 1929. *Med. Mitt.* 2(2): 42–44.

Gottheiner, V. 1971. More than 30 years as physician. Viktor Gottheiner [in German]. *Ther. Ggw.* 110(7): 1069–1084.

Gowers, W. T. 2000. The two cultures of mathematics. In *Mathematics: Frontiers and Perspectives*, eds. V. I. Arnold, M. Atiyah, P. D. Lax, and B. Mazur, 65–78. Providence, RI: American Mathematical Society.

Graff, H. 1897/1898, Oct./Nov.1898. Contribution to the diagnostic value of X-rays [in German]. *Fortschr. Geb. Roentgenstr.* 1(6): 229 & Table XXI, Fig. 1.

Graham, A. K. 1982. Software design: Breaking the bottleneck. *IEEE Spectrum* 19(3): 43–50. doi:10.1109/MSPEC.1982.6366826.

Graham, A. K., and P. Senge. 1980. A long-wave hypothesis of innovation. *Technol. Forecast. Soc. Change* 17(4): 283–311. doi:10.1016/0040-1625(80)90103-1.

Graham, E. A. 1931. The story of the development of cholecystography. *Am. J. Surg.* 12(2): 330–335. doi:10.1016/S0002-9610(31)90065-2.

Graham, E. A., and W. H. Cole. 1924. Roentgenologic examination of the gall bladder. Preliminary report of a new method utilizing the intravenous injection of tetrabromophenolphthalein. *J. Am. Med. Assoc.* 82(8): 613–614. doi: 10.1001/jama.1924.02650340023007.

Grainger, R. G. 1982a. Intravascular contrast media—The past, the present and the future. *Br. J. Radiol.* 55(649): 1–18. doi:10.1259/0007-1285-55-649-1.

Grainger, [G.] R. 1982b. Intravascular contrast media. *Br. J. Radiol.* 55(651): 251. doi:10.1259/0007-1285-55-651-251.

Grainger, R. G. 1982c. Intravascular contrast media. *Br. J. Radiol.* 55(655): 544. doi:10.1259/0007-1285-55-655-544-a.

Grainger, R. G. 1986. Clinical, ethical and economic considerations on contrast media. A personal point of view from Britain [in Italian]. In *Ionici o nonionici, i molteplici aspetti dei mezzi di contrasto*, ed. A. Chiesa, 59–64. Brescia: Clas Editoriale.

Grainger, R. G., Sheffield, UK. 1994, Feb. 9. Letter RGG/MA to Christoph de Haën, Milano, Italy, and related telephone conversation.

Grainger, R. G. 1995. Development of intravascular contrast agents: The first 100 years. *Adv. X-ray Contrast* 3(2): 26–33. doi:10.1007/978-94-011-3959-5_12.

Grainger, R. G., and A. M. K. Thomas. 1999. History of the development of radiological contrast media (1895–1996). In *Textbook of Contrast Media*, eds. P. Dawson, D. O. Cosgrove, and R. G. Grainger, 3–14. Oxford: Isis Medical Media.

Greenbaum, F. R. 1929. Organic addition compounds of calcium chloride and calcium iodide. *J. Am. Pharm. Assoc.* 18(8): 784–789. doi:10.1002/jps.3080180807.

Greitz, T., R. Telenius, and G. Törnell. [1964, 1966] 1967. Influence of methylglucamine and calcium ions in metrizoate (Isopaque) on the bradycardial effect in carotid angiography. Proc. of symposia held in Copenhagen, Denmark, Nov. 1964 and Sandefjord, Norway, 1966. *Acta Radiol. Diagn.* 6(Suppl. 270): 208–215. doi:10.1177/0284185167006S27027.

Grigg, E. R. N. 1965. *The Trail of the Invisible Light: From X-Strahlen to Radio(bio)logy.* Springfield, IL: Charles C. Thomas.

Grönberg, T., T. Almén, K. Golman, S. Mattsson, and S. Sjöberg; *Elementaranalys Almén & Grönberg AB, Malmö.* Swedish Prior. 1982, Jan. 12. Publ. 1986, Feb. 25. X-ray Fluorescence Analyzers. Patent US4573181.

Guttstadt, A., ed. 1891. *Die Wirksamkeit des Koch'schen Heilmittels gegen Tuberkulose.* Berlin: Springer.

GV (Gesamtverzeichnis des deutschsprachigen Schrifttums 1911–1965) 1977. vol. 32, En-Eri, Munich: Verlag Dokumentation.

Haavaldsen, J. 1980. Iohexol. Introduction. *Acta Radiol. Suppl.* 362: 9–11.

Habicht, E., and G. Feth; *Cilag-Chemie AG, Schaffhausen.* Prior. 1962, Aug. 7. Publ. 1966, Apr. 15. Methods of Preparation of New 3-Acylamino-2,4,6-triiodobenzoic Acids and Utilization Thereof as X-ray Contrast Agents [in German]. Patent CH400181.

Habicht, E., and R. Zubiani; *Cilag-Chemie Ltd, Schaffhausen.* Swiss Prior. 1962, Apr. 19, and 1963, Feb. 1. Publ. 1966, Feb. 15. Esters of 3,5-Diiodo-4-pyridone-N-acetic Acid. Patent US3235461 (Incorporates Swiss Patents CH379059 & CH377346).

Habicht, E., and R. Zubiani; *Cilag-Chemie Ltd, Schaffhausen.* Prior. 1963, Oct. 17. Publ. 1966, May 24. Amides and Their Use. Patent US3252985.

Hallouard, F., N. Anton, P. Choquet, A. Constantinesco, and T. Vandamme. 2010. Iodinated blood pool contrast media for preclinical X-ray imaging applications—A review. *Biomaterials* 31(24): 6249–6268. doi:10.1016/j.biomaterials.2010.04.066.

Hamer, W. J., and Y.-C. Wu. 1972. Osmotic coefficients and mean activity coefficients of uni-univalent electrolytes in water at 25°C. *J. Phys. Chem. Ref. Data* 1(4): 1047–1099. doi:10.1063/1.3253108.

Hammarlund, E. R., and K. Pedersen-Bjergaard. 1958. A simplified graphic method for the preparation of isotonic solutions. *J. Am. Pharm. Assoc. Sci. Ed.* 47(2): 107–114. doi:10.1002/jps.3030470211.

Hammerstein, N. 1999. Die Deutsche Forschungsgemeinschaft in der Weimarer Republik und im Dritten Reich. Wissenschaftspolitik in Republik und Diktatur 1920–1945. Munich: C. H. Beck.

Hansen, P.-E., H. Holtermann, and K. Wille; *Nyegaard & Co. A/S, Oslo.* British Prior. 1982, Nov. 8. Publ. 1986, July 16. X-ray Contrast Agents. Patent EP108638.

Harrison, A. J. 1993. Examining science's feminine face. *Chem. Eng. News* 71(51): 40–42 & 66.

Hartman, G. W., R. R. Hattery, D. M. Witten, and B. Williamson Jr. 1982. Mortality during excretory urography: Mayo clinic experience. *AJR Am. J. Roentgenol.* 139(5): 919–922. doi:10.2214/ajr.139.5.919.

Haschek, E., and O. T. Lindenthal. 1896, Jan. 23. A contribution to the practical utilization of the photography according to Roentgen [in German]. *Wien. Klin. Wochenschr.* 9(4): 63–64.

Hausmann, H. 1985. 113 Years of Urology in Berlin: The Development of the Urology in Berlin from 1871–1984 [in German]. Med. diss., Humboldt-Univ. Berlin.

Hausmann, H. 1990. Johannes Volkmann (1889–1982)—His contribution to the development of the excretion urography. Medical history considerations on the occasion of his 100th birthday [in German]. *Z. Urol. Nephrol.* 83(4): 205–209.

Hayes, C. W., J. H. Foster, R. Sewell, and D. A. Killen. 1966. Experimental evaluation of concentrated solutions of iothalamic acid derivatives as angiographic contrast media. *Am. J. Roentgenol. Radium Ther. Nucl. Med.* 97(3): 755–761. doi:10.2214/ajr.97.3.755.

Hecht, G. 1927. The osmotic effects [in German]. Chap. 1 In *Handbuch der experimentellen Pharmakologie*, eds. A. Heffter, and W. Heubner, part 3: 1–39, Berlin: Springer.

Hecht, [G.] Pharmacological Laboratories of I G. Farbenindustrie AG, Werk Leverkusen. 1930. Annual Report 1929 [in German.] Bayer AG, Corporate History & Archives, Leverkusen.

Hecht, [G.] Pharmacological Laboratories of I G. Farbenindustrie AG, Werk Leverkusen. 1931. Annual Report 1930 [in German.] Bayer AG, Corporate History & Archives, Leverkusen.

Hecht, [G.] Pharmacological Laboratories of I G. Farbenindustrie AG, Werk Leverkusen. 1933. Annual Report 1932 [in German.] Bayer AG, Corporate History & Archives, Leverkusen.

Hecht, G. 1939. X-ray contrast agents [in German]. In *Handbuch der experimentellen Pharmakologie, Ergänzungswerk*, eds. W. Heubner, and J. Schüller, 8: 79–163. Berlin: Springer.

Hecht, G., A. Ossenbeck, and E. Tietze; *I. G. Farbenindustrie AG, Frankfurt a. M.* Prior. 1930a, June 11. Publ. 1937, June 30. Roentgenological Method [in German]. Patent DE647244.

Hecht, G., A. Ossenbeck, and E. Tietze; *Winthrop Chemical Comp., Inc., New York.* German Prior. 1930b, July 8. Publ. July 26, 1932. Contrast Media for X-ray Photography. Patent US1868602.

Heckenbach, W. 1930. Functional examinations in the course of excretion of uroselectan [in German]. *Klin. Wochenschr.* 9(15): 684–689. doi:10.1007/BF01726018.

Heilbron, J. L. 1999. From horsehair to lightning rods. Analogy is a powerful tool in Benjamin Franklin's natural philosophy. *Nature* 401(6751): 329. doi:10.1038/43791.

Heidenhain, R. 1883. Physiology of secretory processes [in German]. In *Handbuch der Physiologie*, vol. 5, ed. L. Hermann, part 1:1–420. Leipzig: F. C. W. Vogel.

Hellige, H. D. 1996. Visions in the time-sharing life cycle: From 'Multi-Access' to 'Interactive On-line Community' [in German]. In *Technikleitbilder auf dem Prüfstand. Leitbild-Assessment aus Sicht der Informatik- und Computergeschichte*, ed. H. D. Hellige, 205–234. Berlin: Edition Sigma.

Hesse, M. B. 1966. *Models and Analogies in Science*. Notre Dame, IN: University of Notre Dame Press.

Heubner, W. 1943. Victor Salle [in German]. *Klin. Wochenschr.* 22(6): 131–132. doi:10.1007/BF01771518.

Heubner, W. 1946. In memory of Leopold Lichtwitz [in German]. *Dtsch. Gesundheitsw.* 1: 777–779.

Heuser, C. 1919. Pyelography with potassium iodide and the intravenous injections of potassium iodide in radiography [in Spanish]. *Sem. Med.* 26(17): 424.

Heuser, C. 1921. Chinosol in abdominal typhus [in German]. *Dtsch. Med. Wochenschr.* 47(17): 477.

Heuser, C. 1924. Radiology in gynecology. Radiography of ovaries, oviducts, and uterus [in Spanish]. *Sem. Med.* 31(52): 1496–1501.

Heuser, C. 1925. Lipiodol in diagnosis of pregnancy. *Lancet* [London] (II/5335): 1111–1112.

Heuser, C. 1926a. Table for the x-ray examination in urology [in Spanish]. *Sem. Med.* 33(7): 368–372.

Heuser, C. 1926b. Metrosalpingography [in Spanish]. *Sem. Med.* 33(51): 1667–1671.

Heuser, C. 1927. Uterosalpingography [in Spanish]. *Sem. Med.* 34(40): 904–906.

Heuser, C. 1932. Direct arteriography [in Spanish]. *Sem. Med.* 39(41): 1074.

Heuser, C. 1933. Direct arteriography, aortography by impregnation of the reticuloendothelial system, angiography of the lungs by means of probing the right atrium [in Spanish]. *Rev. Asoc. Med. Argent.* 46: 1119–1128.

Hilal, S. K. [1965] 1966. Hemodynamic changes associated with the intraarterial injection of contrast media. New Toxicity Tests and a New Experimental Contrast Medium. Poster presented at the 51th Annu. Meet. Radiol. Soc. North Am., Chicago, Nov. 28–Dec. 3, 1965. *Radiology* 86(4): 615–633. doi:10.1148/86.4.615.

Hilal, S. K. 1966, Apr. Determination of the blood flow by a radiographic technique. Physical considerations and experimental results. *Am. J. Roentgenol. Radium Ther. Nucl. Med.* 96(4): 896–906. doi:10.2214/ajr.96.4.896.

Hilal, S. K. 1970. Trends in preparation of new angiographic contrast media with special emphasis on polymeric derivatives (with discussion). Presentation at the Symp. Contrast Medium Toxicity, Skytop PA, May 3–6, 1970. *Invest. Radiol.* 5(6): 458–472.

Hilal, S. K., and H. Morgan. 1968. A Trimer of Iothalamate—Preliminary Toxicity Studies. Presentation at the 54th Meet. Radiol. Soc. North Am., Chicago, Dec. 1–6, 1968. Unpublished paper cited in Hilal (1970).

Hildebrand, [H.], [E.] Scholz, and [J.] Wieting [-Pascha]. Preface by T. Rumpf and H. Kümmell. 1901. The human arterial system in the stereoscopic X-ray image [in German]. *Sammlung von stereoskopischen Röntgenbildern aus dem Neuen Allgemeinen Krankenhaus Hamburg-Eppendorf.* no. 1. Wiesbaden: V. J. F. Bergmann.

Hillgruber, K. 1930. About a method of determination of organically bound iodine (uroselectan) in the urine [in German]. *Klin. Wochenschr.* 9(50): 2353–2354.

Hoey, G. B.; *Mallinckrodt Chemical Works, St. Louis MO.* Prior. 1963, Mar. 6. Publ. 1966, Dec. 6. 5-Amino-*N*-alkyl-2,4,6-triiodoisophtalamic Acid Derivatives. Patent US3290366.

Hoey, G. B., P. E. Wiegert, and R. D. Rands. 1971. Organic iodine compounds as X-ray contrast media. Chap. 2 In *International Encyclopedia of Pharmacology and Therapeutics.* Section 76, Radiocontrast Agents, vol. 1, ed. P. K. Knoefel, 23–145. Oxford: Pergamon Press.

Hoey, G. B., K. R. Smith, S. El-Antably, and G. P. Murphy. 1984. Chemistry of X-ray contrast agents. In *Radiocontrast Agents*, ed. M. Sovak, 23–125. Berlin: Springer.

Hoffenberg, R. 2001. Christiaan Barnard: His first transplants and their impact on concepts of death. *Br. Med. J.* 323(7327): 1478–1480.

Hollingsworth, M. D., M. E. Brown, A. C. Hiller, B. D. Santarsiero, and J. D. Chaney. 1996. Superstructure control in the crystal growth and ordering of urea inclusion compounds. *Science* 273(5280): 1355–1359. doi:10.1126/science.273.5280.1355.

Holtermann, H., ed. 1973a. Metrizamide. *Acta Radiol. Diagn.* 14(Suppl. 335): 1–390. doi:10.1177/0284185173014S33503 to doi:10.1177/0284185173014S33547.

Holtermann, H. 1973b. Metrizamide. *Acta Radiol. Diagn.* 14(Suppl. 335): 1–4. doi:10.1177/0 284185173014S33503.

Holtermann, H., H. Baerum, L. G. Haugen, V. Nordal, and J. L. Haavaldsen; *Nyegaard & Co., A/S, Oslo.* Norwegian Prior. 1961, Feb. 28. Publ. 1965, Apr. 13. Process for the N-Alkylation of Acetanilides Halogen Substituted in the Nucleus. Patent US3178473.

Holtzmann Kevles, B. 1997. *Naked to the Bone.* New Brunswick: Rutgers University Press.

Hoppe, J. O. 1959. Some pharmacological aspects of radiopaque compounds. *Ann. N. Y. Acad. Sci.* 78(3): 727–739. doi:10.1111/j.1749–6632.1959.tb56059.x.

Hoppe, J. O. 1963. X-ray contrast media. *Med. Chem.: A Series of Rev.* 6: 290–349.

Hoppe, J. O., A. A. Larsen, and F. Coulston. 1956. Observations on the toxicity of a new urographic contrast medium, sodium 3,5-diacetamido-2,4,6-triiodobenzoate (Hypaque Sodium) and related compounds. *J. Pharmacol. Exp. Ther.* 116(4): 394–403.

Howell, J. D. 1986. Early use of X-ray machines and electrocardiographs at the Pennsylvania hospital. *J. Am. Med. Assoc.* 255(17): 2320–2323. doi: 10.1001/jama.1986.03370170084040.

Howell, J. D. 1989. Machines and medicine: Technology transforms the American hospital. In *The American General Hospital: Communities and Social Context*, eds. D. E. Long, and J. Holden, 109–134. Ithaca, NY: Cornell University Press.

Hryntschak, T. [1928] 1929. Discussion: Roentgenographic rendering of the kidneys and pyelography by the intravenous route [in German]. Information presented at the 8th Congr. German Soc. Urol., Berlin, Sept. 26–29, 1928. *Verh. Dtsch. Ges. Urol./Z. Urol. Sonderband* 8: 434–435.

Hryntschak, T. 1929. Studies of the roentgenological rendering of the renal parenchyma and the renal pelvis by the intravenous route [in German]. *Z. Urol.* 23: 893–904.

Hryntschak, T., and M. Sgalitzer. 1921. The shape of the bladder in different body positions [in German]. *Z. Urol.* 16: 11–16.

Hunold, G. A. 1929. Regarding the Knowledge of Ortho-substituted Derivatives of Pyridines [in German]. PhD diss., Friedrich-Wilhelms-Univ., Berlin.

Hutcheson, I. R., T. M. Griffith, M. R. Pitman et al. 1999. Iodinated radiographic contrast media inhibit shear stress- and agonist-evoked release of NO by the endothelium. *Br. J. Pharmacol.* 128(2): 451–457. doi:10.1038/sj.bjp.070278.

I. G. Farbenindustrie AG, Frankfurt a. M. (Dr. Schmitz & Co., GmbH, Düsseldorf). Prior. 1924, June 13. Publ. 1926, Nov. 18. Methods of Preparation of Halogen-Containing Alcohols from Halogen-Containing Aldehydes [in German]. Patent DE437160.

I. G. Farbenindustrie AG, Frankfurt a. M. (Binz, A.). Prior. 1935, Jan. 28. Publ. 1936, Sept. 16. Methods of Preparations of Halogen methanesulfonic Acids and Their Salts [in German]. Patent DE635242.

Isenman, L. D. 1997. Toward an understanding of intuition and its importance in scientific endeavor. *Perspect. Biol. Med.* 40(3): 395–403. doi:10.1353/pbm.1997.0018.

Iseri, L. T., M. A. Kaplan, M. J. Evans, and E. D. Nickel. 1965. Effect of concentrated contrast media during angiography on plasma volume and plasma osmolality. *Am. Heart J.* 69(2): 154–158.

Jaches, L. 1930. Intravenous urography (Swick method). Presentation at the Annu. Sess. Am. Med. Assoc., Section Urol., Detroit, June 26–27, 1930. *J. Am. Med. Assoc.* 95(19): 1409–1412. doi:10.1001/jama.1930.02720190021005.

Jaches, L., and M. Swick. [1933] 1934. Opaque media in urology, with special reference to a new compound, sodium ortho-iodohippurate. Presentation at the 19th Annu. Meet. Radiol. Soc. North America, Chicago, Sept. 25–30, 1933. *Radiology* 23(2): 216–222. doi:10.1148/23.2.216.

Jacob, F. 1977. Evolution and tinkering. *Science* 196(4295): 1161–1166. doi:10.1126/science.860134.

Jäger, E. On behalf of the Preussische Minister für Wissenschaft, Kunst und Volksbildung, Berlin. 1933, Nov. 21. Letter to Dr. Alexander von Lichtenberg, Berlin, Germany [in German]. Universitätsarchiv der Humboldt Universität, Document UK-L 147, Bd. 1, Bl. 4.

Jaggar, A. M., and S. R. Bordo, eds. 1989. *Gender/Body/Knowledge: Feminist Reconstructions of Being and Knowing*. New Brunswick: Rutgers University Press.

Jasanoff, S., G. E. Markle, J. C. Petersen, and T. Pinch, eds. 1995. *Handbook of Science and Technology Studies*. Thousand Oaks, CA: Sage Publications.

Joerges, B. 1989. Sociology and technology [in German]. In *Technik als sozialer Prozess*, ed. P. Weingart, 44–89. Frankfurt a. M.: Suhrkamp.

Jones, C. B., and J. D. Babb. 2008. Contrast selection in the cardiac catheterization lab. Does choice of contrast agent really matter? *Card. Interv. Today* 2(1): 52–55.

Joseph, E. 1921. A new contrast agent for pyelography [in German]. *Zentralbl. Chir.* 48(20): 707–708.

Juliani, G. 1977. An organo-iodinated, nonionic contrast agent for angiography and urography. First clinical trials [in Italian]. Presentation at the Accad. Med. Torino, Jan. 21, 1977. *G. Accad. Med. Torino* 140(1–12): 22–28.

Jung, C. G. 1923. *Psychological Types: Or the Psychology of Individuation*. Oxford, UK: Harcourt, Brace.

Kahlenborn, W., M. Dierkes, C. Krebsbach-Gnath, S. Mützel, and K. W. Zimmermann. 1995. *Berlin, Zukunft aus eigener Kraft. Ein Leitbild für den Wirtschaftsstandort Berlin*. Berlin: FAB Verlag.

Katayama, H., K. Yamaguchi, T. Kozuka, T. Takashima, P. Seez, and K. Matsuura. 1990. Adverse reactions to ionic and nonionic contrast media. A report from the Japanese Committee on the Safety of Contrast Media. *Radiology* 175(3): 621–628. doi:10.1148/175.3.621.

Kelly, H. A., and R. M. Lewis. 1913. Silver iodide emulsion—A new medium for skiagraphy of the urinary tract. *Surg. Gynec. Obstet.* 16(6): 707–708.

Kielleuthner, [L.]. 1928. VIII. Meeting of the German Society of Urology, Sept. 26–29, 1928, Berlin [in German]. *Munch. Med. Wochenschr.* 75/II(45): 1939–1943.

Kihlström, J.-E. 1952. Osmotic pressure of water-solutions of Methylglucamine-Umbradil (3,5-Diiodine-4-pyridon-N-acetic acid) mixtures. *Acta Radiol.* os-38(5): 399–402. doi:10.1177/028418515203800509.

Kirklin, B. R. 1945. Background and beginning of cholecystography. *Am. J. Roentgenol. Radium Ther.* 54(6): 637–639.

Knie, A. 1994. *Wankel-Mut in der Autoindustrie. Anfang und Ende einer Antriebsalternative*. Berlin: Edition Sigma, Rainer Bohn Verlag.

Knoefel, P. K. 1961. *Radiopaque Diagnostic Agents*. Springfield, IL: Charles C. Thomas.

Knoefel, P. K., ed. 1971a. *International Encyclopedia of Pharmacology and Therapeutics*. Section 76, Radiocontrast Agents, vol. 2. Oxford: Pergamon Press.

Knoefel, P. K. 1971b. Glossary of names of iodine-containing organic compounds used as radiocontrast agents. Chap. 1 In *International Encyclopedia of Pharmacology and Therapeutics*. Section 76, Radiocontrast Agents, vol 1, ed. P. K. Knoefel, 299–333. Oxford: Pergamon Press.

Knorr, K. D. 1977. Producing and reproducing knowledge: Descriptive or constructive? Toward a model of research production. *Soc. Sci. Inf.* 16(6): 669–696. doi:10.1177/053901847701600602.

Kochendoerfer, G.; *Deutsche Gold und Silber Scheideanstalt vormals Roessler AG, Frankfurt a. M.* Prior. 1925, Jan. 25. Publ. 1930, Nov. 25. Method of Preparation of Iodine-Substituted Pyridine Derivatives [in German]. Patent DE513293.

Köhler, H. 1930. The excretion pyelography by means of metufan [in German]. Presentation at the Assoc. Intern. Med. Pediatr., Berlin, June 23, 1930. *Munch. Med. Wochenschr.* 77/II(29): 1252.

Klickstein, H. S. 1966. Wilhelm Conrad Röntgen on a new kind of rays: A bibliographic study. In *Mallinckrodt Classics of Radiology*, vol. 1. St. Louis, MO: Mallinckrodt Chemical Works.

Krause, W., H. Miklautz, U. Kollenkirchen, and G. Heimann. 1994a. Physicochemical parameters of X-ray contrast media. *Invest. Radiol.* 29(1): 72–89. doi:10.1097/00004424-199401000-00015.

Krause, W., U. Speck, and W.-R. Press; *Schering AG, Berlin.* Prior. 1994b, Dec. 9. Publ. 1996, June 13. Use of Additives to Contrast Agents to Improve Imaging [in German]. Patent application WO9617629.

Kretzer, H. 1897. To the knowledge of iodosobenzoic acids [in German]. *Ber. Dtsch. Chem. Ges.* 30(2): 1943–1948.

Krüger, H.-P. 1990. *Kritik der kommunikativen Vernunft. Kommunikationsorientierte Wissenschaftsforschung im Streit mit Sohn-Rethel, Toulmin und Habermas.* Berlin: Akademie Verlag.

Kümmell, [H.] 1896. The diagnosis of bone foci by X-ray rays [in German]. Presentation at the 25th Congr. German Soc. Surgery, Berlin, May 27–30, 1896. *Verh. Dtsch. Ges. Chir.* 25: 25–28.

Kümmell, H. 1901. X-rays in the service of practical medicine [in German]. *Berl. Klin. Wochenschr.* 38(1): 4–7 & (2): 43–45.

Kundt, A., and W. C. Röntgen. 1879. Demonstration of the electromagnetic rotation of the polarization plane of light in carbon disulfide vapor [in German]. *Ann. Phys.*, [Leipzig] 3rd ser. 6(3): 332–336 & 1 ill. doi:10.1002/andp.18792420303.

Kutner, R. 1892. An attempt to stain the urine with methylene blue for diagnostic purposes [in German]. *Dtsch. Med. Wochenschr.* 18(48): 1086–1087. doi:10.1055/s-0029-1199556.

Lachmund, J. 1997. *Der abgehorchte Körper. Zur historischen Soziologie der medizinischen Untersuchung.* Opladen: Westdeutscher Verlag.

Lafay, L. 1893. Clinical-chemical Study on the Urinary Elimination of Iodine After Absorption of Potassium Iodide [in French]. Med. diss., Univ. of Paris.

Lance, E. M., D. A. Killen, and H. W. Scott. 1959. A plea for caution in the use of sodium acetrizoate (urokon) for aortography. *Ann. Surg.* 150(1): 172.

Landow, M. 1903. Radiographic findings after injections of Jodipin [in German]. *Munch. Med. Wochenschr.* 50/II(38): 1634–1636.

Langecker, H., A. Harwart, and K. Junkmann. 1953. 2,4,6-Triiodo-3-acetaminobenzoic acid derivatives as contrast agents [in German]. *Naunyn Schmiedebergs Arch. Exp. Pathol. Pharmakol.* 220(3): 195–206. doi:10.1007/BF00246720.

Langecker, H., A. Harwart, and K. Junkmann. 1954. Diacetylamino-2,4,6-triiodobenzoic acid as X-ray contrast agent [in German]. *Naunyn Schmiedebergs Arch. Exp. Pathol. Pharmakol.* 222(6): 584–590. doi:10.1007/BF00246909.

Larsen, A. A.; *Sterling Drug, Inc., New York.* Corrected US Prior. 1953a, Mar. 27. Publ. 1963, Jan. 29. Acylated 3,5-Diaminopolyhalobenzoic acids. Patent US3076024.

Larsen, A. A.; Schering AG, Berlin. US Prior. 1953b, Mar. 27 and 1954, Feb. 19. Publ. Feb. 8, 1968. Methods of Preparation of *N*-Acylderivatives of 3,5-Diamino-2,4,6-triiodobenzoic Acid [in German]. Examined and published patent application DE1260477.

Larsen, A. A.; *Sterling Drug, Inc., New York.* Prior. 1957, Apr. 17. Publ. 1964, Jan. 28. Iodinated Esters of Substituted Benzoic Acids and Preparation Thereof. Patent US3119858.

Larsen, A. A.; *Sterling Drug, Inc., New York.* Prior. 1963, June 6. Publ. 1968, Nov. 5. 3,3'-Diamino-5,5'-dicarboxy-hexaiodocarbanilides and Derivatives. Patent US3409662.

Lasser, E. C. 2011. X-ray contrast media mechanisms in the release of mast cell contents: Understanding these leads to a treatment of allergies. *J. Allergy* (Cairo, Egypt) 2011: 276258. doi:10.1155/2011/276258.

Latour, B. 1988. How to write 'The Prince' for machines as well as for machinations. In *Technology and Social Change*, ed. B. Elliott, 20–43. Edinburgh: Edinburgh University Press.

Latour, B., and S. Woolgar. 1986. *Laboratory Life: The Construction of Scientific Facts.* Princeton, NJ: Princeton University Press.

Lauteala, L., M. Kormano, and M. R. Violante. 1984. Effect of intravenously administered iodipamide ethyl ester particles on rat liver morphology. *Invest. Radiol.* 19(2): 133–141. doi:10.1097/00004424-198403000-00011.

Lautrou, J., M. Bergin, and A. Cardinal. [1977] 1978. Effect of contrast agent solutions on cardiac hemodynamics during coronagraphy in pigs [in French]. Presentation at the 2nd Congr. Eur. Soc. Cardio-vascular Radiol. Uppsala, Sweden, May 25–27, 1977. *Ann. Radiol.* (Paris) 21(4/5): 261–265.

Leander, P., P. Höglund, A. Børseth, Y. Kloster, and A. Berg. 2001. A new liposomal liver-specific contrast agent for CT: First human phase-I clinical trial assessing efficacy and safety. *Eur. Radiol.* 11(4): 698–704. doi:10.1007/s003300000712.

Lecher, Z. K. 1896. A sensational discovery [in German]. *Die Presse* (Vienna) 49(5) Su., Jan. 5: p. 1, col. 3- p. 2, col. 1–2 & (6) Tu. morn. ed. Jan. 7: p. 1, col. 3- p. 2, col. 1–2.

Legueu, F., B. Fey, and P. Truchot. 1931a. Intravenous urography with Ténébryl: Difficulties of interpretation [in French] *Presse Med.* (Paris) 39(90): 1653.

Legueu, F., B. Fey, and P. Truchot. 1931b. Intravenous urography. I. A new iodine compound: Ténébryl (Guerbet) II. Results interpretation [in French]. Presentation at the 31st French Congr. Urol., Paris Oct. 6–10, 1931. *Congr. Franc. d'Urol./Procès Verbaux, Mémoires et Discussions* 31: 252–258.

Lenarduzzi, G., and R. Pecco. 1927, Sept./Oct. Intravenous injections of sodium iodide (experimental radiographic research) [in Italian]. *Arch. Radiol.* (Naples, Italy) 3(5): 1055–1060.

Lenhard, D. C., H. Pietsch, M. A. Sieber et al. 2012. The osmolality of nonionic, iodinated contrast agents as an important factor for renal safety. *Invest. Radiol.* 47(9): 503–510. doi:10.1097/RLI.0b013e318258502b.

Lerch, J. U. 1884. Study on chelidonic acid [in German]. *Monatsh. Chem.* 5(1): 367–414. doi:10.1002/nadc.200747305.

Lerner, B. H. 1992. The perils of X-ray vision: How radiographic images have historically influenced perception. *Perspect. Biol. Med.* 35(3): 382–397.

Lesch, J. E. 1993. Chemistry and biomedicine in an industrial setting: The invention of sulfa drugs. In *Chemical Sciences in the Modern World*, ed. S. H. Mauskopf, 158–215. Philadelphia: University of Pennsylvania Press.

Lesky, E. 1971. 75 years of Roentgen rays. Viennea and Roentgen's discovery [in German]. *Münch. Med. Wschr.* 113(4): 134–136.

Leszczyński, S. 2000a. The development of world and polish radiology [in Polish]. Chap. 1 In *Historia radiologii polskiej na tle radilogii światowej*, ed. S. Leszczyński, 9–296. Krakow: Wyd. Medycyna Praktyczna.

Leszczyński, S. 2000b. Biography [in Polish]. Chap. 3 In *Historia radiologii polskiej na tle radilogii światowej*, ed. S. Leszczyński, 507–755 & Annex 1–18. Krakow: Wyd. Medycyna Praktyczna.

Li, D., P. Azoulay, and B. N. Sampat. 2017. The applied value of public investments in biomedical research. *Science* 356(6333): 78–81. doi:10.1126/science.aal0010.

von Lichtenberg, A. [1928] 1929. Discussion [in German]. Presentation at the 8th Congr. German Soc. Urol., Berlin, Sept. 26–29, 1928. *Verh. Dtsch. Ges. Urol./Z. Urol. Sonderband* 8: 435.

von Lichtenberg, A. [1929] 1930a. Clinical trial of uroselektan [in German]. Presentation at the 9th Congr. German Soc. Urol., Munich, Sept. 26–28, 1929. *Verh. Dtsch. Ges. Urol./Z. Urol. Sonderband* 9: 332–336.

von Lichtenberg, A. [1929] 1930b. About uroselectan [in Italian]. Presentation at the 8th Congress of the Italian Society of Urology, Genoa, Oct. 26, 1929. *Atti Soc. Ital. Urol. VIII Congresso* (6): 338–339.

von Lichtenberg, A. [1930] 1931. The principles of intravenous urography. Presentation at the Annu. Meet. Am. Urol. Assoc., New York City, June, 1930. *J. Urol.* 25(3): 249–257 and discussion 257–263. doi:10.1016/S0022-5347(17)72842-X.

von Lichtenberg, A. 1931. Principles and new advances in excretion urography. *Br. J. Urol.* 3(2): 119–165.

von Lichtenberg, A. 1932. Basics and progress in excretion urography [in German]. *Arch. Klin. Chir.* 171: 3–28.

von Lichtenberg, A. 1935, Dec. 6. Letter (unsigned copy) to Verwaltungsdirektor, Friedrich-Wilhelms-Univ. Berlin [in German]. Universitätsarchiv der Humboldt Universität, Document UK-L 147, Bd. 1, Bl. 6R.

von Lichtenberg, A., and M. Swick. 1929. Clinical trial of uroselectan [in German]. Presentation at the 9th Congr. German Soc. Urol., Munich, Sept. 26–28, 1929. 1929, Nov. 5. *Klin. Wochenschr.* 8(45): 2089–2091. doi:10.1007/BF01875122.

von Lichtenberg, A., and M. Swick. 1930. Clinical trial of uroselectan [in German]. *Mediz. Mitt.* 2(2): 37–41.

Lichtwitz, L. 1930a. Lecture on X-ray visualization of the kidney and urinary tract by intravenous injection of a new contrast agent, uroselectan, by Dr. M. Swick, New York [in German]. *Med. Mitt.* 2(2): 35–37.

Lichtwitz, L. 1930b. About uroselectan [in German]. *Chirurg* (Heidelberg). 2(8): 357–361.

Lichtwitz, L. 1930c. Abstract of discussion to Jaches (1930). *J. Am. Med. Assoc.* 95(19): 1409. doi:10.1001/jama.1930.02720190021005.

Lichtwitz, L. 1930d. Abstract and discussion. Presented at the 81st Ann. Meet. Am. Med. Assoc., Sect. Urol., Detroit, June 23–27, 1930. *Trans. Sect. Urol. Am. Med. Ass.* (81): 86–87.

Lichtwitz, L., and A. Renner. 1914. About the temperature dependence of the gold number and the viscosity of colloidal solutions [in German]. *Hoppe Seyler's Z. Physiol. Chem.* 92(1): 113–118. doi:10.1515/bchm2.1914.92.1.113.

Lieberman, H. 2007. From whole earth to the whole web. *Science* 315(5817): 1369. doi:10.1126/science.1138361.

Lightman, A. 1996. *Dance for Two: Selected Essays.* New York: Pantheon Books.

Lin, Y.; *Mallinckrodt Inc., St. Louis MO.* Prior. 1982, Jan. 11. Publ. 1983, Aug. 2. Triiodoisophthalamide X-ray Contrast Agent. Patent US4396598.

Lindgren, E., ed. 1980. Iohexol. A non-ionic contrast medium. Pharmacology and toxicology. *Acta Radiol. Suppl.* 362: 5–134.

Lindgren, P., K. Dahlström, and L. Fondberg. [1964, 1966] 1967. Cations in intravascular contrast media and development of specific metrizoate formulas. Pharmacological and clinical studies. Proc. of symposia held in Copenhagen, Denmark, Nov. 1964 and Sandefjord, Norway, 1966. *Acta Radiol. Diagn.* 6(Suppl. 270): 7–243. doi:10.1177/0284185167006S27001 to doi:10.1177/0284185167006S27029.

Lipton, J. 2006. Dustin Hoffmann on Laurence Olivier—'Look at me'. TV-show Inside The Actors Studio, The Actors Studio, Inc., June 18, 2006. www.youtube.com/watch?v=z9xWM-u6XRQ (Accessed Apr. 26, 2018).

Loewe, O. 1931. About the radiographic localization of emboli and other occlusions in the peripheral arteries [in German]. *Zentralbl. Chir.* 58(35): 2182–2185.

Lorenz, K. Z. 1974. Analogy as a source of knowledge. *Science* 185(4147): 229–234. doi:10.1126/science.185.4147.229.

Loughlin, K. R., and C. E. Hawtrey. 2003. Moses Swick, the father of intravenous urography. *Urology* 62(2): 385–389.

Ludmerer, K. M. 2005. *American Medical Education from the Turn of the Century to the Era of Managed Care.* Oxford: Oxford University Press.

Ludwig, J. W. 1995. Heart and coronaries—The pioneering age. In *Radiology in Medical Diagnostics. Evolution in X-ray Applications 1895–1995*, eds. G. Rosenbusch, M. Oudkerk, and E. Ammann, English language ed. P. F. Winter, 213–224. Oxford: Blackwell.

Lusic, H., and M. W. Grinstaff. 2013. X-ray computed tomography contrast agents. *Chem. Rev.* 113(3): 1641–1666.

Luther, M. 2013. *Das grosse Handbuch der Kreativitätsmethoden. Wie Sie in vier Schritten mit Pfiff und Methode Ihre Problemlösungskompetenz entwickeln und zum Ideen-Profi werden.* Bonn: managerSeminare Verlags GmbH.

Lütjens, J. 1896. About the chemical behavior and the oxidation of tetraiodo-terephthalic acid, and about triiodo-diaminobenzoic acid [in German]. *Ber. Dtsch. Chem. Ges.* 29(0): 2833–2839. doi:10.1002/cber.18960290384.

Macintyre, J. 1897. X-ray records for the cinematograph. *Arch. Skiag.* 1(4): 37 and plate XXIV.

Mackaness, G. B., and J. P. Hou; *E. R. Squibb & Sons, Inc., Princeton.* Prior. 1978, Sept. 29. Publ. 1980, Mar. 11. Contrast Media Containing Liposomes as Carriers. Patent US4192859.

Macrakis, K. 1993. *Surviving the Swastika.* New York City: Oxford University Press.

Magidson, O., and G. Menschikoff. 1925. Submitted Nov. 28, 1924. About the iodination of α-amino-pyridine [in German]. *Ber. Dtsch. Chem. Ges.* 58(1): 113–118. doi:10.1002/cber.19250580125.

Magnusson, M. 2008. Angiography—University Hospital Lund: The Mecca of Radiology and Particularly Selective Angiography During the 1960s. www.med.lu.se/english/klinvetlund/cardiology/lhrc/history_of_innovation (Accessed Feb. 2017).

Mallinckrodt Chemical Works, St. Louis. 1931. *The Biography of Iodeikon, Including Technique for Intravenous and Oral Administration for Gall Bladder Diagnosis.* St. Louis, MO: Mosby.

Mallinckrodt Chemical Works, St. Louis MO. Prior. 1954, July 26. Publ. 1957, Sept. 4. Manufacture of New 2:4:6-Triiodobenzoic Acid Compounds. Patent GB782, 313.

Maluf, N. S. R. 1956. Role of roentgenology in the development of urology. *Am. J. Roentgenol. Radium Ther. Nucl. Med.* 75(5): 847–854.

Mann, C. C., and M. L. Plummer. 1991. *The Aspirin Wars.* Boston, MA: Harvard Business School Press.

Manning, P. K. 1989. The limits of knowledge: The role of information in regulation. In *Making of Regulatory Policy*, eds. K. Hawkins, and J. Thomas, 49–87. Pittsburgh, PA: University of Pittsburgh Press.

Marshall, V. F. 1977. The controversial history of excretory urography. In *Clinical Urography*, 4th ed., eds. J. Emmett, D. M. Witten, G. H. Myers, and D. C. Utz., part 1: 2–5. Philadelphia, PA: Saunders.

Martin-Luther-Universität Halle-Wittenberg: Johannes Volkmann [in German]. 2005. Archive UAH PA 6154 Volkmann; Rep. 6 Nr. 1407. www.catalogus-professorum-halensis.de/volkmannjohannes.html (Accessed Apr. 26, 2018).

Massell, T. B., S. M. Greenstone, and E. C. Heringman. 1957. Evaluation of diatrizoate (hypaque) in peripheral angiography and aortography. *J. Am. Med. Assoc.* 164(16): 1749–1752. doi:10.1001/jama.1957.02980160021006.

Mattiuzzi, M., F. Arfelli, R.-H. Menk, L. Rigon, and H.-J. Besch; *Bracco Imaging SpA., Milan.* Prior. 2003, Feb. 10. Publ. 2006, Oct. 19. Contrast-Enhanced X-ray Phase Imaging. Patent App. US 2006/0235296.

Maurer, H.-J., and W. Clauss. 1993. Historical overview of the development of contrast media for radiographic imaging diagnostic procedures [in German]. In *Kontrastmittel in der Praxis*, eds. P. Dawson, and W. Clauss, 1–6. Berlin: Springer.

May, F., and M. Schiller. 1954. Urografin, a new means of excretion urography [in German]. *Med. Klinik* 49(35): 1388 & 1403–1405.

Medawar, P. B. 1963. Is the scientific paper a fraud? In *The Threat and the Glory*, ed. P. B. Medawar, 228–233. New York: Harper Collins.

Melchior, E., and M. Wilimowski. 1916. X-ray imaging of bullet channels, war-surgical fistular tracts and abscess cavities using Jodipin [in German]. *Bruns' Beitr. Klin. Chir.* 103(1–3): 334–353.

Melicow, M. M. 1966. Presentation of the fourth Ferdinand C. Valentine medal and award to Moses Swick. *Bull. N. Y. Acad. Med.* 42(2): 123–127.

Mensch, G. 1979. *Stalemate in Technology: Innovations Overcome Depressions.* Cambridge, MA: Ballinger Publishing.

Merck, E. 1909. *E. Merck's Jahresbericht 1908*, vol. 22. Darmstadt: E. Merck, Chemische Fabrik.

Merck E., and H. Winternitz. 1897, June 24. Contract [in German]. (Company Archives of E. Merck, Darmstadt).

Merck E., and H. Winternitz. 1900, May 29. Contract. [in German]. (Company Archives of E. Merck, Darmstadt).

Merton, K., and E. G. Barber. 2004. *The Travels and Adventures of Serendipity: A Study in Historical Semantics and the Sociology of Science.* Alexandia, VA: Alexander Street Press.

Meslans, M. 1896, Feb. 10. Influence of the chemical nature of bodies on their transparency to Röntgen rays [in French]. *CR. Acad. Sci.* 122(6): 309–311.

Meurer, K., M. Laniado, N. Hosten, B. Kelsch, and B. Hogstrom. 2015. Intraarterial and intra-venous applications of iosimenol 340 injection, a new nonionic, dimeric, iso-osmolar radiographic contrast medium: Phase 2 experience. *Acta Radiol.* 56(6): 702–798. doi:10.1177/0284185114536157.

Meuser, H. 1953. Prof. Dr. Theodor Hryntschak [in German]. *Z. Urol.* 46(1): 1–5.

Meyer, D., and S. Legreneur; *Guerbet S.A., Villepinte.* French Prior. 1991, Dec. 4. and 1992, Mar. 13. Publ. 1999, Feb. 16. Macromolecular Polyamine Iodine-Containing Compound, Process for its Preparation and its Use as a Contrast Agent. Patent US5871713.

Minchey, S. R., C. E. Swenson, A. S. Janoff, L. Boni, K. A. Stewart, and W. Perkins; *Liposome Comp., Inc., Princeton.* Prior. 1988, May 20. Publ. 1995, May 16. High Ratio Active Agent: Lipid Complex. Patent US5415867.

Molinder, H. K. M. 1994. The development of cimetidine: 1964–1976. A human story. *J. Clin. Gastroenterol.* 19(3): 248–254.

Moll, F. 1995. Friedrich Voelcker, Alexander von Lichtenberg. Pyelography [in German]. *Aktuelle Urol.* 26, 70–74.

Moll, F., and P. Rathert. 2012. Development of diagnostic imaging in urology [in German]. In *Illustrierte Geschichte der Urologie*, eds. L. Konert, and H. Dietrich, 206–211. Berlin: Springer.

Moniz, E. 1927a. Arterial encephalography, its importance in the localization of brain tumors [in French]. Presentation at the Soc. Neurol., Paris, July 7, 1927. *Rev. Neurol.* 34/II(1): 72–90.

Moniz, E. 1927b. Carotid injections and injectable substances opaque to X-rays [in French]. *Presse Med.* 35(63): 969–971.

Moore, S. D. 1995. Schering warns year's profit will fall sharply on charges. *The Wall Street Journal Europe* 13: 3.

Moore, T. D., and R. F. Mayer. 1955. Hypaque: An improved medium for excretion urography. A preliminary report of 210 cases, and following discussion. *South. Med. J.* 48(2): 135–141.

Morcos, S. K., P. Dawson, J. D. Pearson et al. 1998. The haemodynamic effects of iodinated water soluble radiographic contrast media: A review. *Eur. J. Radiol.* 29(1): 31–46.

Mosse, M., and C. Neuberg. 1903. About the physiologic degradation of iodo-albumin [in German]. *Hoppe Seylers Z. Physiol. Chem.* 37: 427–441.

Mount Sinai Hospital, New York City. n. d. [Appointment Card of Moses Swick] Archive of the Icahn School of Medicine at Mount Sinai, New York City. Consulted 2017.

Mount Sinai Medical Center, New York City. 1978. Special to the Daily News. Archive of the Icahn School of Medicine at Mount Sinai, New York City. Consulted 2017.

Müller, R.-P., B. I. Wenzel-Hora, and H.-W. Addicks. 1985. First reports on indirect lymphography with iotasul in the head and neck area [in German]. *RöFo* 142(2): 218–221. doi:10.1055/s-2008-1052635.

Muth, H. 1989. History of the German Thorotrast studies. In *Risks From Radium and Thorotrast*, eds. D. M. Taylor, C. W. Mays, G. B. Gerber, and R. G. Thomas, BIR Report 21, 93–97. London: British Institute of Radiology.

Mützel, W., H.-M. Siefert, U. Speck, H. Pfeiffer, P.-E. Schulze, and B. Acksteiner; *Schering AG, Berlin*. Earliest German Prior. 1979, June 28. Publ. 1983, June 29. Dimers of Tri-Iodinated Isophthalic Acid Diamides, Their Preparation, X-ray Contrast Agent Containing Them, and Dimer of Tri-iodinated Isophthalic Acid Chloride [in German]. Patent EP0022056.

Myers, G. 1995. From discovery to invention: The writing and rewriting of two patents. *Soc. Stud. Sci.* 25(1): 57–105. doi:10.1177/030631295025001004.

Nesbit, R. M., and J. Lapides. 1950. Preliminary report on urokon, a new excretory pyelographic medium. *J. Urol.* 63(6): 1109–1112. doi:10.1016/S0022-3547(17)68871-2.

Neudert, W., and H. Röpke. 1954. About the physico-chemical behavior of di-sodium salts of adipic acid bis-(2,4,6-triiodo-3-carboxy-anilide) and other triiodobenzene derivatives [in German]. *Chem. Ber.* 87(5): 659–667. doi:10.1002/cber.19540870507.

Neumann-Kleinpaul, [K.], and [O.] Pessinger. 1929. On the treatment of some equine febrile general illnesses with selectan [in German]. *Arch. Wissenschaftl. Prakt. Tierheilk.* 59: 315–349.

New Specialties. 1930, Jan. 4. Pyelognost [in German] *Klin. Wochenschr.* 9(1): 47. doi:10.1007/BF01740715.

Newton, B., M. Thaning, D.-J. in't Veld, K. Lang-Seth, and P. M. Evans; *GE Healthcare A/S, Oslo*. Eur. Prior. 2012, Jan. 11. Publ. 2013, July 18. X-ray Imaging Contrast Media with Low Iodine Concentration and X-ray Imaging Process. Patent App. WO2013/104690.

Niculescu, S. 1928. About Halogen Derivatives of Pyridine. A Contribution to the Relationship Between Chemical Constitution and Physical Constants [in German]. PhD diss., Vereinigte Friedrichs-Univ. Halle-Wittenberg, Halle a. d. S.

Nobelprize.org. 2014a. *Nomination Database 1938*. Stockholm: Nobel Media. www.nobelprize.org/nomination/archive/show_people.php?id=1026 (Accessed Apr. 26, 2018).

Nobelprize.org. 2014b. *Nomination Database 1931.* Stockholm: Nobel Media. www. nobelprize.org/nomination/archive/show.php?id=13353 (Accessed Apr. 26, 2018).

Nordal, V., and H. Holtermann; *Nyegaard & Co. A/S, Oslo.* Prior. 1976, June 11. Publ. 1979, July 18. Triiodoisophthalic Acid Amides. Patent GB1548594 (German disclosure DE2726196. Publ. 1977, Dec. 22).

Novello, N. J., S. R. Miriam, and C. P. Sherwin. 1929. Comparative metabolism of certain aromatic acids. IX fate of some halogen derivatives of benzoic acid in the animal body. *J. Biol. Chem.* 67(3): 555–566.

Nunn, A. D. 2007. Molecular imaging and personalized medicine: An uncertain future. *Cancer Biother. Radiopharm.* 22(6): 722–739.

Nutrizio, V., V. Ivekovic, J. Paladino, J. Papa, and V. Marinšek-Čičin-Šain, 1981. Radiologic anatomy of the subarachnoid space of the brain displayed by means of contrast materials during computerized tomography. *Acta Med. Iugosl.* 35(3): 157–164.

Nyman, U. 2016. Moses Swick—The father of the iodine contrast media and the battle of its honor [in Swedish]. *Imago Medica: Medlemsforum* (2): 10–18. http://docplayer. se/19907835-Moses-swick-jodkontrastmedlens-fader.html.

Nyman, U., O. Ekberg, and P. Aspelin. 2016a. Torsten Almén (1931–2016): The creator of non-ionic contrast agents—His history [in Swedish]. *Imago Medica: Medlemsforum* (1): 10–15. http://docplayer.se/25341188-Torsten-almen-skaparen-av-icke-joniska-kontrastmedel.html.

Nyman, U., O. Ekberg, and P. Aspelin. 2016b. Torsten Almén (1931–2016): The father of nonionic iodine contrast media. *Acta Radiol.* 57(9): 1072–1078. doi:10.1177/0284185116648504.

Odén, S. 1955. Triurol in cerebral angiography. *Acta Radiol.* os-43(2): 97–103. doi:10.1177/028418515504300201.

Oftedal, S.-I. 1973. Toxicity of water soluble contrast media injected suboccipitally in cats. *Acta Radiol. Diagn.* 14(Suppl. 335): 84–92. doi:10.1177/0284185173014S33510.

Olsson, O. 1954. Contrast media in diagnosis and the attendant risks. The tolerance of the organism to some common contrast media. *Acta Radiol.* os.42(Suppl. 116): 75–83. doi: 10.1177/0284185154042S11610.

Operskalski, J. T., and A. K. Barbey. 2016. Risk literacy in medical decision-making. How to better represent the statistical structure of risk. *Science* 352(6284): 413–414. doi:10.1126/science.aaf7966.

Osborne, E. D. 1921. Iodine in the cerebrospinal fluid, with special reference to iodide therapy. *J. Am. Med. Assoc.* 76(21): 1384–1386. doi:10.1001/jama.1921.02630210008003.

Osborne, E. D. 1922. Contributions to the pharmacology and therapeutics of iodides. *J. Am. Med. Assoc.* 79(8): 615–617. doi:10.1001/jama.1922.02640080017006.

Osborne, E. D., C. G. Sutherland, A. J. Scholl Jr., and L. G. Rowntree. 1923. Roentgenography of urinary tract during excretion of sodium iodide. *J. Am. Med. Assoc.* 80(6): 368–373. doi:10.1001/jama.1923.02640330004002.

Oselladore, G. 1930. In search of the most harmless contrast medium to achieve arteriography [in Italian]. *Minerva Med.* 21/II(38): 410–415.

Oselladore, G. 1937. Mastography with contrast agents and its importance in the diagnosis of breast disease [in Italian]. *Ann. Radiol. Fis. Med.* [Bologna] 11(3): 209–151.

Osol, E., ed.-in-chief. 1950. *The Dispensatory of the United States of America, 1950 ed. New Drug Developments Volume,* vol. 2. 2047–2048. Philadelphia, PA: Lippincott.

Ossenbeck, A., E. Tietze, and G. Hecht; *I. G. Farbenindustrie AG, Frankfurt a. M.* Prior. 1929a, Dec. 25. Publ. 1931, Sept. 3. Methods of Preparation of Iodomethane Sulfonic Acid, Respectively its Salts [in German]. Patent DE532766.

Ossenbeck, A., E. Tietze, and G. Hecht; *Winthrop Chemical Comp. Inc., New York.* German Prior. 1929b. Dec. 24. Publ. 1932, Jan. 26. Iodomethane Sulphonic Acid and Homologues Thereof. Patent US1842626.

Pacey, A. 1999. *Meaning in Technology.* Cambridge, MA: MIT Press.

Pallardy, G., M.-J. Pallardy, and A. Wackenheim. 1989. *Histoire illustré de la radiologie.* Paris: Éditions Roger Dacosta.

Palmer, F. J. 1988. The RACR survey of intravenous contrast media reactions: Final report. *Australas. Radiol.* 32(4): 426–428. doi:10.1111/j.1440-1673.1988.tb02770.x.

Pariser, R. 1998. Creativity at the crossroads between science and technology. *CHEMTECH.* 28(6): 48–54.

Pasteur, L. 1854. Speech given in Douai, France, December 7, 1854. Reprinted 1939. In *Oeuvres de Pasteur Vol. 7. Mélanges scientifiques et littéraires,* ed. L. Vallery-Radot, 129–132. Paris: Masson.

Pasveer, B. 1993. Depiction in medicine as a two-way affair. X-ray pictures and pulmonary tuberculosis in the early twentieth century. In *Medicine and Change: Historical and Sociological Studies of Medical Innovation,* ed. I. Löwy, 85–105. Montrouge: INSERM/John Libbey Eurotext.

Patrias, K.; techn. ed. D. L. Wendling. [2007], updated 2015. *Citing Medicine: The NLM Style Guide for Authors, Editors, and Publishers,* 2nd ed. Bethesda, MD: National Library of Medicine. www.nlm.nih.gov/citingmedicine (Accessed Apr. 26, 2018).

Peiser, F. 1930. Concerning Abrodil [in German]. Company-internal communication from Pharmazeutische Abteilung, I. G. Farbenindustrie AG, Leverkusen to Direktions-Registratur, Elberfeld] Bayer AG, Corporate History & Archives, Leverkusen.

Pérez Castro, E. 1980. Urological antology: Professor Alexander von Lichtenberg (1880–1980) [in Spanish]. *Arch. Esp. Urol.* 33(6): 529–538.

Periasamy, M. P., and B. D. Doty; *Mallinckrodt Inc., St. Louis MO.* Prior. 2004, Mar. 11. Publ. 2012, Oct. 12. Low Osmolar X-ray Contrast Media Formulations. Patent EU2253332.

Perlmann, S. 1931. Experience with Abrodil in intravenous pyelography [in German]. *Munch. Med. Wochenschr.* 78/I(22): 955.

Personalblatt Alexander von Lichtenberg. 1910. Kaiser-Wilhelms-Univ., Strasbourg, Germany [in German]. Document in Arch. Départementales du Bas-Rhin, Strasbourg, France. Document 103 AL 552.

Personalblatt Alexander von Lichtenberg. 1917. Kaiser-Wilhelms-Univ., Strasbourg, Germany [in German]. Document in Arch. Départementales du Bas-Rhin, Strasbourg, France. Document 103 AL 552.

Personalblatt Alexander von Lichtenberg. 1933. Friedrich-Wilhelms-Univ. Berlin [in German]. Universitätsarchiv der Humboldt Universität, Document UK-L 147, Bd. 1, Bl. 1 & 1R

Personalblatt Curt Räth. n.d. Friedrich-Wilhelms-Univ., Berlin [in German]. Universitätsarchiv der Humboldt Universität, Document UK-Pers., R 9, Bd. 1, Bl. 2.

Perutz, A., C. Siebert, and R. Winternitz. 1930. *Pharmakologie der Haut. Arzneimittel. Allgemeine Therapie.* Berlin: Verlag Julius Springer.

Pfeiffer, G. 1887. About halogen derivatives of pyridine bases from pyridine carbonic acids [in German]. *Ber. Dtsch. Chem. Ges.* 20(1): 1343–1353. doi:10.1002/cber.188702001297.

Pfeiffer, H., and U. Speck; *Schering AG, Berlin.* Prior. 1976, June 23. Publ. 1985, Feb. 21. New Dicarbonic Acid Bis(3,5-dicarbamoyl-2,4,6-triiodoanilide) [in German]. Patent DE2628517. (German disclosure DE2628517. Publ. 1978, Jan. 5).

Pharmacist Georg Otto, Berlin. 1957. 70 years old [in German]. *Dtsch. Apoth. Ztg.* 97(36): 310.

Philander [1892]. Electra. A twentieth century physical diagnostic fairy tale [in German]. In *Medizinische Märchen,* ed. Philander, 186–198. Stuttgart: Levy & Müller.

Pinch, T. J., and W. E. Bijker. 1984. The social construction of facts and artefacts: Or how the sociology of science and the sociology of technology might benefit each other. *Soc. Stud. Sci.* 14(3): 399–441. doi:10.1177/030631284014003004.

Pitrè, D., and E. Felder: 1980. Development, Chemistry, and Physical properties of iopamidol and its analogues. *Invest. Radiol.* 15(Suppl. 6): S301–S309. doi:10.1097/00004424-198011001-00065.

Polanyi, M. 1966. *The Tacit Dimension.* Garden City, NY: Doubleday.

Pollack, H. M. 1994. The history of urographic contrast media: the real story. *A. U. A. Today* 7(2): 6 & 9.

Pollack, H. M. 1996. Uroradiology. Chap. 9 In *A History of the Radiological Sciences. Diagnosis*, eds. R. A. Gagliardi, and B. L. McClennan, 195–253. Reston, VA: Radiological Centennial.

Pollack, H. M. 1999. History of iodinated contrast media. Chap. 1 In *Trends in Contrast Media*, eds. H. S. Thomsen, R. N. Muller, and R. F. Mattrey, 3–19. Berlin: Springer.

Pollack, H. M., and M. P. Banner. 1985. Current status of excretory urography: A premature epitaph? *Urol. Clin. North Am.* 12(4): 585–601.

Pool, I. de S. 1990. *Technologies Without Boundaries: On Telecommunication in a Global Age*, ed. E. M. Noam. Cambridge, MA: Harvard University Press.

Porter, R. 2001. *Bodies Politic: Disease, Death and Doctors in Britain, 1650–1900.* Ithaca, NY: Cornell University Press.

Prange, G. 1923. On the Synthesis of Heterocyclic Ring Systems by Intramolecular Condensation of Aminoacetals [in German]. PhD diss., Friedrich-Wilhelms-Univ. Berlin.

Price, D. J. de S. 1984. On sealing wax and string. *Natural Hist.* 93(1): 49–57.

Pribram, B. O. 1926. About a new contrast agent for the radiographic representation of the gallbladder [in German]. *Dtsch. Med. Wochenschr.* 52(31): 1291–1292. doi:10.1055/s-0028-1127554.

Priewe, H., and R. Rutkowski; *Schering AG, Berlin.* Prior. 1952, Aug. 7. Publ. 1956, Feb. 2. Methods of Preparation of *N*-acyl Derivatives of 2,4,6-Triiodo-3-aminobenzoic Acid [in German]. Patent DE936928.

Progress in Therapy. Per-Abrodil [in German]. 1932. *Ther. Ber. d. I. G. Farbenindustrie AG* [Leverkusen, Germany] 9: 224.

Prusiner, S. B. 2002. Discovering the cause of AIDS. *Science* 298(5599): 1726–1727. doi:10.1126/science.1079874.

R. S. 1924. Therapeutic communications. Dominal X as a contrast agent [in German]. *Munch. Med. Wochenschr.* 71/I(15): 489.

Rasor, J. S., and E. G. Tickner; *Schering AG, Berlin.* Prior. 1980, Nov. 17. Publ. 1984, Apr. 17. Microbubble Precursors and Methods for their Production and Use. Patent US4442843.

Ramstetter, M. H. 1966. Eugen de Haën [in German]. *Hannoversche Geschichtsbl.*, n. ser. 20(1/3): 107–190.

Räth, C. 1920. Synthesis of Some New Mydriatics and Anesthetics, as a Contribution to the Clarification of the Relationship between Chemical Constitution and Physiological Action (and Curriculum Vitae) [in German]. PhD diss., Friedrich-Wilhelms-Univ. Berlin.

Räth, C. 1924a. About intramolecular condensation reactions of amino-acetals and amino-aldehydes I: About a synthesis of dihydroquinoline and some homologs thereof [in German]. *Ber. Dtsch. Chem. Ges.* 57(3): 550–553. doi:10.1002/cber.19240570332.

Räth, C. 1924b. About intramolecular condensation reactions of aminoacetals and aminoaldehydes II: About the course of the reaction in the formation of dihydroquinoline and a new preparation of indole derivatives [in German]. *Ber. Dtsch. Chem. Ges.* 57(4): 715–718. doi:10.1002/cber.19240570426.

Räth, C.; *Schering-Kahlbaum AG, Berlin.* Prior. 1924c, July 1. Publ. 1930, Feb. 19. Process for the Preparation of Iodine Substitution Products of Pyridine Derivatives [in German]. Patent DE491681.

Räth, C.; *Schering-Kahlbaum AG, Berlin*. Prior. 1925, Feb. 3. Publ. 1928, Jan. 14. Methods of Preparation of 2-Amino-5-iodopyridine [in German]. Patent DE454695.

Räth, C. 1927a. Pharmacological-biological experiments with preparations of the pyridine series. Meeting report [in German]. Presentation at the Joint Sess. Tierärztl. Ver. Prov. Brandenburg & Tierärztl. Ges. Berlin, Tierärztl. Hochsch. Berlin, Nov. 27. *Z. Angew. Chem.* 40: 1438–1439.

Räth, C.; *Schering-Kahlbaum AG, Berlin*. Prior. 1927b, May 12. Publ. 1930, Sept. 13. Process for Increasing the Solubility of Halogen-Substituted Oxypyridines in Water [in German]. Patent DE506425.

Räth, C. 1930. About the isomerism of derivatives of 2-oxypyridine [in German]. *Liebigs Ann. Chem.* 484(1): 52–64. doi:10.1002/jlac.19304840104.

Räth, C. 1931a. To the knowledge of 3-aminopyridin [in German]. *Liebigs Ann. Chem.* 486(1): 95–106. doi:10.1002/jlac.19314860106.

Räth, C. 1931b. To the knowledge of pyracridone [in German]. *Liebigs Ann. Chem.* 486(1): 284–294. doi:10.1002/jlac.19314860115.

Räth, C. 1931c. Mercaptans and sulfonic acids of pyridine [in German]. *Liebigs Ann. Chem.* 487(1): 105–119. doi:10.1002/jlac.19314870107.

Räth, C. 1931d. On the hydrogenation of pyridine and its N-alkyl derivatives [in German]. *Liebigs Ann. Chem.* 489(1): 107–118. doi:10.1002/jlac.19314890109.

Räth, C., and G. Prange. 1928. About a synthesis of 2-aminonicotinic acid and its behavior against nitric acid [in German]. *Liebigs Ann. Chem.* 467(1): 1–10. doi:10.1002/jlac.19284670102.

Rathert, P. 1992. Excretion urography: Controversies in its historical development [in German]. *Jahrbuch der Urologie* 1: 131–137.

Rathert, P., H. Melchior, and W. Lutzeyer. 1974. Johannes Volkmann, M.D., pioneer in intravenous urography. *Urology* 4(5): 613–616.

Rathert, P., H. Melchior, and W. Lutzeyer. 1975. Contributions to the controversial history of excretion urography [in German]. *Urologe B* 15, 21–28.

Ratschow, M. 1930. Uroselektan in angiography, with special consideration of varicography [in German]. *Fortschr. Geb. Rontgenstr.* 42(1): 37–45.

Rave, [W.] [1929] 1930. Intravenous representation of the urinary tract with uroselektan [in German]. Presentation at the 3rd Conf. North- and East-German Röntgen-Soc., Dresden, Oct 12–13, 1929. *Fortschr. Geb. Rontgenstr.* 41(1): 77.

Raymond, D. P., J. R. Duffield, and D. R. Williams. 1987. Complexation of plutonium and thorium in aqueous environments. *Inorg. Chim. Acta* 140: 309–313. doi:10.1016/S0020-1693(00)81112-3.

Rehn, E. 1924. Demonstration of topical kidney diagnostics on the basis of the acid-base excretion [in German]. Presentation at the 48th Meet. Dtsch. Ges. Chir., Berlin, Apr. 23–26, 1924. *Arch. Klin. Chir.* 133: 263–266.

Reitmann, J.; *I. G. Farbenindustrie AG, Frankfurt a. M.* Prior. 1930, Dec. 21. Publ. 1933, June 22. Methods of Preparation of Iodinated Derivatives of 4-Pyridones [in German]. Patent DE579224.

Reitmann, J.; *I. G. Farbenindustrie AG, Frankfurt a. M.* Prior. Oct. 16, 1931a. Publ. 1933, Feb. 2. Method for Obtaining Highly Soluble Descendants of Pyridine Derivatives [in German]. Patent DE570860.

Reitmann, J.; *Winthrop Chemical Comp., Inc., New York*. German Prior. 1931b, Oct. 15. Publ. 1935, Mar. 5. Aliphatic Amine Salts of Halogenated Pyridones Containing an Acid Group. Patent US1993039.

Report on Lectures Regarding Uroselectan, Held in Genoa, Trieste and Rome, Fall 1929 [in German]. 1930. *Med. Mitt.* 2(2): 41–42.

Richardson, J. F., and D. K. Rose. 1950. Clinical evaluation of Urokon in pyelography. *J. Urol.* 63(6): 1113–1119. doi:10.1016/S0022-5347(17)68872-4.

Riedel, B. 1975. President of the Berlin Urol. Soc.: Appointment of Moses Swick as honorary member.

Rieder, H. 1904. Radiological examinations of the stomach and intestine in living humans [in German]. *Munch. Med. Wochenschr.* 51/II(35): 1548–1551.

Rigler, L. G. 1945. The development of Roentgen diagnosis. *Radiology* 45(5): 467–502. doi:10.1148/45.5.467.

Roentgen contrast agents patentable [in German]. 1935. *Angew. Chem.* 48(14): 217. doi:10.1002/ange.19350481407.

Röntgen, W. C. 1881. On tones, which arise by intermittent irradiation of a gas [in German]. *Ann. Physik Chem.*, 3rd ser. 12(1): 155–159.

Röntgen, W. C. 1885. Experiments on the electromagnetic effect of dielectric polarization [in German]. *Sber. Kgl. Preuss. Akad. Wissensch. Berlin.* I/xi: 195–198.

Röntgen, W. C. 1888. On the electrodynamic force caused by movement of a dielectric in the homogeneous electric field [in German]. *Sber. Kgl. Preuss. Akad. Wissensch. Berlin.* I/ii: 23–28.

Röntgen, W. C. 1895. About a new kind of rays (provisional communication) [in German]. *Sber. Phys.-Med. Ges. Wuerzburg.* 29(9): 132–141.

Röntgen, W. C. 1896. A new kind of rays. II. Communication [in German]. *Sber. Phys.-Med. Ges. Wuerzburg.* 30(1): 11–16 & (2): 17–19.

Röntgen, W. C. 1897. Further observations on the properties of X-rays [in German]. *Sber. Kgl. Preuss. Akad. Wissensch. Berlin.* I/xxvi: 576–592.

Rosati, G., and C. de Haën. [1990] 1991. Considerations on the research and development of new X-ray contrast media: A critical appraisal. Presentation at the 2nd Int. Symp. Contrast Media, Osaka, Nov. 9–10, 1990. In *New Dimensions of Contrast Media*, eds. H. Katayama, and R. C. Brasch, 29–34. Amsterdam: Excerpta Medica.

Rosati, G., A. Morisetti, and P. Tirone. 1992a. Toxicity in animals and safety in humans: The predictive value of animal studies. *Toxicol. Lett.* 64/65: 705–715.

Rosati, G., S. Leto di Priolo, and P. Tirone. 1992b. Serious or fatal complications after inadvertent administration of ionic water-soluble contrast media in myelography. *Eur. J. Radiol.* 15(2): 95–100. doi:10.1016/0720-048X(92)90131-R.

Rosen, F. S. 1997. Transforming the theater of surgery. *Nature* 388(6645): 841.

Rosenbusch, G., M. Oudkerk, and E. Ammann. (1995) English language. In *Radiology in Medical Diagnostics. Evolution of X-ray application 1895–1995*, ed. P. F. Winter. Oxford: Blackwell.

Roseno, A. [1928] 1929. Studies on intravenous pyelography [in German]. Presentation at the 8th Congr. German Soc. Urol., Berlin, Sept. 26–29, 1928. *Verh. Dtsch. Ges. Urol./Z. Urol. Sonderband* 8: 431–433.

Roseno, A. 1929, June 18 & Aug. 27. The intravenous pyelography. II. Communication. Clinical results [in German]. *Klin. Wochenschr.* 8(25): 1165–1170 & (35): 1623. doi:10.1007/BF01737378 & doi:10.1007/BF01847956.

Roseno, A. [1929] 1930. Intravenous pyelography [in German]. Presentation at 9th Congr. German Soc. Urol., Munich, Sept. 26–28, 1929. *Verh. Dtsch. Ges. Urol./Z. Urol. Sonderband* 9: 337–343.

Roseno, A., and H. Jepkens. 1929, May. The intravenous pyelography. I. Communication. Results from animal experiments [in German]. *Fortschr. Geb. Rontgenstr.* 39(5): 859–863.

Rosenstein, P. [1924] 1925. Discussion [in German]. Presentation at the 6th Congr. German Soc. Urol., Berlin, Oct. 1–4, 1924. *Verh. Dtsch. Ges. Urol./Z. Urol. Sonderband.* 4: 320–321.

Rowland, L. P. 2009. *The Legacy of Tracy J. Putnam and H. Houston Merritt: Modern Neurology in the United States.* New York: Oxford University Press.

Rubritius, H. [1919] 1920. Potassium iodide as contrast agent in the roentgenology of the urinary tract [in German]. Presentation at the College of Physicians, Vienna, June 20, 1919. *Z. Urol.* 14(2): 57–60.

Rumpel, T. 1897. The clinical diagnosis of spindle-shaped esophageal dilatation [in German]. Presentation at the Aerztlicher Verein Hamburg, Feb. 2, 1897. *Munch. Med. Wochenschr.* 44/I(15): 383–386, Apr. 13, 1897 and (16): 420–421, Apr. 20.

Sachs, A. P.; *Zonite Products Corp., New York.* Prior. 1933, July 26. Publ. 1938, Nov. 1. Radiographic Substance. Patent US2135474.

Sachse, A. 2008. Iodinated liposomes as contrast agents. In *Nanoparticles in Biomedical Imaging. Emerging Technologies and Applications*, eds. J. W. M. Bulte, and M. M. J. Modo, 371–410. New York: Springer.

Sachse, A., G. Rossling, and J. Leike; *Schering AG, Berlin.* German Prior. 1993, Dec. 2. Publ. 2002, Nov. 2. Process for Increasing the Stability of Liposome Suspensions that Contain Hydrophilic Active Ingredients. Patent US6475515.

Salomon, A.; *N. V. Orgachemia, Oss, The Netherlands.* Prior. 1937, June 3. Publ. 1940, Mar. 25. Improvements In or Relating to the Production of Stable Concentrated Solutions of the Salts of Ortho-iodohippuric Acid [in Dutch]. Patent NL48110.

Salvesen, S. 1973. Acute toxicity tests of metrizamide. *Acta Radiol. Diagn.* 14(Suppl. 335): 5–13. doi:10.1177/0284185173014S33504.

Salvesen, S., P. L. Nilsen, and H. Holtermann. 1967. Effects of calcium and magnesium ions on the systemic and local toxicities of the *N*-methyl-glucamine (meglumine) salt of metrizoic acid (Isopaque). *Acta Radiol. Diagn.* 6(Suppl. 270): 180–193. doi:10.1177/02 84185167006S27023.

Sandström, C. 1953. Contrast media for the kidneys, heart and vessels, and their toxicity. *Acta Radiol.* os-39(4): 281–298. doi:10.1177/028418515303900403.

Sanen, F. J. 1962. Considerations of cholecystographic contrast media. *Am. J. Roentgenol. Radium Ther. Nucl. Med.* 88(4): 797–801.

dos Santos, J. C. 1971. Obituary. In memoriam Reynaldo dos Santos. Profile of a scientist. *J. Cardiovasc. Surg.* [Turin, Italy] 12(1): 78–81.

dos Santos, R., [A.] C. Lamas, and [J.] P. Caldas. 1931a. Recent advances in the technique of arteriography of the abdominal aorta [in French]. *Presse Med.* 39(31): 574–577.

dos Santos, R., A. C. Lamas, and J. P. Caldas. 1931b. *Artériographie des membres et de l'aorte abdominale.* Paris: Masson.

Sarkowski, H. 1996. *Springer-Verlag History of a Scientific Publishing House: Part 1: Foundation 1842–1945, Maturation, Adversity.* Dordrecht: Springer. doi:10.1007/978-3-540-92887-4.

Sartorius, F., and H. Viethen. 1933. Clinical and experimental investigations on the question of contrast agents in retrograde pyelography, in particular on the usability of Thorotrast [in German]. *Z. Urol. Chir.* 36: 312–342.

Schepelmann, [E]. 1910. On blood vessel shadows in X-ray images [in German]. *Munch. Med. Wochenschr.* 57/II(30): 1914–1916.

Schering, AG, Berlin. ≥1941. Hepatoselectan for intravenous hepatolienography. Vasoselectan for Arteriography [in German]. Dokument SchA-S1-121/4 in Bayer AG, Corporate History and Archives, Leverkusen.

Schering, AG, Berlin (Paul Diedrich). Prior. 1953, Feb. 6. Publ. 1958, Aug. 21. Methods of Preparation of N-Acyl Derivatives of 3,5-Diamino-2,4,6-triiodobenzoic Acid [in German]. Patent DE970133.

Schering-Kahlbaum, AG, Berlin. 1930. (UROSELECTAN advertisement). *Med. Mitt.* 2(2): 31.

Schering-Kahlbaum, AG, Berlin. Prior. 1936, Aug. 29. Publ. 1938, Feb. 28. X-ray Contrast Media and the Production of X-ray Photographs. Patent GB4807211.

Schering, AG, and G. J. Wlasich. 1991. *Aus einem Jahrhundert Schering-Forschung Pharma. Scheringianum – Schriftenreihe.* Berlin: Schering Stiftung.

Schich, M., C. Song, Y.-Y. Ahn et al. 2014. A network framework of cultural history. *Science* 345(6196): 558–562. doi:10.1126/science.1240064.

von Schickh, O.; *Schering-Kahlbaum AG, Berlin.* Prior. 1926, May 11. Publ. 1928, Jan. 26. Methods of Preparation of Pure Tetraiodophenolphthalein [in German]. Patent DE454763.

von Schickh, O.; *Schering-Kahlbaum AG, Berlin.* Prior. 1927a, Mar. 10. Publ. 1930, Feb. 26. Methods of Preparation of Bis-ω-halogen Acetylated Diphenyl Ethers [in German]. Patent DE492321.

von Schickh, O.; *Schering-Kahlbaum AG.* Prior. 1927b, Apr. 13. Publ. 1929, Mar. 12. Methods of Preparation of 5-Iodo-2-aminopyridine [in German]. Patent DE473213.

von Schickh, O. 1933. About new synthetic drugs [in German]. *Angew. Chem.* 46(29): 485–490. doi:10.1002/ange.19330462902.

von Schickh, [O.]. 1938. The development of pyridine chemistry. Arthur Binz's 70th birthday [in German]. *Angew. Chem.* 51(45): 779–798. doi:10.1002/ange.19380514502.

Schiffmann, F. 1925. About a New Mode of Formation of δ-Coniceine. On Substitution Products of Pyridine [in German]. PhD diss., Friedrich-Wilhelms-Univ., Berlin.

Schlenk, O. 1934. *Chemische Fabrik von Heyden Aktiengesellschaft, Radebeul-Dresden, 1874–1934: Erinnerungsblätter aus 6 Jahrzehnten.* Radebeul: Kupky & Dietze.

Schlief, E. 1930. About retrograde pyelography with Abrodil [in German]. *Munch. Med. Wochenschr.* 77/II(52): 2220.

Schlottmann, H. 1928. About Isomerisms of Derivatives of α-Pyridon [in German]. PhD diss., Friedrich-Wilhelms-Univ., Berlin.

Schmidt, W. 1930. The angiography with uroselectan, a method for the radiographic representation and functional testing of the peripheral circulation [in German]. *Chirurg* (Heidelberg) 2(14): 652–663.

Schneider, M., H. Tournier, and B. Lamy; *Bracco Industria Chimica SpA, Milano.* Swiss Prior. 1987, May 22. Publ. 1994, May 17. Injectable Opacifying Composition Containing Liposomes of High Encapsulation Capacity for X-ray Examinations. Patent US5312615.

Schneider, M., H. Tournier, R. Hyacinthe, C. Guillot, and B. Lamy; *Bracco International BV, Amsterdam.* Eur. Prior. 1990, Dec. 11. Publ. 1995, Feb. 28. Method for Making Liposomes of Enhanced Entrapping Capacity Toward Foreign Substances to be Encapsulated. Patent US5393530.

Schoeller, W., and M. Dohrn; *Schering-Kahlbaum AG, Berlin.* Prior. 1926, May 22. Publ. 1932, May 18. Methods of Preparation of More Than Two-Fold Halogen-Substituted 2-Arylchinoline-4-carbonic Acids [in German]. Patent DE506349.

Schoeller, W., and M. Gehrke; *Chemische Fabrik auf Actien (vormals E. Schering) Berlin.* German Prior. 1924, July 2. Publ. 1927, Feb. 1. Pharmaceutical Product Containing Arsenic and Process of Making Same. Patent US1616144.

Schoeller, W., and M. Gehrke. 1927. To the knowledge of the effects of thyroxine [in German]. *Klin. Wochenschr.* 6(41): 1938–1939. doi:10.1007/BF01733867.

Schoeller, W., and K. Schmidt; *Chemische Fabrik auf Actien (vormals E. Schering), Berlin.* German Prior. 1924a, Dec. 10. Publ. 1926, June 8. Monoiodo-oxindole and Process for Making Same. Patent US1587866.

Schoeller, W., and K. Schmidt; *Chemische Fabrik auf Actien (vormals E. Schering), Berlin.* German Prior. 1924b, Dec. 10. Publ. 1926, July 13. Poly-Iodinated Isatins and Process Making Same. Patent US1592386.

Schoeller, W., and K. Schmidt; *Schering-Kahlbaum AG, Berlin.* Prior. 1926, Aug. 3. Publ. 1928, Oct. 27. Methods of Preparation of Iodo-Substituted Benzonitriles with Phenolether-Type Bond [in German]. Patent DE467639.

Schulz, W. G. 1996. 20/20 Focus on industry future. Technological vision 2020 aims to keep U.S. Chemical Companies competitive in emerging global business environment. *Chem. Eng. News* 74(40): 12–14.

Schumpeter, J. [1912] 1997. *Theorie der wirtschaftlichen Entwicklung*, 9th ed. Berlin: Duncker & Humblot.

Schumpeter, J. 1939. *Business Cycles: Theoretical, Historical and Statistical Analysis of the Capitalist Process.* New York: McGraw-Hill.

Schwarz, G. S. 1972. Immigrant radiologist of Canadian and non-American origin. *N. Y. State J. Med.* 72(11) 1300–1305.

Seeliger, E., D. C. Lenhard, and P. B. Persson. 2014. Contrast media viscosity versus osmolality in kidney injury: Lessons from animal studies. *Biomed. Res. Int.* 2014: 358136. Doi:10.1155/2014/358136.

Servi, I. S., and D. Turnbull. 1966. Thermodynamics and kinetics of precipitation in the copper-cobalt system. *Acta Metallica* 14(2): 161–169. Doi:10.1016/0001-6160(66)90297-5.

Sewell, W. H. Jr. 1992. A theory of structure: Duality, agency, and transformation. *Am. J. Sociol.* 98(1): 1–29. Doi:10.1086/229967.

Sgalitzer, M. 1921. X-ray examination of the bladder in the axial direction of projection [in German]. *Wien. Med. Wochenschr.* 71(11): 513–517.

Sgalitzer, [M.] 1930. Abstract of discussion on uroselectan in the Viennese Society of Röntgenology [in German]. *Med. Mitt.* 2(3): 84.

Sgalitzer, M., and T. Hryntschak. 1921. X-ray examination of the bladder in the lateral direction [in German]. *Z. Urol.* 16: 399–406.

Sgalitzer, M., and T. Hryntschak. 1924. Questions of technique and indication of pyelography. The 'Oblique' pyelography. Pyelograms in ptotic kidneys. Notifications of a death [in German]. *Fortschr. Geb. Rontgenstr.* 32(1/2): 97–104.

Sgalitzer, M., R. Demel, V. Kollert, and H. Ranzenhofer. 1930. On the representation and treatment of diseases of peripheral arteries [in German]. *Wien. Klin. Wochenschr.* 43(27): 833–837.

Sgalitzer, M., V. Kollert, and R. Demel. 1931. Contrast representation of the veins in the X-ray image [in German]. *Klin. Wochenschr.* 10(36): 1659–1663. doi:10.1007/BF01755385.

Shermer, M. 1995. Exorcising Laplace's Demon: Chaos, antichaos, history and metahistory. *History and Theory* 34(1): 59–83.

Sibum, H. O. 2004. What kind of science is experimental physics? *Science* 306(5693): 60–61. doi:10.1126/science.1093598.

Sicard, J.-A., and J. Forestier. 1921. Radiographic methods of exploration of the epidural cavity by lipiodol [in French]. *Rev. Neurol.* (Paris) 28(12): 1264–1266.

Simmons, R. 2001. Sense and sensibility. *Nature* 411(6835): 243. doi:10.1038/35077175.

Simons, A. 1923/1924. Röntyum, a new X-ray contrast agent for the visualization of the gastrointestinal tract [in German]. *Fortschr. Geb. Rontgenstr.* 31(1): 90.

Šimunić, S., V. Gvozdanović, V. Nutrizio, and J. Papa (corrected from Vedran). 1977. The significance of the intravenous application of contrast medium in computerized tomography of the cranium [in Serbo-Croatian]. Presentation at the 8th Intersectional Meet. Radiologists Bosnia and Herzegovina, Sbrije, Macedonia, Voivodina and Kosovo, Pristina, Yugoslavia, June 22–25, 1977. *Radiol. Iugosl.* (Ljubljana) 11: 387–392.

Singleton, A. O. 1928. Use of intraarterial injections of sodium iodide in determining the condition of circulation in the extremities. *Arch. Surg.* 16(6): 1232–1241. doi:10.1001/archsurg.1928.01140060107007.

Sjögren, S. E. 1957. The Effect on Heart Rate by Different Contrast Media Used in Carotid Angiography. Presentation at the 5th Symp. Neuroradiologicum, Bruxelles, July 21–27. (Not published with Proc.)

Skrepetis, K., G. Paranichiannakis, and N. Antoniou. 2004. Controversies about discovery and development of excretory urography. *J. Pelvic Med. Surg.* 10(2): 71–80. doi:10.1097/01. spv.0000130280.09274.41.

Sobel, D. 1998. *Longitude. The True Story of a Lone Genius Who Solved the Greatest Scientific Problem of His Time.* London: Forth Estate.

Söhnel, O., and P. Novotný. 1985. *Densities of Aqueous Solutions of Inorganic Substances.* Amsterdam: Elsevier.

Sovak, M., ed. 1984. *Radiocontrast Agents.* New York: Springer.

Sovak, M.; *Schering AG, Berlin.* Earliest Prior. 1989, July 5. Publ. 1997, Dec. 16. Carboxamide Nonionic Contrast Media. Patent US5698739.

Sovak, M., and R. Ranganathan; *The Reagents, Univ. California, Berkeley.* Prior. 1980, Jan. 31 and Apr. 17. Publ. July 27, 1982. Novel Amino-dioxepane Intermediates for the Synthesis of New Non-Ionic Contrast Media. Patent US4341756.

Sovak, M., and R. Ranganathan; *Cook Imaging Corp., Bloomington.* Prior. 1985, Aug. 9. Publ. 1990, Sept. 4. Non-Ionic Polyol Contrast Media from Ionic Contrast Media. Patent US4954348.

Sovak, M., B. Nahlovsky, J. Lang, and E. C. Lasser. 1975. Preliminary evaluation of di-iodo-triglucosyl benzene. An approach to the design of nonionic water-soluble radiographic contrast media. *Radiology* 117(3): 717–719. doi:10.1148/117.3.717.

Sovak, M., R. Ranganathan, J. H. Lang, and E. C. Lasser. [1977] 1978. Concepts in design of improved intravascular contrast agents. Presentation at the 2nd Congr. Eur. Soc. Cardio-vascular Radiol., May 25–27, Uppsala, Sweden, 1977. *Ann. Radiol.* (Paris) 21(4–5): 283–289.

Sovak, M., R. Terry, C. Abramjuk et al. 2004. Iosimenol, a low-viscosity nonionic dimer: Preclinical physicochemistry, pharmacology, and pharmacokinetics. *Invest. Radiol.* 39(3): 171–181. doi:10.1097/01.rli.0000115332.25954.a3.

Sovak, M., R. Terry, O. Masner, C. Abramjuk, B. Adolph, and A. Seligson. 2005. Iosimenol: The second generation of nonionic dimers. *Acad. Radiol.* 12(5): S52–S53. doi:10.1016/j. acra.2005.03.010.

Spann, M., M. Adams, and W. Sounder. 1995. Measures of technology transfer effectiveness. Key dimensions and differences in their use of sponsors, developers and adopters. *IEEE T. Eng. Manage.* 42(1): 19–29. doi:10.1109/17.366400.

Speck, U. 1995. Development of intravascular contrast media. In *Radiology in Medical Diagnostics. Evolution of X-ray Applications 1895–1995*, eds. G. Rosenbusch, M. Oudkerk, and E. Ammann, English language ed. P. F. Winter, 121–130. Oxford: Blackwell.

Speck, U., P. P. Blaszkiewicz, D. Seidelmann, and E. Klieger; *Schering AG, Berlin.* Prior. 1979, Mar. 8. Publ. 1982, May 19. Triodinated Isophthalic Acid Diamides, their Preparation and X-ray Contrast Media Containing Them [in German]. Patent EP0015867. (German disclosure DE2909439. Publ. 1980, Sept. 18).

Speicher, M. E. 1956. A report on hypaque, a new intravenous urographic medium. *Am. J. Roentgenol. Radium Ther. Nucl. Med.* 75(5): 855–869.

Spence, H. M. 1990. The life and times of Alexander von Lichtenberg. *Urology* 35(5): 464–469. doi:10.1016/0090-4295(90)80096-6.

Spinazzi, A., S. Ceriati, P. Panezzola et al. 2000. Safety and pharmacokinetics of a new liposomal liver-specific contrast agent for CT: Results of clinical testing in non-patient volunteers. *Invest. Radiol.* 35(1): 1–7. doi:10.1097/00004424-200001000-00001.

Spinazzi, A. 2005. Late adverse events following administration of iodinated contrast media. An update. In *Multidetector-row Computed Tomography. Scanning and Contrast Protocols*, eds. G. Marchal, T. J. Vogl, J. P. Heiken, and G. D. Rubin, 121–131. Berlin: Springer.

Staab, E. V., D. B. Fraser, P. Fritsche et al. (RSNA ad hoc Committee for the Centennial Celebration). 1994. Highlights on Our Progress and Growth. *Exhibit presented at the 80th Annu. Meet. Radiol. Soc.*, North Am., Chicago, Nov. 27–Dec. 2.

Standen, J. R., M. B. Nogrady, J. S. Dunbar, and R. B. Goldbloom. 1965. The osmotic effects of methylglucamine diatrizoate (Renografin 60) in intravenous urography in infants. *Am. J. Roentgenol. Radium Ther. Nucl. Med.* 93(2): 473–479.

Staudenmaier, J. 1985. *Technology's Storytellers: Reweaving the Human Fabric.* Cambridge, MA: MIT Press.

Steiger, [H.] Secretary of Agriculture, Berlin. 1929, Dec. 31. Letter to the Rektor der Landwirtschaftlichen Hochschule, Berlin [in German]. Universitätsarchiv der Humboldt Universität, Document UK-Pers. R 9, Bd. 2, Bl. 18.

Stork, D., and A. Sapienza. 1992. Task and human messages over the project life cycle: Matching media to messages. *Proj. Manag. J.* 23(4): 44–49.

Strain, W. H. 1971. Radiocontrast agents for neuroradiology. Chap. 9 In *International Encyclopedia of Pharmacology and Therapeutics.* Section 76, Radiocontrast Agents, vol. 2, ed. P. K. Knoefel, 365–393. Oxford: Pergamon Press.

Strain, W. H. 1987. Historical development of contrast media for medical imaging. In *Contrast Media: Biologic Effects and Clinical Application*, vol. 1, eds. Z. Parvez, R. Moncada, and M. Sovak, 3–23. Boca Raton, FL: CRC Press.

Strain, W. H., J. T. Plati, and S. L. Warren; *Noned Corp. & Eastman Kodak Comp., Rochester.* Prior. 1940, June 11. Publ. 1944, May 9. Compounds for Use in Radiography. Patent US2348231.

Strain, W. H., S. M. Rogoff, R. H. Greenlaw, R. M. Johnston, F. Huegin, and W. P. Berliner. 1964. Radiologic diagnostic agents: A compilation. *Med. Radiogr. Photogr.* 40(Suppl.): iii–v & 1–110.

Stürzbecher, M. 1971. Cholera, Dieffenbach and the catheterization of the heart in 1831 [in German]. *Dtsch. Med. J.* 22: 470–471.

Sturken, M., C. Thomas, and S. J. Ball-Rokeach, eds. 2004. *Technological Visions. The Hopes and Fears that Shape New Technologies.* Philadelphia, PA: Temple University Press.

Sundhedsstyrelsen. 1993. Changes in the indication range of high-osmolar X-ray contrast agents [in Danish]. *Ugeskr. Laeger* 155(6): 414.

Suter, H., and H. Zutter. 1975. Derivatives of diiodofumaric acid as potential X-ray contrast agents [in German]. *Pharm. Acta Helv.* 50(5): 151–152.

Suter, H., H. Zutter, and J. Conti; *Eprova AG, Schaffhausen.* Prior. 1957, July 8. Publ. 1962, Feb. 28. Utilization of Tetraiodoterephthalic Acid Diamides as X-ray Contrast Agents [in German]. Patent CH359446.

Suter, H., H. Zutter, and H. R. Müller; *Eprova AG, Schaffhausen.* Swiss Prior. 1971, Nov. 17. Publ. Dec. 27, 1973. X-ray Contrast Media [in German]. Patent AT312162.

Swarm, R. L. 1971. Colloidal thorium dioxide. Chap. 12 In *International Encyclopedia of Pharmacology and Therapeutics.* Section 76, Radiocontrast Agents, vol. 2, ed. P. K. Knoefel, 431–441. Oxford: Pergamon Press.

Swick, M. 1929 [1930]. Representation of the kidney and the urinary tract in the X-ray image through intravenous injection of the new contrast agent uroselectan [in German]. Presentation at the 9th Congr. German Soc. Urol., Munich, Sept. 26–28, 1929; Nov. 5, 1929. *Klin. Wochenschr.* 8(45): 2087–2089. doi:10.1007/BF01875121; 1930. *Verh. Dtsch. Ges. Urol./Z. Urol. Sonderband* 9: 328–331.

Swick, M. [1929] 1930. About uroselectan [in Italian]. Presentation at the 8th Congress of the Italian Society of Urology, Genoa, Oct. 26, 1929. *Atti Soc. Ital. Urol. VIII Congresso* 6: 339.

Swick, M. 1930a. Intravenous urography by means of uroselectan. Presentation at the Sect. Genito-Urin. Surg., New York Acad. Med., New York City, Jan. 15, 1930. *Am. J. Surg.*, n. ser. 8(2): 405–414.

Swick, M. 1930b. Intravenous urography by means of the sodium salt of 5-iodo-2-pyridine-*N*-acetic acid. Presentation at the 81st Ann. Meet. Am. Med. Assoc., Sect. Urol., Detroit, June 23–27, 1930. *J. Am. Med. Assoc.* 95(19): 1403–1409. doi:10.1001/jama.1930.02720190015004; extended version in: *Trans. Sect. Urol. Am. Med. Assoc.* 81: 64–90 (discussion included).

Swick, M. 1930c. Abstract of discussion to Jaches (1930). *J. Am. Med. Assoc.* 95(19): 1412. doi:10.1001/jama.1930.02720190021005.

Swick, M. 1933a. Excretion urography by means of the intravenous and oral administration of sodium ortho-iodohippurate with some physiological considerations. *Surg. Gynecol. Obstet.* 56(1): 62–65.

Swick, M. 1933b. Some principles of excretion urography with a report of experimental work on new compounds, particularly sodium iodohippurate for oral and intravenous urography. Presentation at the 84th Ann. Sess. Am. Med. Assoc., Section Urol., Milwaukee, June 12–16, 1933. *J. Am. Med. Assoc.* 100(19): 1524.

Swick, M. 1933c. Intravenous and oral urograms demonstrating various urologic conditions. Poster presented at the 84th Ann. Sess. Am. Med. Assoc., Section Urol., Milwaukee, June 12–16, 1933. *J. Am. Med. Assoc.* 100(19): 1531.

Swick, M. 1933d. Excretion urography, with particular reference to a newly developed compound: Sodium ortho-iodohippurate. *J. Am. Med. Assoc.* 101(24): 1853–1855. doi:10.1001/jama.1933.02740490013003.

Swick, M. [1965] 1966. The discovery of intravenous urography: Historical aspects of the urographic media and their role in other diagnostic and therapeutic areas. The Fourth Ferdinand C. Valentine Memorial Lecture. *Presented at the Section Urol., New York Acad. Med.*, Mar. 17, 1965. *Bull. N. Y. Acad. Med.*, 2nd ser. 42(2): 128–151.

Swick, M. New York. 1974a, Feb. 28. Handwritten letter to Peter Rathert, Aachen, Germany. Typed transcription prepared by C. de Haën. Document in possession of Prof. Dr. Peter Rathert, Düsseldorf.

Swick, M. New York. 1974b, Apr. 25. Handwritten letter to Peter Rathert, Aachen, Germany. Typed transcription prepared by C. de Haën. Document in possession of Prof. Dr. Peter Rathert, Düsseldorf.

Swick, M. 1974c. Uroradiographic media. *Urology* 4(6): 750–757. doi:10.1016/0090-4295(74)90266-0.

Swick, M. 1975. The historical development of organic iodine preparations for excretion urography and angiography [in German]. (Lecture on occasion of the award of the Dr. Honoris Causa, Freie Univ. Berlin). *Urologe B* 15: 196–201.

Swick, M. 1978. Radiographic media in urology. The discovery of excretion urography: Historical and developmental aspects of the organically bound urographic media and their role in the varied diagnostic angiographic areas. *Surg. Clin. North Am.* 58(5): 977–994. doi:10.1016/S0039-6109(16)41638-5.

Swick, M. New York 1980a. Fall. Handwritten letter to Ronald G. Grainger, Sheffield, UK. Copy provided by Grainger to C. de Haën, who in return prepared a typed transcription.

Swick, M. New York. 1980b, Dec. 15. Handwritten letter to Ronald G. Grainger, Sheffield, UK. Copy provided by Grainger to C. de Haën, who in return prepared a typed transcription.

Swick, M. New York. 1981, Feb. 2. Handwritten letter to Ronald G. Grainger, Sheffield, UK. Copy provided by Grainger to C. de Haën, who in return prepared a typed transcription.

Swick, M. 1982. The development of iodinated organic contrast media. In *Pioneers in Angiography. The Portuguese School of Angiography*, eds. J. A. Veiga-Pires, and R. G. Grainger, 97–99. Lancaster: MTP Press.

Teichmann, L. 1880. My newly invented method of using glazier's putty for injecting blood vessels [in Polish]. *Rozpr. Spraw. Pos. Wydz. Mat.-Przyr. AU.* 7: 108–157.

Thaning, M.; *GE Healthcare AS, Oslo*. Prior. 2007, July 1. Publ. 2009, Jan. 15. Contrast Agents. Patent App. WO2009008734.

The clichés of R&D day. 2001. *VIVO Bus. Med. Rep.* [Norwalk, CT] 19(1): 13.

The Roentgen Rays – In a fairy tail [in German]. 1896. Neues Dtsch. Familienbl. 25(7): 60.

Theander, G., and L. Wehlin. 1962. Non-ultrafiltrable contrast medium for renal angiography. *Acta Radiol.* os-57(2): 139–144. doi:10.1177/028418516205700208.

Thomas, P. D., and D. Gilbert. 2002. Beyond serendipity. *Scientist* 16(23): 12 & 14.

Thomas, S. F., G. W. Henry, and H. S. Kaplan. 1951. Hepatolienography: Past, present and future. *Radiology* 57(5): 669–683. doi:10.1148/57.5.669.

Thomas, S. M., J. E. Williams, and E. J. Adam. 1997. Intravascular contrast media: Can we justify the continued use of ionic contrast agents? *Clin. Radiol.* 52(1): 59–61.

Thoms, H., ed. 1928. Jodfortan. In Handbuch der praktischen und wissenschaftlichen Pharmazie, vol. VI, 2nd half, 1st part, p. 1145. Berlin: Urban & Schwarzenberg.

Thomsen, H. S., J. W. Archer, L. Schiermer, and P. W. Radensky. 1997. A case study of the decision to restrict use of high-osmolar contrast media in intravascular radiographic procedures. *Acad. Radiol.* 4(6): 446–450. doi:10.1016/S1076-6332(97)80053-0.

Tilly, C. 1981. *As Sociology Meets History*. New York: Academic Press.

Tilly, G., M. J.-C. Hardouin, and J. Lautrou; Laboratoires André Guerbet SA, Aulnay-sous-Bois, *France*. Prior. 1973, Dec. 7. Publ. 1976, May 19. Tri-iodo Benzene Derivatives and Their Use as X-ray Contrast Media. Patent GB1436357. (German disclosure DE2456685. Publ. 1975, June 12).

Tilly, G., M. J.-C. Hardouin, and J. La[u]trou; *Guerbet SA, Aulnay-sous-Bois, France*. Prior. 1974, May 31 and July 31. Publ. 1977, Oct. 19. X-ray Contrast Media. Patent GB1488903. (German disclosure DE2524059. Publ. 1975, Dec. 18).

Tirone, P. 1974. Short Report on Phase I [in Italian]. Company-internal document PLT 3/74, Archives of Bracco Imaging SpA, Milan, Italy.

Tirone, P., and E. Boldrini. 1981a. Systemic and local tolerability of iopamidol. An experimental study. *Rays* 6(Suppl. 2): 11–17.

Tirone, P., and E. Boldrini. 1981b. Cardiovascular and hemodynamic effects of iopamidol. An experimental study. *Rays* 6(Suppl. 2): 19–32.

Tirone, P., and E. Boldrini. 1982. Effects of iopamidol on the nervous system. An experimental study. *Rays* 7(Suppl. 3): 61–71.

Törnell, G. 1968. Bradycardial reactions in cerebral angiography induced by sodium chloride and methylglucamine iothalamate (Conray). Comparison with Urografin in a controlled study in man. *Acta Radiol. Diagn.* 7(6): 489–501. doi:10.1177/028418516800700604.

Tournier, H., and B. Lamy; *Bracco Research SA, Carouge, Switzerland*. Eur. Prior. 1995, Feb. 24. Publ. 2001, Apr. 17. Liposome Suspensions as Blood Pool Imaging Contrast Agents. Patent US6217849.

Trendelenburg, P. 1926. *Grundlagen der allgemeinen und speziellen Arzneimittelverordnung*. Leipzig: F. C. W. Vogel.

Trevan, J. W. 1927. The error of determination of toxicity. *Proc. Roy. Soc. London B Biol. Sci.* 101(712): 483–514. doi:10.1098/rspb.1927.0030.

Tucker, A. S., and G. Di Bagno. 1956. Intravenous urography. A comparative study of Neo-Iopax and Urokon. *Am. J. Roentgenol. Radium Ther. Nucl. Med.* 75(5): 855–864.

Tufte, E. R. 1983. *The Visual Display of Quantitative Information*. Cheshire: Connecticut Graphics Press.

Turner, B. M. 2001. *Chromatin and Gene Regulation. Molecular Mechanisms in Epigenetics*. Oxford: Blackwell.

Tweedle, M. F. 2016. The father of nonionic contrast media Professor Torsten Almén, MD, PhD (1931–2016). *Invest. Radiol.* 51(12): 1. doi:10.1097/RLI.0000000000000305.

Umbral, F. 2001. *Un ser de lejanías*. Barcelona: Editorial Planeta.

Unger, H. 1928. Jodfortan [in German]. *Medizinische Welt.* 2(32): 1205–1206.

United States Pharmacopeia. 1989. General notes. In *USP 22-NF 17 (United States Pharmacopeia, 22th revision—National Formulary,* 17th ed.) 11. Rockville, MD: The United States Pharmacopeial Convention.

Urich, K. 1995. *Success and Failures in the Development of Contrast Media from Wilhelm C. Röntgen to the Present.* Berlin: Blackwell.

Vahlen, T. On behalf of the Preussische Minister für Wissenschaft, Kunst und Volksbildung, Berlin. 1934, July 19. Letter to Dr. Alexander von Lichtenberg, Berlin [in German]. Universitätsarchiv der Humboldt Universität, Document UK-L 147, Bd. 1, Bl. 5.

Vahlen, T. On behalf of the Preussische Minister für Wissenschaft, Kunst und Volksbildung, Berlin. 1936, Nov. 30. Letter to the Rektor der Universität Berlin, Berlin [in German]. Universitätsarchiv der Humboldt Universität, Document UK-Pers., R 9, Bd. 1, Bl. 5.

Verwaltungsdirektor, Friedrich-Wilhelms-Univ. Berlin. 1935, Nov. 23. Letter to Dr. Alexander von Lichtenberg, Berlin [in German]. Universitätsarchiv der Humboldt Universität, Document UK- L 147, Bd. 1, Bl. 6.

Vicentini, G., and G. Pacher. 1896. Experiences with Roentgen rays [in Italian]. Presentation at the Univ. of Padova, Jan. 26, 1896. *Mem. R. Ist. Veneto Sci. Lett. Arti.* 25(7): 1–18 and true photographic plates III and IV and post-script.

Vieweg, K.; *Deutsche Gold-und Silber-Scheidenanstalt vormals Roessler AG, Frankfurt a. M.* Prior. 1926, Dec. 2. Publ. 1931, June 10. Process for the Iodination of 2-Aminopyridine [in German]. Patent DE526803.

Viviani, G. 1986. The radiologist: A doctor between science and art. *Rays* 11(1): 41–50.

Viviani, G. 1989. The artistic and radiological image in the light of today's aesthetic concepts [in Italian]. *Radiol. Med.* 78(3): 145–152.

Voelcker, F., and E. Joseph. 1903. Functional kidney diagnostics without ureteral catheter [in German]. *Munch. Med. Wochenschr.* 50/II(48): 2081–2089.

Voelcker, F., and A. von Lichtenberg. 1905. The shape of the human bladder in the X-ray image [in German]. *Munch. Med. Wochenschr.* 52/II(33): 1576–1578.

Voelcker, F., and A. von Lichtenberg. 1906. Pyelography (X-ray imaging of the renal pelvis after Kollargol filling) [in German]. *Munch. Med. Wochenschr.* 53/I(3): 105–107.

Volkmann, J. 1923, Nov. 16. About experiments on direct inspection of the brain chambers [in German]. *Munch. Med. Wochenschr.* 70/II(46): 1382.

Volkmann, J. 1924a. Laboratory notes of Mar. 7 to Mar. 26 [in German]. Document in possession of Prof. Dr. Peter Rathert, Düsseldorf.

Volkmann, J. 1924b, Nov. 24. Comment to 9. Jüngling—Tübingen: Progress in the field of localization brain tumors by means of ventriculography according to dandy [in German]. Presentation at the 48th Congr. German Soc. Surg., Berlin, Apr. 23–26, 1924. *Arch. Klin. Chir.* 133: 66.

Volkmann, J. 1924c. Comment to Presentation 13. Goetze. Pyelogram of the operatively freed kidney [in German]. Presentation at the 15th Meet. German Röntgen-Soc., Berlin, Apr. 27–29, 1924. *Verh. Dtsch. Rontgen Ges.* 15(1): 35.

Volkmann, J. (minutes taken by Grote). 1924d. Aug. 1. Mr. Volkmann shows X-ray images of the bladder and renal pelvis [in German]. Presentation at the Meet. Soc. Physicians, Halle a. d. S., June 25, 1924. *Munch. Med. Wochenschr.* 71/II(31): 1080.

Volkmann, J. (anonymous minute taker). 1924e, Oct. 25. About the radiographic representation of the urinary tract by administration of shadowing substances [in German]. Presentation at the 5th Middle German Meet. Surgeons, Dresden, June 1924. *Zentralbl. Chir.* 51/II(43): 2376–2377.

Volkmann, J. 1924f, Oct. 19. About the radiographic representation of the urinary tract by administration of shadowing substances [in German]. *Dtsch. Med. Wochenschr.* 50(41): 1413–1414. doi:10.1055/s-0028-1133918.

Volkmann, J. 1928, Jan. Prospects and views in urological surgery [in German]. *Allg. Med. Zentral-Ztg.* 96(3): 25–26.

Volkmann, J. 1931. Experiments with potassium iodide and other compounds [in German]. Presentation at the Med. Naturwiss. Ges. Münster, Jan. 1, 1931. *Med. Klin.* 28(11): 415.

Volkmann, J. 1955. [Lecture before Students of the University of Greifswald about his Life, in German] Archive of the Dtsch. Akad. Naturforsch. Leopoldina, Halle a. d. S. Document in Matrikel-Mappe Nr. 4836.

Volkmann, J. 1957. Contrast representation of the cardiac pouch. Experimental and anatomical investigations [in German]. *Bruns Beitr. Klin. Chir.* 194(1): 87–105.

Volkmann, J. Hannover. 1966, Aug. 31. Unsigned typed letter to one of the authors of Rathert, Melchior, and Lutzeyer (1974). Document in possession of Prof. Dr. Peter Rathert, Düsseldorf.

Volkmann, J. Hannover. 1974. Memorandum concerning the manuscript later published by Rathert, Melchior, and Lutzeyer (1975) [in German]. Document in possession of Prof. Dr. Peter Rathert, Düsseldorf.

Wackenheim, A. 1982. *Les radiologistes: imagiers de la médicine.* Milan: Centro Studi Bracco.

Wall, B., and D. K. Rose. 1951. The clinical intravenous nephrograms: Preliminary report. *J. Urol.* 66(2): 305–314. doi:10.1016/S0022-5347(17)74344-3.

Wallingford, V. H.; *Mallinckrodt Chemical Works, St. Louis.* Prior. 1950, May 31. Publ. 1952, Sept. 23. 3-Carboxylic Acylamino-2,4,6-triiodobenzoic Acids and Their Ethylester and Non-Toxic Salts. Patent US2611786 (plus Certificate of Correction of 1952, Sept. 23).

Wallingford, V. H. 1953. The development of organic iodine compounds as X-ray contrast media. *J. Am. Pharm. Assoc. Sci. Ed.* 42(12): 721–728. doi:10.1002/jps.3030421206.

Wallingford, V. H. 1959. General aspects of contrast media research. *Ann. N. Y. Acad. Sci.* 78(3): 707–719. doi:10.1111/j.1749-6632.1959.tb56057.x.

Wallingford, V. H., H. G. Decker, and M. Kruty. 1952. X-ray contrast media. I. Iodinated acylaminobenzoic acids. *J. Am. Chem. Soc.* 74(17): 4365–4368. doi:10.1021/ja01137a035.

Wallingford, V. H. (Documents in relation to the 5th Iodine Research Award) 1953. The State Historical Society of Missouri Research Center-Saint Louis. Document S0452 Edward Mallinckrodt Jr. Papers, Box 21, Folder 1179.

Walter, E. 1922. Aminoaldehydes and the Intramolecular Condensation of the Latter to Nitrogen-Containing Bicyclic Ring Systems. About a Synthesis of Dihydroquinoline [in German]. PhD diss., Friedrich-Wilhelms-Univ., Berlin.

Waters, C. A., S. Bayne-Jones, and L. G. Rowntree. 1917. Roentgenography of the lungs. Roentgenographic studies in living animals after intratracheal injection of iodoform emulsion. *Arch. Int. Med.* 19(4): 538–549. doi:10.1001/archinte.1917.00080230041005.

Webb, S. 1990. *From the Watching of Shadows. The Origins of Radiological Tomography.* Bristol: Adam Hilger.

Weitl, F. L., M. Sovak, and M. Ohno. 1976. Synthesis of a potential water-soluble radiographic contrast medium, 2,4,6-triiodo-3-acetamido-5-N-methylcarboxamidophenyl-β-D-glucopyranoside. *J. Med. Chem.* 19(3): 353–356. doi:10.1021/jm00225a001.

Weitl, F. L., M. Sovak, T. M. Williams, and J. H. Lang. 1976. Studies in the design of X-ray contrast agents. Synthesis, hydrophobicity, and solubility of some iodoresorcyl bis(β-glucosides). *J. Med. Chem.* 19(12): 1359–1362. doi:10.1021/jm00234a001.

Weld, E. H. 1918. The use of sodium bromide in roentgenography. *J. Am. Med. Assoc.* 71(14): 1111–1112. doi:10.1001/jama.1918.02600400011003.

West, R. C., ed. 1988. *CRC Handbook of Chemistry and Physics,* 68th ed., D219-D269 and D245-D246. Boca Raton, FL: CRC Press.

Westermann, B. 1978. Alexander von Lichtenberg (1880–1949). Biobibliography of a Urologist [in German]. Med. diss., Freie Univ., Berlin.

Wiedeman, M. P. 1964. Influence of low molecular weight dextran on vascular and intravascular responses to contrast media. *Am. J. Roentgenol. Radium Ther. Nucl. Med.* 92(3): 682–687.

Wiegert, P. E.; *Mallinckrodt Chemical Works, St. Louis*. Prior. 1968, Nov. 29 and 1971, Feb. 8. Publ. 1973, Dec. 25. Nitrilotriacyltriimino-tris-(2,4,6-triiodobenzoic acid) Compounds. Patent US3781338.

Wild, C., and S. Puig. 2004. Analogue preparations—Market strategies of drug manufacturers, as well as drug buyers, exemplified with nonionic (monomeric) X-ray contrast media [in German]. *Gesundheitswesen* 66(11): 716–722. doi:10.1055/s-2004-813745.

Wille, K.; *Nyegaard & Co. A/S, Oslo*. Brit. Prior. 1982, Oct. 1. Publ. 1986, May 7. X-ray Contrast Agents. Patent EP0105752.

Wilmot, A., N. Mehta, and S. Jha. 2012. The adoptation of low-osmolar contrast agents in the United States: Historical analysis of health policy and clinical practice. *AJR Am. J. Roentgenol.* 199(5): 1049–1053. doi:10.2214/AJR.11.8426.

Wingler, A.; *I. G. Farbenindustrie AG, Frankfurt a. M.* Prior. 1925, May 12. Publ. 1928, Nov. 1. Methods of Preparation of Diiodobehenolic Acid [in German]. Patent DE468021.

Winner, L. 1986. *The Whale and the Reactor: A Search for Limits in an Age of High Technology*. Chicago, IL: University of Chicago Press.

Winternitz, H. 1897. On the behavior of iodo-fats in the organism and their therapeutic utilization [in German]. *Dtsch. Med. Wochenschr.* 23, Therapeutische Beilage No. 5: 33–34.

Wissenschaftliches Labor. Gehe & Co. AG, Dresden. 1924. Handwritten company-internal memorandum to the scientific division of Gehe & Co. AG, Chemische Fabriken, Dresden-N.: Regarding letter from Privatdozent Dr. Volkmann, Halle a. S [in German]. Document in possession of Prof. Dr. Peter Rathert, Düsseldorf.

Wistrand, L. G., A. Rogstad, G. Hagelin et al. 2010. GE-145, a new low-osmolar dimeric radiographic contrast medium. *Acta Radiol.* 51(9): 1014–1020. doi:10.3109/02841851. 2010.509739.

Wlasich, G. J. 2011. *Schering AG in the Time of National Socialism: Contributions to the Corporate Culture of a Berlin-based Concern* [in German]. Berlin: Kalwang & Eis.

Wlasich, G. J., and C. Berghausen, eds. 1996. *Schering AG: From a Chemist's Workshop to a Multinational Enterprise: A Curriculum Vitae*. Berlin: Schering AG Press and Public Affairs Dept.

Wolf, M., and F. Remenovsky. 1931. The practical application of varicography [in German]. *Wien. Klin. Wochenschr.* 44(11): 353–355.

Wysowski, D. K., and P. Nourjah. 2006. Deaths attributed to X-ray contrast media on U.S. death certificates. *AJR Am. J. Roentgenol.* 186(3): 613–615. doi:10.2214/ AJR.04.1790.

Yu, S.-B., and A. D. Watson. 1999. Metal-based X-ray contrast media. *Chem. Rev.* 99(9): 2352–2377. doi:10.1021/cr980441p.

Zaunick, R. 1954. Carl Hermann von Hoessle, a pioneer of technical pharmaceutical colloid chemistry [in German]. *Chem. Tech.* (Leipzig) 6(12): 693.

Zeitler, E. 1995. Peripheral vessels. In *Radiology in Medical Diagnosis. Evolution of X-ray Applications 1895–1995*, eds. G. Rosenbusch, M. Oudkerk, and E. Ammann, English language ed. P. F. Winter, 236–247. Oxford: Blackwell Science.

Ziegler, J., and H. Köhler. 1930. Peroral pyelography [in German]. *Med. Klin.* 26(1): 10–11.

Zschintzsch, W. On behalf of the Reichs- und Preussische Minister für Wissenschaft, Erziehung und Volksbildung, Berlin. 1936, July 30. Letter to Dr. Alexander von Lichtenberg, Berlin [in German]. Universitätsarchiv der Humboldt Universität, Document UK-L 147, Bd. 1, Bl. 8.

Endnotes

1 The antipode of Leitbild (vision) is Schreckbild (dread, bugbear), the latter signaling a situation to avoid, formulates it in a concise format, and attaches to it so many warning signals as to impede all activities that favor the undesirable situation. Examples of Schreckbilder are the red danger, the human skull in front of two crossed bones, silent spring, global warming, vivisection, etc. While visions (Leitbilder) tend to spur action and technological progress, Schreckbilder tend to favor inactivity and resistance to such progress, even when both address similar issues, e.g., green industrial production versus acid rain. Occasionally the term trend has been used with a similar meaning to the term vision as it is used here. But trend is not considered satisfactory, because it refers to a *post-facto* observation. Furthermore, the term vision is not to be confounded with a simple business objective, for which it is sometimes used in the management literature.

2 Not only the history of X-rays *per se* but also that of contrast agents offers illustrations of ill-advised behavior by users, producers, and government regulators. In the case of contrast agents, the most clamorous ill behavior of industry and government regulatory bodies was the continued sale of thorium-oxide-based products decades after the concrete and deadly danger from their radioactivity had been recognized and radiologists had been warned by some experts (Council A. M. A. 1932; Swarm 1971; Muth 1989). The exquisite images that could be obtained with these agents may have blinded a visually oriented profession to the price in health to pay in the long run. Worth pondering is the possibility that in this case the vision guiding contrast agent technology that focused on short-term safety, may have blurred the perception of reality.

Examples of individual ill-advised behavior span the range from daredevil self-experimentation to reckless risk taking with patients. For instance, the overzealous and deadly pursuit of cerebral angiography with LIPIODOL™ in man, before adequate testing in animals by Jean-Athanase Sicard and Jacques Forestier, and driven by jealousy toward Egas Moniz (Doby 1976), falls in this category. The recommendation by Kaznelson and Reimann (1925) to use for the preparation of solutions of tetrabromophthalein sodium for intravenous injection tap water instead of pharmaceutical grade water, presumably purely because it was cheaper, constitutes another illustration. Fuchs (1928) declared that since the so-called Albrecht-Ulzer halogen solution, a fluoride and hypochlorite containing antiseptic, when injected intraperitonealy or subcutaneously did not cause tissue damage, it could be tested for retrograde pyelography in man without animal experiments. Iofendylate remained in myelographic use for much time after its deleterious effects had been recognized. But such individual misjudgments are distinct from the described problematic behavior of a whole category of users, which are alluded to here.

3 History of contrast agent technology texts: Cole Collaborators 1932; Goldstein and Abeshouse 1935; Hecht 1939; Rigler 1945; Dorn 1957; Knoefel 1961, 1971a; Hoppe 1963; Strain et al. 1964; Bruwer 1964; Barke 1970; Strain 1971, 1987; Rathert, Melchior, and Lutzeyer 1975; Doby 1976; Marshall 1977; Swick 1978; Grainger 1982a; Sovak 1984; Felder 1986; Fischer 1987; Pallardy, Pallardy, and Wackenheim 1989; Rathert 1992; Elke 1992; Maurer and Clauss 1993; Pollack 1994, 1996, 1999; Almén 1995; Rosenbusch, Oudkerk, and Ammann 1995; Speck 1995; Bonati 1995; Urich 1995; Gagliardi and McClennan 1996; Holtzmann Kevles 1997; Grainger and Thomas 1999; Leszczyński 2000a; Skrepetis, Paranichiannakis, and Antoniou 2004.

4 Dierkes, Hoffmann, and Marz (1996) use "Technology genesis and shaping" in their translation of the widely used German term of Technikgenese (Joerges 1989). Referring to work by Latour (1988), the same concept has been circumscribed by socio-technical networking of humans and artifacts (Bauer 1995).

5 One is talking about empirical knowledge cultures, practical ones more often than theoretical ones, and excluding esoteric ones.

6 Dierkes, Hoffmann, and Marz (1996) equate interference between knowledge cultures with technological knowledge. Here interference is considered to give rise to knowledge.

7 Laudable exceptions are Barke (1970), Strain (1971), Grainger (1982a), and Hoey et al. (1984).

8 In contrast, a remarkably well-balanced history of radiological tomography has been published (Webb 1990). Also, the histories of various components of X-ray equipment in the book edited by Rosenbusch, Oudkerk, and Ammann (1995) provide references to patents, while in the same book the chapter on contrast agents does not.

9 In the historical brochure of Schering AG and Wlasich (1991, 54), an anachronistic photomontage superimposing an ampoule of UROSELECTAN B™ in solution on the launch advertisement for UROSELECTAN™ in February 1930 gives the false impression that already the earliest product was available in a form ready for injection. In a promotional brochure by Schering AG, referring to galactose-based ultrasound contrast agents, it is stated that "The worldwide first ultrasound contrast agent was developed in the Schering-Laboratories" (Eberhard-Metzger et al. 1993). There is no mention of the independent inventors, whose patents (Rasor and Tickner 1980) the company acquired. The otherwise beautiful reviews of the history of LIPIODOL™ by authors from Guerbet SA (Cabanis and Iba-Zizena Cabanis 1994; Bonnemain and Guerbet 1995a; Bonnemain 2014) conveniently underestimates contributions by earlier investigators not associated with their company, as discussed in detail here in the main text. These examples all involve companies that are competitors of Bracco Imaging SpA, the one in the past I have been working for. I do not exclude that I have noticed the splinter in competitor's eyes without realizing the wooden beam in my own.

10 The historiographic utility of timelines has recently received constructive analytical attention (Champagne 2016).

11 Stanley B. Prusiner (1942–) (Nobel prize for physiology and medicine 1997) in a historical article states: "I have taken care to determine which of my views about priority are supported by the dates of submission and subsequent publication of manuscripts in refereed journals. After all, this is the currency by which scientific discoveries have been and will be judged for the foreseeable future. I did not consider abstracts, communications at meetings, patent applications, or hearsay" (Prusiner 2002). This statement captures well the prevailing approach to history of science and technology taken by most present-day academic experimental scientists. It differs radically from the one taken here and in de Haën (2001), because this approach inadmissibly limits crucial data. The limitation is most evident in the neglect of scientific and technological disclosures in form of patent applications, disclosures that for legal reasons need to precede publication in journals.

12 A counterexample is the discovery of the 4 K background radiation of the universe that, though capable of explaining a certain technical problem, was not suited as basis for technological innovation in the sense of Schumpeter.

13 A few months after his monumental discovery and initial characterization of X-rays, Röntgen left the field of endeavor. For lack of his prolonged accompaniment of the technology genesis I hesitate to apply to him the term father of the technology and prefer to qualify him as its instigator.

14 Alternative social realizations of a technology are possible, as well illustrated for the case of the Internet (Lieberman 2007).

15 The concept of interference of knowledge cultures is similar to the concept of "trading zone" introduced by Galison (1997).

16 In the history examined here, merely three women participated visibly, and just two during one of the periods covered in depth. To my knowledge none of them left traces illuminating the gender issue. Thus, regrettably nothing of relevance could be learned in this important area.

17 For a critical assessment of Werner Forssmann's personality and past as an early Nazi party member (1932–1945), see Bröer (2002).

18 Röntgen's contributions to physics outside the discovery of X-rays have often been underestimated (e.g., Holtzmann Kevles 1997). Before the discovery of X-rays, he already had made numerous important contributions to physics (Klickstein 1966; Rosenbusch, Oudkerk, and Ammann 1995). Special mention deserves the demonstration of electromagnetically induced optical rotation in molecular gases (Kundt and Röntgen 1879), the extension to gases of Alexander Graham Bell's discovery of the photo-acoustic effect in solids (Bell 1880; Röntgen 1881), and the conclusive demonstration of the displacement current that had first been postulated by James Maxwell and detected by Henry Rowland (Röntgen 1885, 1888; Dawson 1997). In part, the underappreciation of Röntgen's early contributions may be explained by his aversion to theory, the favored vehicle for gaining recognition in physics.

19 According to Röntgen's godchild, protégé and biographer Margret Boveri he also did not much care for theory, neither for those of his colleagues Einstein and Sommerfeld nor for philosophical ones (Boveri 1996).

20 They adhere to the imperative of Lucius Cary, Viscount Falkland: If it is not necessary to change, it is necessary not to change.

21 Dierkes, Hoffmann, and Marz (1996) instead uses the expression "modes of representation" that rely on signs with three functions, as subsequently defined. The role of one of the functions, namely that denominated symbol, is "description". In the belief that none of the intended meaning is changed but perhaps clarity is gained, here the expression "modes of knowledge documentation and conveyance" was substituted for "modes of representation". "Representation" was instead reserved together with description for the role of the function symbol.

22 A similar situation is also encountered in cases where predictions about the same types of events are sought with the help of models based on cause–effect relationships on the one hand, and purely statistical means (logistic regression analysis, artificial neural networks, genetic programming, etc.) on the other.

23 Good examples for how some industrial chemists have weighed the literature and incorporated their perception into reviews of a field, reviews that cover patents and journal publications on an equal footing, are given by von Schickh (1938) or Hoey, Wiegert, and Rands (1971).

24 "Interpretative flexibility" was introduced by Pinch and Bijker (1984) as attribute of technical artifacts and is here used with the same meaning to characterize visions.

25 For rigorous discussions of the word, serendipity, see van Andel (1994) and Merton and Barber (2004). The word was imported from the literary world to the natural sciences by the early X-ray contrast agent researcher, Walter B. Cannon.

26 Translation into English and bracketed specifications are the responsibility of the present author.

27 Ludwig Hopf (pseudonym Philander) (1838–1924) was a physician with a colorful and imaginative personality, who published in the fields of medicine, medical history, ethnology, ornithology and prehistory, in formats that included manuals, school books, and fairy tales.

28 The original radiograph of Haschek and Lindenthal (1896) in the form of a glass plate (Negative) has resurfaced only recently. A modern photomechanical reproduction is

presented here for the first time, in fact as positive mode print (Figure 3.1). The radiograph was initially reproduced in the *Wiener Klinische Wochenschrift* in a slightly cropped form. The lack of signs disclosing a reproduction process involving an engraving step suggests that collotype print was used, a process allowing continuous-tone rendering. Curiously all subsequent reproductions of the radiograph (e.g., in Glasser [1931] 1993; Ellegast and Thurnher 1962; Lesky 1971; Doby 1976; Zeitler 1995; Abrams 1996) show better edge contrast than the two mentioned versions. Through engraving signs they reveal themselves as line-block prints, production of which involved manual translation by engravers of gray tones into line structures. The first of such reproductions is found in the journal *Scientific American* (X-ray Photography, 1896). It was clearly generated having in hand an uncropped positive print. Through serial photomechanical half-tone reproduction of such line-block prints and accompanying Moiré effects, in later reproductions the lines progressively altered their appearance until eventually loosing their conspicuity.

29 Today conventions specify that the naked % symbol for the concentration of a solution of solid and semisolid substances in liquids are to be understood as gram of substance per 100 mL of solution, i.e., % (w/v), for a solution of a liquid in a liquid as milliliter of one liquid per 100 mL of solution, i.e., % (v/v) and for a solution of gases in a liquid as gram gas per 100 g of solution, i.e., % (w/w) (e.g., General Notes 1989). In addition, for a solution of a solid the specification % (w/w) is to be understood as gram solid per 100 g solution. In this book, the naked %-sign signals adherence to these conventions. This book cites numerous papers written in times before the advent of clear conventions, occasions in which authors did not adhere to them or did not define their use of %. In these cases, the % sign is left naked. However, whenever possible, the kind of % is specified based on information provided in the publications, being it direct (e.g., recipe) or indirect (e.g., informative cited sources). Here is an illustration of possible confusions. Brooks (1924) gave a recipe for sodium iodide as contrast agent that in modern terms is a 50% (w/w) solution. Singleton (1928) cites Brooks (1924), calling the product a 100% solution, i.e., he thought in terms of grams of sodium iodide per 100 g of water, without specifying so.

30 Company chronology: The pharmacist Ernst Schering in Berlin opened the pharmacy Grüne Apotheke in 1851 and let it evolve into the private company Ernst Schering, which, on October 23, 1871, was incorporated as Chemische Fabrik auf Actien (vormals E. Schering). Chemische-Fabrik C.A.F. Kahlbaum GmbH, Berlin-Adlershof was founded on April 1, 1818. It initiated as an alcohol distilling and liquor producer and evolved into a chemical manufacturer. It entered the field of radiology by producing X-ray fluorescent screens for photofluorography already in 1896 (Buka 1896; Gaedicke 1896). The fusion of the two companies on April 5, 1927 brought forth Schering-Kahlbaum AG. When on March 31, 1937, Kokswerke & Chemische Fabriken AG took over Schering-Kahlbaum AG, the resulting company assumed the name Schering AG. As a result of its acquisition by Bayer AG (see note 57), on December 29, 2006, the company Bayer Schering Pharma AG, Berlin, was formed. On July 1, 2011, the company was renamed Bayer Pharma AG.

31 When the generic name of a commercial product is excessively long and/or characterizes it insufficiently, its trade name is used instead.

32 Today iodized oil or *oleum iodisatum* (rarely *oleum iodatum*) designates any vegetable oil bearing as halogen only iodine., e.g., products of addition of iodine or hydrogen iodide to unsaturated bonds. At least up to 1910 also chloroiodized oils, including some that contained free iodine fell under that designation.

33 Company chronology: E. Merck [oHG], Chemische Fabrik, Darmstadt originated in 1668 as a pharmacy, the Engel Apotheke, in Darmstadt, owned by Friedrich Jacob Merck. Since then it remained always in the larger Merck family. In 1827 it began

chemical manufacturing for trade with others and became known as E. Merck, Darmstadt, sometimes with the qualification chemical factory, the name still in use these days. For most of the time it was a family owned business, or ordentliche Handelsgesellschaft (oHG). In 1995, it changed societal form to become a partially public company, i.e., Kommanditgesellschaft auf Aktien (KGaA). For a history of this oldest pharmaceutical company worldwide and in particular its involvement in research and development, see Burhop (2009) and references therein.

34 Company chronology: In 1901 Marcel Guerbet and Laurent Lafay created the partnership (société en participation) Société Guerbet & Cie in Paris. Upon the death of Lafay in 1926 Marcel Guerbet's son, André Guerbet acquired shares, leading to the foundation of Laboratoires André Guerbet & Cie, Saint-Ouen, France (Societé en commandite simple). In 1964, the acquisition of complete ownership by the Guerbet family led to the societal structure and name, i.e., SA Laboratoires André Guerbet. In 1968, the company moved to Aulnay-sous-Bois and in 1977 it changed its name to Guerbet SA. Since 1987 headquarters of the Guerbet group are in Villepinte, France (Bonnemain 2014).

35 Breslau, at the time belonging to Germany, is today Wrocław, Poland.

36 The product developed by Burns has frequently and incorrectly been referred to a thorium nitrate. However, already Burns (1915) himself realized that citrate had formed a strong complex with the thorium ion, subsequently describing it as double citrate of thorium and sodium together with an excess of sodium citrate and some sodium nitrate (Burns 1917). Actually, the complex has the chelate structure $[Th(C_6H_5O_7)_2(OH)_2]Na_4$ at neutral pH (Raymond, Duffield and Williams 1987). The mode of preparation of the contrast agent of Burns (1915) guaranteed that the citrate chelate was quantitatively formed and sodium nitrate remained as accompanying salt.

37 Chronology of contrast agent companies using the industry brand name Mallinckrodt: G. Mallinckrodt and Company, Manufacturing Chemists, St. Louis, was founded in 1867 by Edward Mallinckrodt (see note 123) and his two brothers Gustav and Otto. The company was incorporated as Mallinckrodt Chemical Works in 1882. On April 23, 1974, it simplified its name to Mallinckrodt, Inc. On September 3, 1982 the company became part of Avon Products, Inc., and passed 1986 to IMCERA Group., Inc., Chicago. In 1988, the imaging sector of the company became Mallinckrodt Medical, Inc., and in 1996 again Mallinckrodt, Inc. (Diagnostic Products Division, Hazelwood, MO, USA). In October 2000 the company, keeping its name and place of operations, merged with Tyco International Ltd, Pembroke, Bermuda. In 2007 Tyco's healthcare business units were spun off as Convidien plc, Dublin, Ireland. This company's activities regarded, besides other items, specialty pharmaceuticals and imaging agents, and both were sold under Mallinckrodt as industry brand name. With the intent of eventually spinning off these activities, in 2013 Mallinckrodt plc, Dublin, Ireland, was created. June 28, 2013 Mallinckrodt plc., Dublin, spun off from Convidien plc to become Mallinckrodt plc., Chesterfield, UK. On November 27, 2015, the contrast media and delivery systems portion of Mallinckrodt plc. was acquired by Guerbet Group SA, Villepinte, France.

38 In Berberich and Hirsch (1923), the contrast agent in DOMINAL X™ is initially correctly indicated in Latin as *sodium bromatum*, which is sodium bromide. Further down in the same article it is wrongly identified as sodium bromate (in Latin *sodium bromicum*), a very toxic compound.

39 Heuser's idiosyncratic publication behavior extended even to not giving specific references to his own pioneering work in angiography (Heuser 1919) in follow-up papers and reviews (Heuser 1932, 1933), although it has to be recognized that in one review of the field he exceptionally granted limited references to other researchers (Heuser 1933). Moreover, he wrote his publications almost exclusively in Spanish. A number of these were translated by friends and republished in English. But his work received attention mostly in France, where one of his numerous radiology textbooks appeared in translated

version. In German-speaking countries, his work in the field of abdominal typhus therapy (Heuser 1921), earned him in 1923 an award by the Austrian Government, namely
the Ehrenkreuz I. Klasse vom Roten Kreuz mit der Kriegsdekoration.

40 Based on a specification in Osborne (1922), in publications he co-authored the concentrations are given in % (w/v).

41 The radiological literature often uses the terms osmolality and osmolarity interchangi-
bly, which is scientifically incorrect. There are three equivalent ways of describing the
osmotic characteristics of the same solution but with different units, and consequently,
different numerical values. They are the osmolality (ξ_m in osmol/kg solvent or mosmol/
kg solvent), osmolarity (by IUPAC definitively termed osmotic concentration) (ξ_c in
osmol/L solution), and osmotic pressure (Π in atm). A related property is the molal
osmotic coefficient, $\Phi_m = \xi_m/Pm$, wherein P is the number of particles the contrast
agent molecule in solution may dissociate, and m is the molality (mol/kg$_{solvent}$). The
osmotic property called tonicity is defined only in the context of a specified membrane
and its permeability to the solutes in the solution. In practice, it is used loosely for the
semiquantitative classification of contrast agents into hypo-, iso-, and hypertonic types.
Even otherwise authorative sources are often marred by erroneous definitions and measure conversion formulas.

Osmolality has become the preferred one in the field of commercial preparations of
contrast media, since today it is directly measurable by a vapor pressure osmometer at
a relevant temperature. The osmolarity, in the past required by some health authorities
in product labeling, is not directly measurable but can be calculated from the osmolality through, $\xi_c = \xi_m C_1$, where C_1 (in kg/L) is the mass concentration of solvent in
the solution. C_1 can be calculated from the solution density and the concentrations and
molecular masses of the solutes. Exact calculation of the osmotic pressure requires in
addition knowledge of the partial molal volume of the solvent in the solution, whereas
approximate calculations use the inverse density of pure solvent instead.

For most historical contrast agents osmolalities are not available. However, values for
25 °C could be obtained using the following sources and calculations: Densities of aqueous solutions necessary for the transformation of various concentration units into molalities were taken from Söhnel and Novotný (1985), CRC Handbook (1988), Akhundov
et al. (1989), and Blascow and Wade (1972). If necessary, molalities were calculated
from molarities and densities. Unknown densities were estimated based on known
densities at other concentrations and the approximately linear increase in density with
molarity observed with various contrast agents. This yielded a molar density increase.
In the absence of any density data, the unknown molar density increase was estimated
assuming it to be different from those obtained from data on iodopyracet diethanolamine
(Blascow and Wade 1972) and diatrizoate sodium (CRC Handbook 1988) in proportion to the molecular weights. Osmolalities were interpolated from data tables in CRC
Handbook (1988), Robinson (1942) and Hamer and Wu (1972), or they were calculated
from molalities through the empirical equations of Jakli and Van Hook (1972), Pitzer
and Mayorga (1973), Richardson and Kurtz (1984), and Rard and Archer (1995). They
were further checked by an internally generated version of extended Debye-Hückel theory, developed in collaboration with Roberto La Ferla.

In the case of ionic iodinated contrast agents the assumption was made that their
solutions of equal molal ionic strength had equal molal osmotic coefficients. When
the molality, m (mol/kg$_{water}$), and thus the molal ionic strength, I_m, of such a product
was known, osmotic coefficients at 37 °C, Φ_m, were read off the upper curve of Figure
2 in Børdalen, Wang, and Holtermann (1970). It should be remembered that osmolalities of commercial contrast agent preparations include the contributions of both
active ingredients (contrast agent) and pharmaceutical excipients (e.g., partially salified
trometamol, EDTA).

42 In Osborne et al. (1923) Rowntree is specified as the one who had the idea of using
     intravenous sodium iodide for excretion urography. Hence, the interpretation of
     Grainger, which assigns a more ideative role to Osborne (Grainger 1995), cannot be
     upheld. Similarly, there exists no evidence for the interpretation of Volkmann (1957),
     according to which Rowntree obtained the suggestion from Cameron. Some claims by
     Ritter and Rattner (1932) about the use of intravenous sodium iodide in 1921 are appar-
     ently misquotations of their own earlier work. At least the claims are unsupported.

43 Among the equivalent terms, "intravenous urography" is not used and "intravenous
     pyelography" is only found in verbatim quotations. The product of urography is a urogram.

44 Volkmann originally attributed the idea to Voelcker (Volkmann 1924c) but later
     claimed that Voelcker in 10 years never offered suggestions to his assistants (Volkmann
     1955). The claim that the studies in Voelcker's clinic in 1922 with sodium bromide
     involved intravenous administration (Volkmann 1924e) is contradicted by his own
     communications (Volkmann 1924c,f) and must have its origin in a misunderstanding
     by the anonymous minute taker.

45 Could 10 mL in the research notebook be a recording error? 100 mL would be more
     plausible.

46 My efforts to retrieve what Volkmann in German called "primäres Einladungs-
     verzeichnis", and is here translated as "preliminary congressional program", have
     failed. Moreover it is not known whether it would have given titles of presentations or
     only presenter's name.

47 His laboratory notebook shows that he initiated planned experimentation in man only
     on March 7, 1924, in his first experiment using sodium iodide and only in his second
     one strontium bromide (Volkmann 1924a). Based on this evidence one must conclude
     that the patient with the restricted urethra had obtained strontium bromide and not
     sodium iodide as mentioned in Rathert, Melchior, and Lutzeyer (1974).

48 Examination of the dates of Volkmann's experiments and those of the meetings,
     together with his own meeting presentations and meeting presidents he cited, reveals
     that whatever did happen caused him a time delay in communication of no more than
     a few days, i.e., a negligible time in comparison to his 1 year delay on Osborne et al.
     (1923). Moreover, his comment was treated on a par with two other authors' comments
     to undelivered lectures. His story thus reflects more the emotional roller coaster ride on
     which he had been taken by the various vicissitudes than to any objective handicap he
     suffered. The date of 1923 given by Hausmann (1990) for the relevant Congress of the
     German Society of Surgery is in error.

49 Puzzling is the fact that a few years after the described events, Volkmann discussed
     excretion urography with the help of potassium iodide instead of sodium iodide, as if
     this choice of contrast agents was quite common (Volkmann 1928). Did he simply try
     to distinguish himself from Rowntree and collaborators?

50 Company chronology: Gehe & Co. AG, Chemische Fabriken Dresden-Nord has its
     origin in the company founded in 1835 by the merchant Franz Ludwig Gehe, called
     Drogerie- und Farbwarenhandlung Gehe & Comp. The founder was joined in 1859 by
     the pharmacist and chemist R. August Luboldt, and in 1903 the company was incor-
     porated under the first-mentioned name. Later the company became in part Gehe AG,
     Stuttgart, 2003 renamed Celesio AG, and in another part Arzneimittelwerk Dresden
     GmbH, Dresden. 2017 Celesio AG became McKesson Europe AG.

51 This approach was once more addressed, this time successfully but without achieving
     a product of acceptable toxicity (de Haën et al. 1993).

52 At the time the authors talked about an addition complex. Today the nature of such
     products, called inclusion compounds or supramolecular crystalline solids, are better
     understood (Hollingsworth et al. 1996). But even today most chemists have never heard
     of such products.

53 The company's research staff anonymously edited Gehes Codex, a famous encyclope-
    dia of commercially available drugs that appeared first in 1910 and was updated about
    every 6 years. This fostered cultivation of communication networks of investigators and
    companies.
54 At the time the role of chemical companies in inventing pharmaceuticals was taboo.
    Companies therefore felt it advantageous to stay out of the limelight to avoid accusa-
    tions of profiteering on the sick patient. Frequently they chose to convey the impression
    that they solely made available what a leading physician had discovered to be useful.
    This behavior of pharmaceutical companies continued to prevail for many decades and
    did not change until a dominant concern of company management became impressing
    the stock market with products in the research and development pipeline. Physicians
    responsible for initial clinical evaluation only too readily accepted the traditional role
    reserved for them and tried to preserve it until long after the situation had changed.
55 It is interesting that the idea of adding urea to contrast agents would reappear again in
    connection with the latest generation of products, i.e., so-called dimers (Krause, Speck,
    and Press 1994b), although without finding commercial implementation.
56 In the literature, uroselectan and its relatives, selectan and selectan neutral, were some-
    times spelled with k instead of c, even in papers authored by certain protagonists of
    their historical origin. The image of the commercial package of UROSELECTAN™
    (Figure 4.22) establishes the spelling with c as the only correct one.
57 Company history of relevance for contrast agents: Farbenfabriken vormals Friedrich
    Bayer Co. AG, Leverkusen, Germany, on January 1 (retro dated), 1925 became Lower
    Rhine Division of I. G. Farbenindustrie AG, Frankfurt a. M., Werk Leverkusen; in
    1930 Pharmazeutische Abteilung Bayer-Meister Lucius der I. G. Farbenindustrie AG;
    after 1934 »BAYER« I. G. Farbenindustrie AG; after 1947 I. G. Farbenindustrie AG in
    dissolution (under British control), Leverkusen; after 1951 Farbenfabriken Bayer AG;
    after 1974 Bayer AG. In 2006 Bayer AG acquired Schering AG (see note 30).
58 Before World War I pharmaceutical companies generally acquired new drug candi-
    dates from academic institutions and focused their technology innovation activities
    on production process optimization. After the war Farbenfabriken vormals Friedrich
    Bayer Co. AG pioneered professionally managed, centralized in-house drug discovery
    research and eventually other companies followed their example (Burhop 2009).
59 Landwirtschaftliche Hochschule, Berlin is now Landwirtschaftliche Abteilung der
    Humboldt-Universität, Berlin.
60 Here the designation as freelance university lecturer is used as the approximate
    anglosaxon equivalent to Privatdozent in Germany, Austria, Switzerland, and The
    Netherlands. It refers to a person who has written an advanced postdoctoral dissertation
    (Habilitationsschrift) and through this and/or other activities has earned the rights to
    teach extracurricular courses and to be charged with limited teaching assignments, both
    usually against a minimal compensation. Most importantly the person becomes eligible
    for a professorial position in a specific field at the same university (Habilitation = *Venia
    legendi*). In a process called nostrification, another university can transfer the eligibility
    to its own institution, and also the applicable field can be changed. Maintenance of the
    position required continuous research and teaching involvement. After 10 years in the
    position and continued activity without election to a professorship, the title of titular
    professor can be granted. Some of the described traditions are changing.
61 Actually, in the case of Räth (1924c) the sole inventorship of Räth holds only for the
    primary German patent, assigned to Schering-Kahlbaum AG. The British equiva-
    lent instead names both Binz and Räth as inventors and the Chemisches Institut der
    Landwirtschaftlichen Hochschule Berlin as assignee (*Binz and Räth* 1924b).
62 Some compounds were subsequently reported in doctoral theses supervised by Räth.
    Selectan (Figure 4.9, **1**) was described by Schiffmann (1925) and selectan neutral

(Figure 4.9, **2**) by Schlottmann (1928). Compound **5** (Figure 4.11) was described by Garthe (1929), who also prepared compound **4** (Figure 4.11), a compound whose synthesis was already documented in the literature (Pfeiffer 1887). None of the theses dealt with uroselectan. Thus, Räth must have synthesized it himself.

63 Determination of the subcutaneous and intravenous *Dosis toxica* and *Dosis tolerata* in mice, and the therapeutic experiments on rabbits, were performed by Binz's assistant, Gerda Wilke (later Nossak) (Binz [1930] 1931, 1937; Binz and Maier-Bode 1933). Actual toxicological observations in rats are described in Binz, Räth, and Junkmann (1930).

64 The agreement as outlined is not documented, but it can be abduced from a number of documents.

65 After the 5th communication, the subtitle was reduced to "On derivatives of pyridines by A. Binz and C. Räth".

66 Toward the end of the 20th century in certain academic disciplines the situation has begun to change. Increasingly professors are involved in companies and own patents.

67 Actually, less than 3 years after the launch by Schering-Kahlbaum AG of UROSELECTAN™ and UROSELECTAN B™, i.e., in the spring of 1933, Räth, was in a leading position at Chemische Fabrik von Heyden AG, Dresden. There he successfully recruited as academic collaborators Binz, as well as von Schick, who after June 1933 had become Binz's assistant (see note 76). The legal dispute of von Schickh with Schering-Kahlbaum AG over his role in the advent of uroselectan, mentioned in the main text, could explain his passage from Schering-Kahlbaum AG to Binz. It took place around the time that the collaboration with Binz was terminated.

68 Such 2-year trial periods were apparently common at the time at I. G. Farbenindustrie AG (Lesch 1993).

69 Notgemeinschaft der Deutschen Wissenschaft, after 1951 was incorporated into the Deutsche Forschungsgemeinschaft. e. V.

70 This conclusion is supported by a detailed analysis of pertinent international patent families, i.e., equivalent patents in various national jurisdictions having identical or very closely similar application dates and Räth as inventor or co-inventor. Depending on the country, a national patent did or did not list inventors and did or did not include Binz as co-inventor. The assignee was always indicated and varied among national patents. In most cases, the assignee reflected the situation at the date of publication, not that of the date of application, dates that were years apart. Several national patents from applications predating the switch of research support from Deutsche Gold- und Silber-Scheideanstalt vormals Roessler AG to Schering-Kahlbaum AG in 1926 were assigned to the new sponsor (Binz and Räth 1923, 1924a, 1925; Räth 1925). Helpful for the identification of historical industrial sponsors of research was that the German patent by Räth (1924c) was assigned to Schering-Kahlbaum AG, whereas its Swiss equivalent kept the assignment to the former sponsor (*Deutsche Gold- und Silber-Scheideanstalt* 1924). In a letter dated May 8, 1995, Mechthild Wolf from the archives of its present-day successor of the latter company, Degussa AG, informed me, that no documents about the collaboration and the patents is available. In contradiction with the present conclusion, Urich (1995) identified the company that financially helped Binz and Räth as Bayer but did not support the claim by any evidence.

71 The conclusion about the date of termination is derived from the appearance of Binz's name as inventor on patents connected to different applicant companies (Binz and von Schickh 1933)

72 The full chemical name is 2,6-bis(*p*-iodophenyl)-quinoline-4-carboxylic acid sodium salt. After abandoning this product, the trade name, BILOPTIN™, was reused for the new cholecystographic agent sodium ipodate.

73 Today the trade name, SELECTAN™, is used for the veterinary antibiotic florfenicol.

74 It is hard to imagine that Schering-Kahlbaum AG was not also involved.

75 In publications and patents, he called himself Otto von Schickh, but his hand signature on the curriculum vitae accompanying his doctoral thesis spells only Otto Schickh. Here the name used in his publications is adopted.

76 Otto von Schickh worked from 1933 to 1936 for Arthur Binz (*curriculum vitae*). On occasion of the latter's 70th birthday he published a historical overview of the development of the chemistry of pyridine, in the journal *Angewandte Chemie*, the journal of which Binz had been a longtime chief editor (von Schickh 1938). Therein he diligently bypassed identification of any specific person as responsible for the crucial realization that iodinated pyridin derivatives with both elevated water solubility and elevated renal excretion could be used as contrast agents. He also avoided mentioning any industry involvement in the history of uroselectan sodium (see note 54). The birthday was understandably not the opportune moment to contradict Binz's already publicized version of history (Binz 1937a). Of importance for current purposes is the fact that von Schickh's account remains compatible with the one provided here.

77 Krankenanstalt Erzherzogin Sophienstiftung, Wien.

78 Tribromoethanol (AVERTIN™) also came to be recognized as having interesting X-ray contrasting properties (Coulston and Hoppe 1959).

79 Company chronology: An extremely complex history of foundations, mergers and acquisitions is here reduced to events that conditioned a particular American group of companies and units thereof that have dealt with contrast agents. In the 1930s Winthrop Chemical Company, Inc., owned 50% by Sterling Products, Inc., and 50% by American I. G. Chemical Corp., subsidiary of I. G. Farbenindustrie AG, Germany (see note 57), entered the business of contrast agents with products developed by the latter in Germany (see main text). In 1943, the Sterling-Winthrop Research Institute, Division of the in 1942 renamed Sterling Drug, Inc., was founded. In 1944 the German assets in United States were seized by the government and traded to Sterling Drug, Inc., whereupon the Winthrop Chemical Company, Inc., assumed the name, Winthrop Laboratories, Division of Sterling Drug, Inc. In 1944 Winthrop Laboratories merged with the Frederick Stearns & Co. to become Winthrop Stearns, Inc. (1948–1961). While Winthrop Stearns, Inc., was majority-owned by Sterling Drug, Inc., it changed name several times, i.e., Winthrop Laboratories, Division of Sterling Drug, Inc. (1961–1984), Winthrop-Breon Laboratories (1984–1986) and Winthrop Pharmaceuticals, Inc. (1987–1991). In February, 1988, Eastman Kodak Company, Inc., acquired Sterling Drug, Inc., and with it the imaging activities of Winthrop Pharmaceuticals, Inc., as well as those of the Sterling-Winthrop Research Institute. While the latter was closed, the radiological goods business continued under its old company name, Winthrop Pharmaceuticals, Inc., until in 1991, when it changed name to Sterling Winthrop, Inc. In 1991 Sanofi, subsidiary of Elf Aquitaine SA, Paris, and Sterling Winthrop, Inc., created a joint venture, which sold contrast agents under the latter company's name. Eastman Kodak Company in 1994 divested the business of prescription drugs of Sterling Winthrop, Inc., to Sanofi. The freshly-formed Sanofi Winthrop SA, Paris, still in 1994, passed the radiological business to Hafslund Nycomed A/S, Oslo, Norway. For later developments see Endnote 118..

80 The translation attempts to retain the awkwardness in style of the original. The numbers identifying the compounds in Hryntschak's work are put in quotation marks in order to distinguish them from the numbers for compounds in the present work.

81 For example, various process patent applications bearing on iodomethanesulfonic acid were only filed beginning in 1929 (Ossenbeck, Tietze, and Hecht 1929a).

82 It was Hryntschak's practice to use roman numerals to indicate the month.

83 According to Swick (1929) Erbach already possessed the doctoral title. At the time there existed just one Dr. Erbach in Germany young enough to be an assistant of

Lichtwitz, namely Kurt Erbach (GV 1977). This Erbach had got his doctoral degree in medicine from the University of Heidelberg in 1923 (K. Erbach 1923a,b). In the late 1920's his older brother, Otto Erbach, a licensed physician who got his doctoral degree only in 1932, was senior medical officer of the police in Altona (Hamburg) (O. Erbach 1932). Apparently, Kurt Erbach pursued his postdoctoral training in Altona, where his brother lived, and thus ended up in the laboratories of Leopold Lichtwitz. The erroneous name Herbach has been sometimes propagated (Rathert, Melchior, and Lutzeyer 1975; Rathert 1992).

84   Moses Swick was born as Moses (also Morris) Goldstein but assumed the maiden name of his mother, Swick, with court order of April 17, 1924. Most likely he presumed this change to a type of name not associated with Jewishness to help him find a suitable internship position and subsequent hospital employment in a job market, in which Jewish names formed the basis for monitoring antisemitic quotas (Ludmerer 2005; Rowland 2009).

85   Besides Paris, Vienna, and London, Berlin in the second half of the 19th century became a center of excellence for urology, although a chair for this specialty was only created in 1937 (Hausmann 1985).

86   Probably Räth was no longer part of the group, but perhaps Kurt Hillgruber participated.

87   Binz always refers merely to a single meeting in March 1929.

88   The description in 1930 is chemically not consistent with the much later recollection of Swick in 1966 (Swick [1965] 1966). Also, in 1966 only dog experiments are mentioned, whereas in 1930 a study on a male patient is described. The earlier description must be considered more reliable.

89   The trip Portland—Berlin required minimally 10 days.

90   The name "von Salle" used by Grainger (1982), Rathert, Melchior, and Lutzeyer (1975, 1992) and Marshall (1977), is erroneous (Heubner 1943).

91   In Binz (1937a) and in the interview with Moses Swick, conducted by Ronald G. Grainger (1982a,b), only the family name Renner is given. Binz refers to Renner as Leopold Lichtwitz's substitute, while Grainger (1982) and Marshall (1977) describe Renner as Lichtwitz's assistant. Actually, no assistant by that name ever worked with Lichtwitz in Hamburg. But Lichtwitz did publish together with the urological surgeon and specialist in sports medicine, Alfred Renner from the University of Breslau (Lichtwitz and Renner 1914) (see note 35). It is concluded that Lichtwitz asked the chief physician and university professor Alfred Renner to substitute for him in the mediation.

92   Grainger, who in 1981 had personally interviewed Swick, describes the donation by Schering-Kahlbaum AG on one occasion as an annual contribution to the Libman Scholarship Fund (Grainger 1982a,b) and on another occasion as a $1500 donation to Swick, which the latter turned over to the foundation (Grainger 1995). Although Schering-Kahlbaum AG and its successors behaved commendably with respect to Jews under the Nazi regime (Wlasich 2011), it could not have afforded politically to continue an annual contribution to the Libman fellowship fund for any time after 1933.

93   Schering Corp., Bloomfield, New Jersey, USA, was a subsidiary of Schering AG, Berlin, which in the context of the trade war preceding World War II, in 1941 was blocked by the US government. As part of war reparations after the war it became an independent American corporation. In 1971 Schering Corp. merged into Schering-Plough Corp., New York. In 1993 Schering Corp. became a wholly owned subsidiary of Schering-Plough Corp., Kenilworth, NJ. Schering AG, Berlin, is forbidden to use its name in any form in the USA, and its subsidiary dealing with contrast agents thus was called Berlex Laboratories, Inc. Today it is a Division of Bayer AG.

94   Sometimes doses of up to 2.7 times the one described were used (Hecht 1939).

95   Beginning a few months thereafter the chemist Hans-Georg Allardt, from Schering-Kahlbaum AG, requested similar process patents for methiodal sodium

(Allardt 1930, 1931). Based on the timing one would conclude that Allardt arrived at methiodal sodium as contrast agent independently of Ossenbeck, Tietze, and Hecht. But the possibility that news about the product had been leaked from the many clinical testing sites to the people at Schering-Kahlbaum AG cannot be excluded.

96 Clinical trials were organized by the Pharmazeutische Abteilung Bayer-Meister Lucius of I. G. Farbenindustrie AG in Leverkusen (Bronner, Hecht, and Schüller 1930). Köhler (1930) cites the Agfa division of I. G. Farbenindustrie AG as his source of the product, but probably it acted only as facilitator. During the clinical trials, i.e., before June 1930, methiodal sodium carried the provisory name METUFAN™ but was then given the definitive brand name of ABRODIL™ (Peiser 1930).

97 For a history of angiography see Doby (1976), Ludwig (1995), and Abrams (1996).

98 Forssmann's description of doses and concentrations is poor. Here my interpretation of his report is given.

99 Initial attempts to produce X-ray-based moving pictures (cineradiology) were made in April 1896 by J. Mount Bleyer (1896), the otorhinolaryngologist with an honorary law degree, inventor of execution by lethal injection, early promoter of the electric chair and later opponent of capital punishment, in his private practice in New York. He produced them with an Edison kinetograph, fed with serially taken static radiographs of objects moved between image acquisitions. Only little time after Bleyer, John Macintyre in Glasgow began his efforts in the same direction (Macintyre 1897). Taking a succession of static radiographs of an amputated frog leg being progressively stretched and rephotographing them onto a strip of motion picture film, he succeeded creating the illusion of a true movie. In 1934 he is said to have built the first practical cineradiological camera (Burrows 1986). But by 1931 Viktor Gottheiner (Forssmann 1931b; Gottheiner 1971) and by 1932 Robert Janker, lecturer at the Surgical Clinic of the Friedrich-Wilhelms University of Bonn (Cartwright 1995, and ref. therein), already had succeeded.

100 Since the mid-1940s, primarily in the United States, lumbar myelography was performed mainly with the new oily product iofendylate (INN). Only in Europe did the water-soluble contrast agent ABRODIL™ continue to play some role. Iofendylate that was a mixture of 10- and 11-(p-iodophenyl)undecanoic acid ethylester, was invented in 1940 at the University of Rochester, Rochester, NY and the patent was assigned to two companies, one of them the Eastman Kodak Company in the same city (Strain, Plati, and Warren 1940; Strain 1971). The product was developed and manufactured by this company and co-developed by Lafayette Pharmaceutical, Inc., Lafayette, Indiana. Numerous licensees, both pharmaceutical companies and X-ray equipment manufacturers, handled the international distribution. It was known under names such as PANTOPAQUE™ (Lafayette), MYODIL™ (Glaxo Laboratories Ltd., Greenford, Middlesex, UK), ETHIODAN™ (British Drug Houses Ltd., Poole, Dorset, UK), to name just the most important ones. Heavy controversies over the safety of the product led to lawsuits by patients allegedly developing late adhesive arachnoiditis due to it. Scientifically the controversy was never resolved and out-of-court settlements without admission of liability only mudded the issue. In any case, lawsuits were the initial reasons for the product to be withdrawn in some countries. The product in the end disappeared altogether because Eastman Kodak Company stopped producing the active ingredient in 1986.

101 At the University of Strassburg von Lichtenberg declared himself without religious denomination (Personalblatt Alexander von Lichtenberg 1910). When returning after a war-related absence for a short time to the same university in June 1917, he changed his declaration to Roman Catholic (Personalblatt Alexander von Lichtenberg 1917), apparently as a consequence of his marriage on March 4, 1917. When in 1933 asked by the Friedrich-Wilhelms-University in Berlin to fill in a form of particulars requested by

the Nazis, he declared himself of Roman Catholic denomination and of non-Arian race (Personalblatt Alexander von Lichtenberg 1933). However, neither the Diözesenarchiv Berlin, which has fairly complete records, nor the catholic St. Hedwig Krankenhaus at which he worked, has evidence that von Lichtenberg formally joined the Roman Catholic faith. To a more specific inquiry by the university administration of November 23, 1935 about racial and religious affiliations of his grandparents (Verwaltungsdirektor 1935) von Lichtenberg responded that they were Jewish (von Lichtenberg 1935). On November 21, 1933 the Friedrich-Wilhelms-University in Berlin fired von Lichtenberg (Jäger 1933), obviously for racial reasons. Under pressure of the Hungarian government, he was reinstated (Vahlen 1934), but in 1936 he was definitively deprived of his teaching authorization, while being allowed to continue his research until the end of the year (Zschintzsch 1936). This forced him to return to Hungary. After the Hungarian government in 1939 too instituted racial laws, von Lichtenberg fled with his family to Mexico, where further research was not practical.

Von Lichtenberg is just one of a large number of investigators in territories that came under the Nazi regime, who turned up in the present context and who were forced by racial persecution to leave their positions and research career. Apart from the much larger human tragedy, the loss to contrast agent science and medicine of so many highly qualified scientists in those territories could not remain without consequences. Germany lost its dominance in the field of endeavor. The situation in this specific area of science supports a widely held perception for science in general. Admittedly, some recent serious studies have offered evidence to the contrary (Macrakis 1993; Hammerstein 1999).

102 According to the United States Department of Justice Immigration and Naturalization Service, Alexander von Lichtenberg did not apply for a visa in the late 1930s (Butler 1995).

103 A telltale sign, probably a subconscious one, of Swick's pronounced eagerness to be recognized, is found in the citation of the article Jaches and Swick ([1933] 1934) with inverted order of authors (Swick 1978).

104 In Marshall (1977) several historical protagonists, even Swick, are given wrong first or family names. Erroneously von Lichtenberg's presentation at the Annu. Meet. Am. Urol. Assoc., New York City, June, 1930 (von Lichtenberg [1930] 1931), is declared the "American premiere" on intravenous urography, when actually Swick had presented uroselectan already in January of 1930 at the Sect. Genito-Urin. Surg., New York Acad. Med., New York City, and published (Swick 1930a).

105 Oral administration of uroselectan sodium yielded good radiological results, but its terrible taste required administration with the help of a sound (Ziegler and Köhler 1930).

106 This author believes Swick merited the Valentine award, based on being one of the fathers of excretion urography, who had not been appreciated sufficiently before that occasion.

107 Actually, an earlier catheterization of the human heart was performed in 1831 by the surgeon Johann [Friedrich] Dieffenbach at the Charité Hospital in Berlin (Dieffenbach 1834; Stürzbecher 1971). Dieffenbach catheterized the left ventricle through the brachial artery of a cholera patient in desperate condition, in a vain attempt to drain the heart of thickened blood. Unfortunately the patient on whom the procedure was performed was so close to death as to allow conclusions about tolerability of the procedure. In terms of self-catheterization, in general, although not of the heart, Forssmann was not the first one. Fritz Bleichröder, as Director of the City Hospital of Pankow (today Berlin) in 1912 belatedly reported that about in 1905, while still being assistant in the Pathology Institute of the Charité Hospital in Berlin, he had begun experimenting with intravascular catheterization to bring COLLARGOL™ as an antiseptic in higher doses to the site of infection with less exposure of the whole body (Bleichröder 1912). After some experiments in dogs he performed venous self-catheterization. Lastly, he

proceeded to cure four patients suffering from severe puerperal sepsis by administering COLLARGOL, a product about to become used as pyelographic contrast agent, through a catheter inserted into the femoral artery and advanced to the aortic bifurcation. Using a ureteric catheter lubricated inside and outside with liquid paraffin, the procedure went smoothly. Bleichröder thus pioneered intraarterial catheterization, a procedure that would become of preeminent importance in the radiodiagnostic context. In any case, these deeds do not detract from the breakthrough with its practical consequences achieved by Forssmann.

108 It seems more than accident that among radiologists sophisticated visual art appreciation is a frequent trait (Viviani 1989). The eminent radiologist Reynaldo dos Santos (dos Santos 1971), through his writings on the subject, may even be taken as a full-fledged art critic.

109 Excellence in a specific area is not incompatible with the role of research director. For example, Schoeller was not only the organizer who put various academics on projects, e.g., Adolf Butenandt on the project of female sex hormones, but also he made crucial contributions to their chemistry, e.g., the synthesis of progesterone (Bettendorf 1995).

110 Elzen, Enserink, and Smit (1996) use the term resilience to characterize the state of protection against threats to a technological project.

111 An exception is a thyrotoxic crisis days after administration of iodinated contrast agents in certain patients with thyroid anomalies.

112 Systemic toxicity was assessed by intravenous administration of dissolved compounds to rats and mice and determining the lowest dose producing clearly discomforting or deadly effects. In 1927 estimation of the acute median lethal dose, $LD_{50, i.v.}$ was introduced into toxicology (Trevan 1927). Gradually this new measure took hold also in the field of contrast agents and it was expressed preferably as g(Iodine)/kg body weight.

113 Up to the 1960s cryoscopy was the principle means for estimating osmolalities (e.g., Boriani and Boriani 1940; Broman and Olsson 1949; Cotrim 1954; Neudert and Röpke 1954; Bernstein, Reller, and Grage 1962; Almén and Wiedeman 1968). Introduction in the 1950s of the thermistor-based Fiske osmometer (Fiske Associates, Inc., Waterford, PA) rendered such measurements very convenient. It was later realized that the crystallization behavior of concentrated contrast agent solutions did not meet the validity criteria for the method and yielded therefore seriously underestimated osmolalities (Kihlström 1952). In the 1960s, the first commercial thermoelectric vapor pressure osmometer appeared on the market (Mechrolab Inc., Mountain View, CA, USA). It allowed reliable osmolality measurements at ambient and physiological temperatures (Kihlström 1953; Børdalen, Wang, and Holtermann 1970; Krause et al. 1994a).

114 This value corresponds to an osmotic coefficient of roughly 1.4 (see note 41). In highly concentrated solutions such elevated values are possible, as illustrated by the data on electrolytes in Hamer and Wu (1972).

115 Hecht reported the sodium bromide solution in use to have an osmolality 16 times that of tissue fluid. The correct value should be 24 times, as derivable from the values reported in the main text.

116 In von Lichtenberg (1932), dimethiodal bears the code B 1015. The letter B could stand for Bayer (see note 57), the predecessor company of I. G. Farbenindustrie AG, or for Backer, who first published its synthesis (Backer 1926).

117 Company chronology: The company was started 1874 by Morton Nyegaard in Kristiania [after 1924 called Oslo], Norway, as agency for the importation of pharmaceuticals and pharmacy tools, called Morton Nyegaard, cand. pharm. In 1890, Theodor Hafslund joined Nyegaard and the company assumed the name of Nyegaard & Co. (nicknamed Nyco), a name that survived the departure of the founders and the evolution from agency through wholesaler to producer and developer of pharmaceuticals. After 1918,

the company assumed over time the following names: 1918–1981, Nyegaard & Co. A/S; 1981–1982, Nyegaard & Co., Division of Norgas A/S (subsequently Actinor A/S); 1983–1985, Nyegaard & Co. A/S; 1986–1990, Nycomed A/S; 1990–1992, Nycomed A/S (Imaging), separate company within Hafslund Nycomed A/S; 1992–1996, Nycomed Imaging A/S; 1996–1997, Nycomed ASA. In October 1997, the company merged into Nycomed Amersham plc, Little Chalfont, Buckinghamshire, UK, whose diagnostic business then on October 15, 2001, assumed the name of Amersham Health, Division of Amersham plc. On April 8, 2004, the company became a subdivision of General Electric Company and assumed the name GE Healthcare Medical Diagnostics, without change of location.

118 Strain et al. (1964) describe UROTRAST™ as being iodomethamate disodium based. Contradicting this description, Amdam and Sogner (1994) show a picture of a vial from 1934 clearly labeled sodium diiodomethanesulfonate.

119 The duration of patent protection varied as a function of the country and time period. For example, in Germany in 1877 the duration for patents of the kind relevant here was fixed at 15 years, in 1936 at 18 years and in 1976 at 20 years.

120 Company chronology: Zonite Products Corp., founded in 1922, was a Delaware corporation with headquarters in New York. In the 1930s, it became a major producer of pharmaceutical products. In the mid-1950s, the company moved to New Brunswick N. J. and in February 1956 it changed its name to Chemway Corp. In 1971, the company was sold to Cooper Laboratories, Inc., which since 1987 is called Cooper Companies, Inc.

121 Note that the earlier cholecystographic contrast agents, iodophthalein sodium, and diiodophenylcinchonic acid also contained iodinated benzene rings.

122 The co-founder of Mallinckrodt Chemical Works, Edward Mallinckrodt, was trained to open his own fine chemical manufacturing business in the USA in the years 1866/67, by Eugen de Haën, founder and owner of E. de Haën & Cie, List near Hannover, Germany. Mallinckrodt was accepted under a contract as paying trainee and a convention that forbid him to become commercially active in Europe for at least 5 years (Ramstetter 1966). It is possible that the tradition of the company not to extend patents to Europe had its origin in this early history.

123 Actually, Mallinckrodt Chemical Works more than a year later than the two other companies also filed a competing patent application but was granted one only in the UK (*Mallinckrodt Chemical Works* 1954)

124 Although the patent of Larsen (1953a) indicates a posterior filing date, the legal priority date is the one cited here and in patent extensions to other countries.

125 The inventor was Paul Diedrich (U. Speck, personal communication), in agreement with the fact that in a later publication (Langecker, Hartmann, and Junkmann 1954) he is quoted as being the one that synthesized the product.

126 This complexity is reflected in the preliminary publication of a German patent on the same technology, which cites Larsen as inventor and Schering AG as applicant (Larsen 1953b).

127 Company chronology: In 1858 Edward Robinson Squibb founded a pharmaceutical company in Brooklyn, New York. When management responsibility passed to his sons, the company assumed the name E. R. Squibb & Sons. In 1905, the company was divested and became incorporated without name change. In 1952, the company was acquired by Mathieson Chemical Corp. and after the merger in 1954 of the latter with Olin Industries Inc., assumed the name E. R. Squibb & Sons, Inc., Division of Olin Mathieson Chemical Corp., New Brunswick, New Jersey. In 1968, it was spun off as a separate company with the old name E. R. Squibb & Sons, Inc. In 1989, it merged with the Bristol-Myers Corp. into Bristol-Myers Squibb Company, Princeton, NJ, USA. In 1994, its subsidiary Squibb Diagnostics was acquired by Bracco SpA (see note 130),

and two companies were formed with the names Bracco Research, Inc., and Bracco Diagnostics, Inc., Princeton. Recently the former was assumed by the latter.

128 ISOPAQUE™ was subsequently reformulated with other cations (e.g., Lindgren, Dahlström, and Fondberg [1964, 1966] 1967).

129 Company chronology: Elio Bracco and Wilhelm Merck in 1927 founded La Società Italiana Prodotti E. Merck, S.A., Milan, Italy. From 1930 to 1936, it was called Italmerck SpA and from 1936 to 1958 SA Bracco già Italmerck. A separate line of descent started as Cilag Italiana SpA (1940–1955), co-founded and co-owned by Fulvio Bracco and Cilag-Chemie AG, Schaffhausen, Switzerland. From 1955 to 1958 that company became called Industria Chimica Dr. Fulvio Bracco SpA. After fusion of the two lines the company was called, from 1958 to 1992, Bracco Industria Chimica SpA, and beginning in 1992 Bracco SpA. The part of the company involved in contrast agents was split off in 2001 and received the name Bracco Imaging SpA, Milan, Italy.

130 The solutions of the diacidic dimer, iodipamide sodium, at equal iodine concentrations had cryoscopically estimated osmolalities roughly 2/3 of those of the monoacidic mono-mer, acetrizoate sodium (Neudert and Röpke 1954). Erroneous derivations of apparent molecular weights from osmotic properties (see note 114) hampered forward-looking interpretation of the data.

131 Founded in 1857 and commonly named Falckens Hospital, Malmö, Sweden, in 1896 it became Malmö General Hospital (Malmö Allmänna Sjukhus), in 1994 Malmö University Hospital (MAS) and in 2010 Skåne University Hospital in Malmö. Since 1948 the Malmö institution is affiliated with the University of Lund, Sweden. The Department of Diagnostic Radiology was headed in 1952–1970 by Sölve Welin, and in 1970–1977 by Erik Boijsen.

132 Company chronology: The company called Pharmacia AB was founded in 1911 in Stockholm and moved 1951 to Uppsala. Beginning in 1990 the company participated in a turbulent series of mergers and divestitures, with changes in headquarter locations. Pharmacia is today a wholly owned subsidiary of Pfizer, Inc., New York.

133 In Sweden, a medical dissertation involves an extensive individual research project by an already licensed physician and leads to the title of MD Frequently this career step immediately precedes the appointment to docent that is equivalent to the American associate professorship.

134 In concomitance with the described efforts it was found that some linear copolymers experimented with in animals distinguished themselves from classical ionic contrast agents, e.g., diatrizoate sodium/meglumine (GASTROGRAFIN™, Schering AG), in that they did not precipitate at the acidity of the stomac. Owing to this property they became destined at first as gastrointestinal contrast agents (Björk, Erikson, and Ingelman 1970). This very limited clinical indication could hardly justify the invest-ments necessary for a development and commercialization.

135 Eprova AG, Schaffhausen, Switzerland was founded in 1952 by Fulvio Bracco and Hans Suter. Through participation of E. Merck, Darmstadt, in some of Fulvio Bracco's enterprises, it also participated in Eprova AG. After 1973 Eprova AG belonged fully to the E. Merck, Darmstadt concern. In August 2002 it became Merck Eprova AG. In 2011 it was liquidated as separate company and became the Schaffhausen subsidiary of Merck & Cie, headquartered in Altdorf, Switzerland, itself under the corporate owner-ship of Merck KGaA in Darmstadt, Germany.

136 Fondberg told me that Torgny Greitz did some of these studies. In the letter to me men-tioned in the main text, Greitz has denied it and suggested other possibilities. Among those suggested and contacted, none confirmed having been involved in such activities. Only the pharmacologist Lindgren at the Karolinska Institute in Stockholm could not be asked, since he has died, and so has Fondberg.

137 In the 1960s early self-administration of experimental drugs by their champions was still a tolerated and not unusual practice. For a company such an experiment, especially by a non-employee, carried great risks.

138 As described in the text, this patent application was abandoned very early in the process of evaluation. Thus, it has not been possible to trace it at the Swedish Patent Office. Accordingly information on inventorship, assignment and priority dates are unknown.

139 Today part of Akzo Nobel Nederland BV, Arnheim, The Netherlands.

140 In 1983 Nyegaard & Co. A/S, Oslo, Norway, participated at the 51% level in the foundation of the company A/B Varilab, Gothenburg, Sweden, of which Lars Fondberg was a co-founder and director of clinical studies. Thus, trust between former employer and employee had at that point not been irreversibly lost.

141 Torsten Ekstrand worked in sales at Erco-Läkemedel AG, Stockholm, and Ansger Munksgaard was a chemist in the company's Danish subsidiary.

142 Patents are not peer reviewed in the same way as journal articles, which the decisive authorities consider a major reason for treating them differently from publications in serious scientific journal.

143 In Almén (1993b) Astra AB is excluded, but in Almén (2004) it replaces Leo Läkemedel AB.

144 Founded as Chemisches Industrielles Laboratorium AG, Schaffhausen, Switzerland, in May 1936, it became subsequently known as Cilag AG. From 1958 to 1981 it was called Cilag-Chemie AG, whereupon it returned to its previous name. In October 1959 Cilag AG became a subsidiary of Johnson & Johnson, Inc., New Brunswick, NJ, and is today integrated into the Janssen Pharmaceutical Companies of Johnson & Johnson.

145 True to the reigning design principles for contrast agents at the time, initial efforts were directed at an increased iodine content. A patent for the first contrast agent molecules with a tetraiodinated benzene ring, namely tetraiodoterephthaldiamides was deposited (Suter, Zutter, and Conti 1957). No commercial product came of it.

146 Contrary to the impression given by Amdam and Sogner (1994) that Almén's dissertation dealt with effects of a plurality of established contrast agents, including ISOPAQUE™, it was focused on the construction of a new catheter. ISOPAQUE was the sole established contrast agent used in Almén's own experimentation, and then mostly as a tool.

147 The compound could have been synthesized by Giertz. But the description fits also one of the metrizoate-derived molecules synthesized by Wickberg's group in fulfillment of its contract with Erco-Läkemedel AB (Ekstrand, Munksgaard, and Wickberg 1970). This fact, together with the timing, would allow the possibility that Fondberg in 1967, while he was still an employee of the company, had provided Almén with a trimer from Wickberg's group. His denial of knowledge about a collaboration of Erco-Läkemedel AB with Wickberg, expressed to me, argues against this possibility.

148 Hilal contends that it was he who introduced Mallinckrodt Chemical Works, to the idea of dimeric and trimeric contrast agents (Hilal [1965] 1966, 1966, 1970). The evidence here supports instead the described scenario.

149 Molecules of the size of current contrast agents and small oligomers thereof do not experience water as a continuum, whereas large macromolecules do. This renders the continuum-based hydrodynamic theories called upon by Almén (Almén 1966a) invalid for oligomeric contrast agents.

150 The restored version of the affidavit shows **Almén**'s signature that was cropped in the first published version (Almén 1985) and the H_3 of the acetyl groups in structural formulas, which was cropped elsewhere (Amdam and Sogner 1994; Almén [1994] 2001). It preserves the visibility of the embossed seal of the notary public wiped out in a printed version of Almén ([1994] 2001). It does not suffer from modifications blemishing

various published versions, modifications that Almén declared to me to have made at later times, i.e., added underlining of hydroxyl groups and the indication Figure 9 (Amdam and Sogner 1994; Almén [1994] 2001), as well as arrows pointing at hydroxyl groups, and the correction for a missing methylene group (Almén [1994] 2001; Nyman, Ekberg, and Aspelin 2016a,b).

151   The numbering of compounds at Nyegaard & Co. A/S is somewhat confusing. Newly synthesized compounds were numbered consecutively, numbers becoming large. Compounds selected for biological testing were renumbered with smaller numbers, e.g., cpd 543 ≡ C29 or simply 29 (Haavaldsen 1980; Holtermann 1973b). In patents yet another numbering was used, metrizamide becoming number 11 (Almén, Nordal, and Haavaldsen 1969). Finally, in a publication Holtermann additionally labeled the chemical structure of metrizamide with a Roman number V (Holtermann 1973b).

152   Almén was a polymath. Other than what is described in the main text, he was co-inventor of a novel X-ray fluorescence analyzer. The patent was assigned to the newly created company Elementaranalys Almén & Grönberg AB in Malmö (Grönberg et al. 1982). Founded in 1983, the company soon changed name to Elementaranalys AB and after substantial growth was sold in 1989. Almén also collaborated in a research group working on rechargeable lithium batteries, in one on an artificial kidney based on porous capillary tubes and in one dealing with the therapeutic use of the magnetic resonance contrast agent, mangafodipir.

153   In 1972 Schering AG hired Ulrich Speck and this event accompanied a renewed commitment to contrast agents.

154   B 6590 is shown here in Figure 4.33, **24**.

155   The English version read at the symposium is conserved in the Archive of Bracco Imaging SpA, Milan. It corresponds exactly to the published Italian version (Felder 1970), except that slides instead of figures are cited.

156   Italy was in the midst of social turmoil and prices for pharmaceutical products were dictated at a very low level by government.

157   The German patent application was laid open on January 7, 1971 (Almén, Nordal, and Haavaldsen 1969) and was cited in a list of major German patent claims regularly published by the Wila-Verlag, a publication that at the time was subscribed to by Eprova AG.

158   Also, a research project on L-amino acid production in part financed by a government-supported institution needed to be diligently pursued until its termination in 1973.

159   By an error of the Swiss patent office the patents Felder and Pitrè (1974b,c) do not report the legal priority date, available instead on all other related patent documents.

160   Readers not familiar with commercial contrast agents may appreciate the extraordinarily elevated concentrations often needed, considering the following example: ISOVUE®-370 is 755 g of solid iopamidol dissolved in 650 g of water.

161   Amdam and Sogner (1994) give a vibrant description of how the surprising problem of crystallization, which manifested itself very late in the development of one of their candidate products, was lived through at Nyegaard & Co. A/S. The described consternation closely mirrors the reactions occurring at Bracco Industria Chimica SpA in view of the same observations with their candidate product sometime later.

162   Iomeprol to date has never spontaneously crystallized.

163   Cook Imaging Corp., Bloomington, Indiana, was established in 1982 as a contrast agent developing company, part of the Cook Group of companies. In 2002 the assets of OXILAN™ passed to the US subsidiary of Guerbet SA (see note 34), Guerbet LLC, Bloomington, IN, which continued to offer the product.

164   Milos Sovak was a polyglot polymath, a professor of radiology, scientist, inventor, artist, patron of the arts and entrepreneur, all at the same time. His extraacademic fields

of activity included art etching presses, design and edition of bibliophilic editions, toys and hobbies, chains for firefighter's motor saws, diagnostic and therapeutic pharmaceuticals, phytochemicals, functional food, cosmetics, as well as hotels and motels. Beginning in 1974 he created a web of organizations and companies, including Ettan Press Company, Biophysica, Inc., Biophysica Foundation, Sovak Corp. and Roosevelt Corp. At the Czechoslovakian first postcommunist government's request, he helped restructure the pharmaceutical industry and formulate national health care laws. He participated 1992 in the reprivatization of Interpharma Praha AS, Prague, Czechia (see note 173) (Deaths 2009).

165 The overwhelming number of patients was treated with iopamidol, the rest with iohexol.

166 Sir James W. Black is another famous example. He is the inventor and developer of the β-adrenergic blocking drug propranolol and the histamine receptor type 2 antagonist cimetidine. His project aimed at the latter product was several times closed by top management of the company he worked for, but he and his immediate superior found surreptitious ways of keeping it alive until success. Nobel Prize for physiology and medicine 1988 (Molinder 1994).

167 The recognition of such an organizational champion in an academic environment has been debated in the connection with the question of who should have won the Nobel Prize for the discovery of gravitational waves (Cho 2016).

168 The even more recent trends toward combinatorial chemistry and novel receptor-driven drug discovery research again deviates from this pattern.

169 The fatal outcome rates, which for new contrast agents at early times remain always poorly determined, were definitively too small for patients to comprehend, and were never presented to them as an argument.

170 Henceforth osmotic properties will consistently be discussed in terms of osmolality, even when the cited authors used other parameters. Osmolality (ξ_m in osmol/kg solvent or mosmol/kg solvent), osmolarity (by IUPAC definitively termed osmotic concentration) (ξ_c in osmol/L solution), and osmotic pressure (Π in atm), are three equivalent ways of describing the osmotic characteristics of the same solution but with different units, and consequently, different numerical values. Osmolality has become the preferred one in the field of commercial preparations of contrast media, since it is reliably measurable by a vapor pressure osmometer and includes the contributions of both active ingredients (contrast agent) and pharmaceutical excipients (e.g., partially salified trometamol, EDTA) at a relevant temperature. The osmolarity, in the past required by some health authorities in product labeling, is not directly measurable but can be calculated from the osmolality through, $\xi_c = \xi_m C_1$, where C_1 (in kg/L) is the mass concentration of solvent in the solution. C_1 can be calculated from the solution density and the concentrations and molecular masses of the solutes. Exact calculation of the osmotic pressure requires in addition knowledge of the partial molal volume of the solvent in the solution, whereas approximate calculations use the density of pure solvent instead. Even otherwise authorative sources are often marred by erroneous definitions and unit conversion formulas.

The osmotic property called tonicity is defined only in the context of a specified membrane and its permeability to the solutes in the solution. In practice, it is used loosely for the semiquantitative classification of contrast agents into hypo-, iso- and hypertonic types.

171 It is too early to know whether X-ray phase contrast imaging (Mattiuzzi et al. 2003) will have a future and what role contrast agents may play therein.

172 Interpharma Praga AS, Prague, Czechia, founded in 1932, nationalized in 1946, and reprivatized with the participation of Milos Sovak in 1992, is specialized in the chemical manufacturing of generic drug substances. The contrast agents offered are iohexol and iodipamide. In 2008, the company was acquired by Otsuka Pharmaceutical Co., Ltd, Tokyo, Japan (see note 174).

173 Otsuka Pharmaceutical Co., Ltd, Tokyo, Japan, is a very large pharmaceutical company, thus far present with its own name in the field of medical imaging only with the oral gastrointestinal contrast agent for magnetic resonance imaging, ferric ammonium citrate (FERRISELTZ™) and since 2008 as owner of the generic contrast agent drug substance producer Interpharma Praha SA, Prague, Czechia (see note 173).

174 On the Japanese market in 2003 generic nonionic contrast agents have reached the 8% level. The stronger impact of generic products in Japan than elsewhere is best explained by the fact that there the prices are still higher than elsewhere in the world.

175 Under the first-to-invent regime, applicable for example in the USA before 2013, the effective priority may be earlier than the one declared on the published patent.

176 Patents are commonly issued and published in the final version only years after they have been applied for and have become public in the form of applications.

Person Index

Well-known personalities having no direct relationship with contrast agent history, but serve to illustrate aspects of the sociological analysis, are not listed. Portions of names not usually used in scientific communications are given within brackets. Titles referring to nobility or special merit are placed in parenthesis, if regularly used, and in brackets, if not. Nicknames, pseudonyms, replaced names, names with alternative spellings in the relevant language of the person, which are used in documents of present interest, are provided behind an equal sign within a parenthesis. Occasionally, only the last reported place of residence or the last date of mention as alive (me.) may be provided. Where association of historical names of localities with present day names and national affiliation are not obvious, modern short names are provided in brackets. (OS) designates the Julian calendar, which was generally used in most territories of the Russian Emp. until 1918. The Julian calendar is behind the modern day Gregorian calendar (NS) by 12 days up to Feb. 28(NS), 1900, and 13 days thereafter.

A

Allardt, Hans-Georg (b. Aug. 30, 1898, Gnesen, Prov. of Posen, Kingd. of Prussia, German Reich [today, Gniezno, Poland] // me. 1940, Philippstal near Potsdam, Prov. of Brandenburg, State of Prussia, German Reich), 131

Almén, Torsten [Håkan Oscar] (b. Sept. 2, 1931, Lund, Malmöhus Län, Kingd. of Sweden // d. Jan. 8, 2016, Falsterbo, Skåne Län, Kingd. of Sweden), 145, 156–169, 175–195, 199, 204, 207, 214, 217–219, 221–226, 162, 303–304

Alwens, [Edmund] Walter [Johannes Daniel] (b. June 24, 1880 Stuttgart, Kingd. of Württemberg, German Reich // d. July 27, 1966, Frankfurt a. M., State of Hesse, Fed. Rep. of Germany), 40

Amundsen, Per (b. July 1, 1912, Øystre Slidre, Fylke of Oppland, Kingd. of Norway // d. Feb. 5, 1983, Oslo, Kingd. of Norway), 187

B

Barke, Reinhard (b. Sept. 21, 1920, Dresden, Rep. of Saxony, German Reich // d. Nov. 13, 2003, Dresden, State of Saxony, Fed. Rep. of Germany), 145

Becher, Wolf (b. May 6, 1862 Filehne, Prov. of Posen, Kingd. of Prussia [today Wieluń, Poland] // d. Apr. 29, 1906, Schlachtensee, Prov. of Brandenburg, Kingd. of Prussia, German Reich [today Berlin]), 36

Belt, Elmer (b. Apr. 10, 1893, Chicago, IL, USA // d. May 17, 1980, Los Angeles, CA, USA), 104

Berberich, Joseph (b. May 20, 1897, Grosskrotzenburg a. M., Prov. of Hesse-Nassau, Kingd. of Prussia, German Reich // d. June 3, 1969, New York City, NY, USA), 40, 49, 51

Bier, August [Karl Gustav] (b. Nov. 24, 1861, Helsen, Principality of Waldeck and Pyrmont [today Bad Arolsen, State of Hesse, Germany] // d. Mar. 12, 1949, Sauen near Beeskow, State of Brandenburg, Soviet Zone of Allied-occupied Germany), 87

Binz, Arthur [Heinrich] (b. Nov. 12, 1868, Bonn, Rhine Prov., Kingd. of Prussia // d. Jan.25, 1943, Berlin, State of Prussia, German Reich), 59, 60, 61, 63, 68–72, 74–76, 78, 83, 86–87, 89–93, 96–97, 104–109, 111–115, 117–121, 125, 184, 223, 294–297

Bjäringer, Erik H[arald] (b. Nov. 16, 1917, Laholm, Hallands Län, Kingd. of Sweden // d. July 2, 2009, Djursholm/Danderyd, Stockholms Län, Kingd. of Sweden), 156, 157, 158, 212

Björk, Lars (b. Dec. 20, 1924, Ludvika, Kopparbergs Län, Kingd. of Sweden // d. May 27, 2009, Uppsala, Uppsala Län, Kingd. of Sweden), 149–152, 155, 156, 165, 168–169, 174–175, 183, 190, 212, 215

Bjørnson, Olav (b. Jan. 14, 1909, Loge Ekli, Nes sokn, Lister og Mandals amt, Kingd. of Norway [today Flekkefjord, Fylke of Vest-Agder] // d. Mar. 12, 1996, Oslo, Fylke of Oslo, Kingd. of Norway), 158, 221

Bleichröder, Fritz (b. Jan. 12, 1875, Berlin, Prov. of Brandenburg, Kingd. of Prussia, German Reich // d. Nov. 8, 1938, Berlin (-Pankow), State of Prussia, German Reich), 300

Subject Index

317